NEW ROOT FORMATION IN PLANTS AND CUTTINGS

Developments in Plant and Soil Sciences
Volume 20

New Root Formation in Plants and Cuttings

Edited by

MICHAEL B. JACKSON
Long Ashton Research Station
University of Bristol
UK

1986 **MARTINUS NIJHOFF PUBLISHERS**
a member of the KLUWER ACADEMIC PUBLISHERS GROUP
DORDRECHT / BOSTON / LANCASTER

Distributors

for the United States and Canada: Kluwer Academic Publishers, 190 Old Derby Street, Hingham, MA 02043, USA
for the UK and Ireland: Kluwer Academic Publishers, MTP Press Limited, Falcon House, Queen Square, Lancaster LA1 1RN, UK
for all other countries: Kluwer Academic Publishers Group, Distribution Center, P.O. Box 322, 3300 AH Dordrecht, The Netherlands

Library of Congress Cataloging in Publication Data

Main entry under title:

New root formation in plants and cuttings.

  (Developments in plant and soil sciences ; 20)
  Includes indexes.
  Bibliography: p.
  1. Roots (Botany) 2. Plant cuttings. 3. Plant
regulators. I. Jackson, Michael B. II. Title:
Root formation in plants and cuttings. III. Series:
Developments in plant and soil sciences ; v. 20.
QK644.N48 1986   581.3   85-25902
ISBN 90-247-3260-3

ISBN 90-247-3260-3 (this volume)
ISBN 90-247-2405-8 (series)

Copyright

PRINTED IN THE NETHERLANDS

In memory of The AFRC Letcombe Laboratory
(Deceased, 31st March, 1985)

Introduction

The formation of roots is in some respects one of the least fundamentally understood of all plant functions. Propagation by cuttings is the aspect that will occur first to most gardeners and horticulturists, and it is certainly the most useful application. But any observant traveller in the tropics can notice that some trees have the habit of forming roots in the air. Climbers like *Cissus* bear long fine strings of roots hanging down. *Pandanus* trees tend to have stout aerial roots issuing from the bases of the long branches, while the tangle of roots around the trunk of many of the *Ficus* species is characteristic. In *Ficus bengalensis,* in particular, stout cylindrical roots firmly embedded in the ground from a height of 3 to 5 meters give support to the long horizontal branches, enabling them to spread still further. In the big old specimen at Adyar near Madras, the spread of these branches all around the tree, each with a strong root growing out every few meters, makes a shaded area under which meetings of almost 5000 people are sometimes held.

The history of how the formation of roots on stem cuttings was found to be under hormonal control is worth repeating here. It was the demonstration by van der Lek in Holland in 1924 that grape cuttings tended to form roots directly below a young bud or shoot, and on the same side of the stem, that led to the modern study of the physiology of root formation. Frits Went, who had always been impressed by Julius Sachs' idea of a root-forming substance, went to work on the subject in 1929 when he left Utrecht to work for a while in Java. Together with Raymond Bouillenne, he found that 'Dedek', the extract of rice polishings that had been prepared for Jansen and Donath's famous isolation of thiamine, promoted the rooting of *Acalypha* stem cuttings. Although the active factor in this complex extract was not identified, the simple fact that there must be a root-forming or root-promoting substance in the extract did give reality to the Sachsian idea. Thus when Went came to the California Institute of Technology in 1933, where I had been working on the extraction and identification of auxin from cultures of *Rhizopus,* it was natural that we should join forces to look for a root-forming substance, or 'rhizocaline', which evidently had a real existence. The auxin bioassay was, of course, for coleoptile elongation. It quickly became apparent, however, that the highly concentrated and partially purified preparations of auxin had powerful rhizocaline action, and Went's ingenious and highly repeatable bioassay with the stems of etiolated pea seedlings showed that, surprisingly, the root-forming ability went parallel with the auxin action as purification proceeded. To the last it seemed improbable that auxin, then viewed as perhaps identical with Sachs' stem-forming substance, could be identical with the proposed root-forming substance, but the crystallization of indole-acetic acid, and the testing of analogues, clinched the matter. The induction of splendid root systems on commercial cuttings – first on citrus by Bill Cooper – left no doubt of the potential horticultural utility of the discovery. The story is a good example of the highly practical and down-to-earth results that come from pure basic research, driven only by curiosity.

It was almost a quarter of a century later when Skoog and Miller discovered the cytokinins and showed that they acted as 'shoot-forming substances', at least in tissue cultures. As is well known, the cytokinins antagonize the inhibiting effect of auxin on the development of lateral buds, and it is now beginning to seem probable that this so-called 'apical dominance' exerted by auxin is actually a balance between the polarly downward moving auxin that inhibits bud growth, and an upward stream of cytokinin coming mainly from the roots that promotes bud growth. What is less well known is that there is a corresponding balance in the formation of lateral roots on a main root. Although auxin does inhibit root elongation it promotes the formation of lateral roots (in line with its root-forming action on stems) and here the cytokinins clearly act as inhibitors. Thus the hormonal balance in roots is a sort of mirror image of that in shoots, with the roles of auxin and cytokinin as inhibitor and promoter simply reversed. In this way the organization and control of the plant's behaviour has remarkable symmetry. However, comprehension of the biochemistry that underlies this symmetry will be for the future. Some aspects of the hormonal balance within the root are taken up in detail later in this book. Indeed, the book may represent the first occasion when lateral root formation on roots is presented in the context of the more extensively studied formation on roots on stem tissue.

As was said at the outset, certain mysteries remain. The basic difference between plants that are easy to root and those that are difficult or refractive is still elusive. So is the evident relationship between ease of rooting and juvenility of form. So also, unfortunately, is the practical problem of inducing the difficult plants to change their behaviour. In some cases there is apparently an inverse relation between ease of rooting and rate of elongation. But then all the more strange is the behaviour of such plants as the humble blackberry, which manages to combine rapid elongation of internodes with profuse root formation at nodes. It is said that some six million acres of Chile is occupied by blackberries that, following innocent introduction by some settler, have entrenched themselves in this way. Thus although the present book is most valuable in presenting the quantity of knowledge and experience that has been accumulated about the theory and practice of root formation, it also makes evident that generalizations about the behaviour of 'plants' have still to be made with caution.

I am honoured to have the opportunity to introduce a book by such an array of expert and knowledgeable authors, and I am certain that it will prove of great and immediate value to both pure and applied researchers.

Santa Cruz   KENNETH V. THIMANN
California

Preface

The study of plant roots is currently enjoying something of a resurgence. Perhaps we are coming to realize more keenly than before the benefits to agriculture and horticulture that can flow from such work. Furthermore, the essential nature of root research for any understanding of plants as whole organisms that interact in complex ways with their environment is self evident. The key position of root systems in the broader biological context is abundantly clear from a realization that their evolution was a pre-requisite for the extensive colonization by plants of 'dry-land' some four hundred million years ago. The workings of roots are thus central to the existance and survival of plants on the land and indeed indirectly to the survival of almost all terrestrial creatures. Thus, one cannot at all easily overestimate the importance of plant root systems and the need to understand them.

The attributes of roots that assured the success of plants as terrestrial colonizers are equally vital to the success of today's crops and forests. Their cultivation is largely a matter of tending the root environment and thus of manipulating roots and root-based processes. Recognition of this has led to the publication in recent years of several books about roots. They concentrate principally upon root system morphology, root growth and more especially on root functioning and its interaction with environment (see the book list on page XIII). There is also evidence that some authors, concerned with more natural environments and with conservation matters, are also aware of the importance of plant roots for the survival or successful management of vegetation in uncultivated areas. Surprisingly, very little is written in these publications about the *formation* of roots, the underlying physiological processes that guide this formation nor of how this knowledge can be of practical help, particularly in horticulture. One aim of the present book is to begin to rectify this shortcoming.

In seven chapters, the formation of new roots on the existing framework of older roots (lateral rooting), and the formation of adventitious roots from other plant organs are considered. Lateral rooting and the branched structures it forms are essential if plants are to extract water and inorganic nutrients efficiently from the soil. Lateral roots are also thought to be a source of hormones needed by the shoot system. On the other hand, adventitious rooting lies at the heart of a huge industry concerned with the vegetative propagation of desirable plant material, mostly from cuttings. But adventitious roots have another dimension, as highly diverse but functional organs of intact plants. This much neglected subject is also considered in this volume.

The book is not addressed exclusively to any one group. I hope that in the reading, research workers, lecturers and students in the basic or applied plant sciences, agriculturists, horticulturists, plant propagators and indeed anyone whose professional life or natural curiosity draws them to an interest in plant roots will learn much that is useful and worthwhile.

Long Ashton, August 1985   MICHAEL B. JACKSON

Some other books about roots

WEAVER, J.E. (1926) Root Development of Field Crops. McGraw-Hill, New York.
WEBER, H. (1950) Die Bewurzelungsverhältnisse der Pflanzen. Verlag Herder, Freiburg.
KUTSCHERA, L. (1960) Wurzelatlas Mitteleuropäischer Ackerunkräuter und Kulturpflanzen. DLG-Verlags-GmbH, Frankfurt-am-Main.
SHUURMAN, J.J. and GOEDEWAAGEN, M.A.J. (1965) Methods for the Examination of Root Systems and Roots. Purdoc, Wageningen.
WHITTINGTON, W.J., ed. (1968) Root Growth. Butterworths, London.
GHILAROV, M.S., KOVDA V.A., NOVICHKOVA-IVANOVA, L.N., RODIN, L.E. and SVESHNIKOVA, V.M. (1968) Methods of Productivity Studies in Root Systems and Rhizosphere Organisms (First International Root Symposium. Leningrad, USSR, 1968). Nauka, Leningrad.
KOLEK, J., ed. (1971) Structure and Function of Primary Root Tissues. Veda, Bratislava.
HOFFMANN, G., ed. (1974) Ökologie und Physiologie der Wurzelwachstums (Second International Root Symposium, Potsdam, DDR, 1971). Akademie Verlag, Berlin.
CARSON, E.W., ed. (1974) The Plant Root System and its Environment. University Press of Virginia, Charlottesville.
TORREY, J.G. and CLARKSON, D.T., eds. (1975) The Development and Function of Roots. Academic Press, London.
RUSSELL, R.S. (1977) Plant Root Systems, Their Function and Interaction With the Soil. McGraw-Hill, London.
BÖHM, W. (1979) Methods for Studying Root Systems. Springer-Verlag, Berlin.
HARLEY, J.L. and RUSSELL, R.S., eds. (1979) The Soil-Root Interface. Academic Press, London.
SEN, D.N., ed. (1980) Environment and Root Behaviour. Geobios International, Jodhpur.
RUSSELL, R.S., IGUE, K. and MEHTA, Y.R. (1981) The Soil-Root System in

XIV

Relation to Brazilian Agriculture. Fundação Instituto Agronômica do Paraná.
BROUWER, R., GASPARIKOVÁ, O., KOLEK, J. and LOUGHMAN B.C., eds. (1981) Structure and Function of Plant Roots. Martinus Nijhoff/Dr W. Junk, The Hague.
ARKIN, G.F. and TAYLOR, H.M., eds. (1981) Modifying the Root Environment to Reduce Crop Stress. American Society of Agricultural Engineers, St. Joseph.
KUTSCHERA, L. and LICHTENEGGER, E. (1982) Wurzelatlas Mitteleuropäischer Grünlandpflanzen Band I (Monocotyledonae) Band II (Dicotyledoneae). Gustav Fischer Verlag, Stuttgart.
BÖHM, W., LICHTENEGGER E. and KUTSCHERA, L. eds. (1983) Root Ecology and its Practical Application. Bundesanstalt für Alpenländische Landwirtschaft, Irdning.
JACKSON, M.B. and STEAD, A.D., eds. (1983) Growth Regulators in Root Development, Monograph 10. British Plant Growth Regulator Group, Wantage.
SUTTON, R.F. and TINUS, R.W. (1983) Root and Root System Terminology. Society of American Foresters, Washington, D.C.

Contents

List of contributors

ANDERSEN, A.S., Royal Veterinary and Agricultural University, Department of Horticulture, Rolighedsvej 23, Frederiksberg, DK1958 Denmark.

BARLOW, P.W., Long Ashton Research Station, University of Bristol, BS18 9AF/U.K.

HAISSIG, B.E., USDA – Forest Service, North Central Forest Experiment Station, Rhinelander, Wisconsin 54501, U.S.A.

JARVIS, B.C., Department of Botany, The University, Sheffield, S10 2TN U.K.

LOVELL, P.H., Department of Botany, Private Bag University of Auckland, Auckland, New Zealand.

PETERSON, C.A., Department of Biology, University of Waterloo, Waterloo, Ontario N2L 3G1, Canada.

PETERSON, R.L., Department of Botany, University of Guelph, Guelph, Ontario N1G 2W1, Canada.

TORREY, J.G., Harvard Forest, Harvard University, Petersham, Massachusetts O1366, U.S.A.

WHITE, J., Department of Botany, Private Bag University of Auckland, Auckland, New Zealand.

1. Ontogeny and anatomy of lateral roots

R L PETERSON and CAROL A PETERSON[1]

Department of Botany, University of Guelph, Guelph, Ontario, Canada, N1G 2W1

[1] Department of Biology, University of Waterloo, Waterloo, Ontario, Canada, N2L 3G1

1. Introduction

Lateral roots arise endogenously from other roots and increase the absorptive surface of the plant. Sometimes referred to as secondary roots, lateral roots ultimately mimic to a large extent, the structure of the root from which they originate. Since most lateral roots are initiated some distance basipetal to the apical meristem, differentiated cells must become reprogrammed to give rise to the initials of lateral root primordia. Subsequently, these initials divide and enlarge in very precise patterns to organize a new organ recognizable as a root. This chapter will consider the structural aspects of lateral root initiation as well as tissue involvement and reorganization in the parent root associated with this process. In addition, the development of some specialized lateral roots will be considered. The most thorough previous treatments of the structural aspects of lateral root formation can be found in Van Tieghem and Douliot [1], Von Guttenberg [2] and McCully [3].

2. Involvement of parent root tissues in the initiation of lateral root primordia

2.1. Pericycle

Although the initiation site of lateral root primordia is commonly assigned, in general terms, to the pericycle in angiosperms and gymnosperms and to the endodermis in many pteridophytes [4] other tissues may be involved. In angiosperms, however, the pericycle is the most important contributing layer to lateral root primordium initiation [5, 6, 7, 8] and several investigations have documented the early changes in cells of this layer. Pericyclic cells from which lateral root primordia are initiated are usually vacuolated, differentiated cells [9, 10]. In *Raphanus sativus,* these cells have nuclei with a 4C DNA level indicating that they have synthesized DNA, are in the G_2 phase of the cell cycle and this may be the reason these cells proceed rapidly into mitosis following auxin treatment [11].

In *Pisum sativum* [7] and *Convolvulus arvensis* [12], radial enlargement of pericyclic cells is the first indication of lateral root primordium initiation while in many species, increases in the amount and staining intensity of the cytoplasm are the first indications of primordium initiation. In *Zea mays,* early events involve a marked increase in cytoplasmic basophilia and cytoplasmic volume in a few pericyclic and stelar parenchyma cells and what appears to be delignification of these same cells [9, 13]. In *Fagopyrum esculentum* [5], the aquatic plant *Pontederia cordata* [14], *Ipomoea purpurea* [15] and *Calystegia soldanella* [16], radial enlargement and changes in staining response of pericycle cell cytoplasm precede cell divisions leading to lateral root primordium initiation. In *Ipomoea,* pronounced nucleolar enlargement in these cells is also evident and the first cell divisions are anticlinal [15] in contrast to being periclinal in other systems. Many species show changes in nuclear staining of pericycle cells prior to primordium initiation [2]. The above observations, and those of Bayer *et al.* [17] on trifluralin-treated cotton roots, Foard and Haber [18] on gamma irradiated wheat, and Foard *et al.* [19] on colchicine-treated wheat roots, all treatments that permit the formation of enlarged primordiomorphs at sites of lateral root primordia, clearly indicate that pericyclic cells begin their reprogramming prior to the initial cell divisions.

Initial periclinal divisions in the pericycle are sometimes asymmetric, resulting in the larger of the two daughter cells being positioned next to the stele [10, 13]; but this apparent asymmetry may be due to the plane of section [8]. In *Zea mays*, the first periclinal divisions in the pericycle cells are accompanied by a number of wall changes [20]. The anticlinal walls elongate, become considerably thinner, lose much of their birefringence and cease to give a positive lignin reaction. The thinned areas of the wall have more fluorescence when stained with aniline blue and more plasmodesmata than the thicker areas. The mechanisms by which all these wall changes occur have not been determined [20].

Some authors [21, 22] simply describe the initiation of lateral roots as involving periclinal divisions of pericycle cells. Following the initial periclinal divisions, a series of divisions both periclinal and anticlinal, lead to the formation of a protuberance at the periphery of the vascular cylinder in which a complex series of events occurs involving pericyclic derivatives, the endodermis, stelar parenchyma, and the cortex [2].

2.2. Endodermis

It is inevitable that the endodermis, a layer of parenchyma cells contiguous to the pericycle, will be affected by lateral root initiation. In spite of this, it is surprizing how few detailed studies exist concerning the early changes in this layer. Most studies describe changes at the light microscope level. For example, in *Fagopyrum esculentum* [5], *Daucus carota* [6], *Convolvulus arvensis* [12] and *Ipomoea purpurea* [15] endodermal cells adjacent to pericyclic derivatives increase their cytoplasmic content and in *Zea mays* [13] there is increased basophilia. In *I. purpurea*, nucleoli of endodermal cells adjacent to early lateral root primordia enlarge [15]. Byrne [22] has shown that in *Malva sylvestris*, endodermal nuclei incorporate ^{3}H-thymidine prior to radial enlargement and anticlinal divisions. All the above authors report anticlinal divisions in these cells after early cytological changes while other authors report anticlinal divisions in endodermal cells as the first event associated with lateral root initiation [2, 21, 23, 24, 25, 26]. Popham [7] claims that in *Pisum sativum*, the endodermis divides periclinally forming what he refers to as a transversal meristem.

The most detailed accounts of endodermal changes associated with lateral root primordia are those of Bonnett [27] for *Convolvulus arvensis* and Karas and McCully [9] for *Zea mays*. In *C. arvensis*, endodermal cells change from being highly vacuolated to cytoplasmic cells with small vacuoles. In early stages of endodermis reactivation, cell wall synthesis occurs not only where new walls form during division, but in preexisting regions of the walls, including the region of the Casparian strip. In the latter region, fragments of old plasmalemma are embedded in newly synthesized cell wall and additional plasmalemma forms to maintain the continuity of the membrane. New endodermal cells formed by anticlinal divisions lack the ability to form Casparian strips, an observation made earlier by Berthon [28] and Von Guttenberg [2]. In *Z. mays* early changes in endodermal cells include elongation along the periclinal axis, disappearance of suberin lamellae from tangential walls of enlarged cells and their persistence in anticlinal walls. The Casparian strip is retained in these cells. The thickened secondary cell walls become thinner and lose their affinity for lignin stains. Cytoplasmic changes involve an increase in basophilia along with an increased volume of cytoplasm, and several changes in ultrastructural features. Activated endodermal cells display an increase in rough endoplasmic reticulum cisternae, ribosomes and clearly defined dictyosomes with close-packed cisternae. Dictyosome vesicles lack the dense core characteristic of those in endodermal cells prior to lateral root primordium initiation [9].

2.3. Stelar parenchyma

Most studies of lateral root initiation have not included early changes in stelar parenchyma near the primordium. In *Ipomoea purpurea*, some of the cells in the position of incipient protoxylem elements revert to a meristematic state, becoming densely cytoplasmic and therefore easily distinguishable from other stelar parenchyma [15]. Byrne *et al.* [29, 30] have documented early changes in stelar parenchyma in *Glycine max* and *Lycopersicon esculentum* and have shown the importance of these cells in forming vascular connections between the main and lateral root, as discussed in detail in Sections 3.6.1. and 3.6.2. In *Zea mays*, stelar parenchyma cells adjacent to lateral root primordia show an increase in acid phosphatase activity [31] but do not show β-glucosidase activity [32]. They also lose their thickened walls, staining reaction for lignin, and birefringence [13]; they become basophilic and subsequently divide [3, 13].

2.4. Cortex

The most dramatic changes in the cortex of the main root occur during lateral root outgrowth; these will be discussed in Section 4. Changes in the cortex associated with lateral root initiation have been considered for only a few species. Janczewski [5] claimed that cells of the inner cortex of several species of angiosperms become protoplasmic and divide. Berthon [28] separated angiosperm roots into two broad categories – the pea type and the maize type – the former involving the pericycle, endodermis and cortex in the initiation of lateral root primordia, and the latter involving only the pericycle. However, his observations, particularly on maize, have been questioned [9]. Byrne *et al.* [29] showed that in *Glycine max* the inner cortex adjacent to pericyclic and endodermal cells that are dividing periclinally also undergoes periclinal divisions. Mallory *et al.* [24] make the general statement that in *Cucurbita maxima*, the cortex is involved in the formation of the lateral root primordium although the exact contribution is not stated. Mitoses in cortical cells are frequently associated with older stages of lateral root primordia [33].

3. Formation of primary tissues of lateral roots

3.1. Apical meristem

One of the most important events in the development of lateral root primordia is the establishment of an apical meristem, which in most primordia is responsible for a period of indeterminant growth. The establishment of this meristem is directly related to the organization of the initials for the various primary tissues considered in subsequent sections. The apical meristem of lateral roots has been studied from three main perspectives; the heterogeneous nature of the population of cells comprising the meristem, the development of initials of the primary tissues and the organization of a quiescent centre. Methods of analysis of root meristem cell populations and the main results obtained by the use of these techniques are reviewed by Webster and MacLeod [34].

The population of cells comprising a lateral root primordium has its origin from rather few parent root cells [12,35], and consequently these must proliferate and organize into various sized primordia which ultimately emerge as lateral roots [36]. Lateral root primordia, although having potential as developmental systems to study cell population dynamics, are complicated by the fact that they do not behave uniformly throughout their development [34].

In *Vicia faba*, two populations of cells differing in mean cycling time are established in small primordia of 1000–1500 cells [37]; early growth of primordia occurs primarily through proliferation of cells occupying a central core, while subsequent growth occurs through proliferation of peripheral cells [38]. Detailed analyses of various parameters of cell proliferation have shown a number of general features of primordium growth in *V. faba*. Most of the cells in developing primordia become quiescent at some time, usually just prior to lateral root emergence [38, 39, 40]. Considering the primordium as a whole, the length of the mitotic cycle and the period of DNA synthesis increase as the primordium increases in cell number [39, 41] while the size of the proliferating population of cells (the growth fraction) decreases [39, 42]. Subsequently the growth fraction increases as laterals increase from 2 to 10 mm in length [43]. The overall increase in cell number, meristem length and volume is linear as primordia are displaced basipetally along the main root [44]. MacLeod and McLachlan [45, 46] have shown quite clearly that during lateral root elongation there is considerable variation in the labelling index (% of cells with nuclei which incorporate ^{3}H-thymidine) and mitotic index (% of cells in mitosis) among primordia and lateral roots of differing lengths, and among tissues within any one length class. Both labelling index and mitotic index are low in just-emerged laterals but increase as the lateral root elongates. The initial emergence of lateral roots is therefore accomplished through cell elongation with cell proliferation resuming later [38, 39, 40, 44, 47]. The planes of cell division vary in the developing primary tissues of primordia and within any one tissue along the length of the primordium as lateral roots elongate [48].

A feature of most root meristems, whether primary or lateral, is the development of a population of slowly cycling cells, the quiescent centre, at the pole of the cortex and stele [49]. Clowes [50] was the first to show that quiescent centres do develop in lateral root meristems, specifically in the roots of two aquatic floaters, *Pistia* and *Eichhornia*. In both genera, the quiescent centre is formed before lateral root emergence. In *Malva sylvestris* [22] and *Vicia faba* [47, 51] lateral roots do not develop a quiescent centre until after emergence from the parent root, a situation which is implied as well for *Ipomoea purpurea* [15]. The results with *V. faba* may be explained, in part, by poor penetration of ^{3}H-thymidine into unemerged primordia [40]. In *Malva sylvestris* [22] the quiescent centre of both secondary and tertiary roots appears at a time when distal root cap cells no longer incorporate ^{3}H-thymidine and the endodermal cover is being sloughed.

In *Zea mays* [52], the majority of lateral root primordia lack a quiescent centre before emergence but almost all fully-developed laterals have a quiescent centre. The first quiescent centre appears to develop as soon as a large enough root cap forms from endodermal derivatives. This quiescent centre disappears after the first set of cap initials stops cycling and a second quiescent centre is established after a new set of cap initials of pericyclic origin begins division. This usually occurs after lateral root emergence. Lateral roots of very narrow diameter may not develop the first quiescent centre. The difference in timing of quiescent centre organization may be related to the organization of the apical meristem either as an 'open' or 'closed' system [52].

3.2. Root cap

There has been and continues to be some ambiguity in the literature concerning the source and nature of the tissue which forms the interface between the lateral root primordium and the cortex of the parent root. Although there is likely much variation between species, some of the confusion no doubt arises from a lack of detailed analysis of the ontogeny of lateral roots in all but a few species. In his comprehensive treatment of roots, Von Guttenberg [2] describes the formation of a 'Tasche' formed by the parent root endodermis which in all but a few aquatic species, is subsequently shed as the lateral root emerges from the parent root. A similar structure, the 'poche digestive', a presumed source of degradative enzymes responsible for the lysis of cortical cells has been described by Van Tieghem and Douliot [1]. The 'poche digestive' is also sloughed as the lateral root penetrates the epidermis of the parent root. Byrne [22] described an ephemeral 'endodermal cover' in lateral root primordia of *Malva sylvestris*. He agrees with Seago [15], that in the absence of consistent terminology for this tissue in the literature that 'endodermal cover' be used for all cases in which there is no permanent histogenic involvement of cells derived from the endodermis. Curiously, cells of the endodermal cover in *Malva sylvestris* continue to incorporate ^{3}H-thymidine until the lateral root is 5 mm long, and have cytological features such as amyloplasts with starch grains comparable to mature root cap cells.

A root cap, distinct from an endodermal cover, can originate from pericyclic derivatives in some species [6, 12, 21, 22], from the endodermis [5, 9, 13, 16] or from the cortex [5, 7]. The most detailed studies of root cap origin during lateral root development are those of Bell and McCully [13] and Karas and McCully [9] who elaborated on the earlier observations of Janczewski [5] on *Zea mays*. In this species, the endodermis at first divides anticlinally to form a single layer of cuboidal epidermal cells. A few of these cells at the summit of the lateral root primordium then divide periclinally, forming an outer layer of root cap cells and an inner layer of root cap initials. Further periclinal divisions in the initials result in a small, multiseriate root cap. In this species, therefore, the root cap which is of endodermal origin, is not sloughed from the lateral root. In fact, individual root cap cells are not sloughed until the lateral emerges from the parent root. Karas and McCully [9] have described the cytological changes which occur as the cuboidal epidermal cells are transformed into either root cap cells or root cap initials. Some of these features are the secretion of a weakly periodic acid Schiff-positive but metachromatic substance, the early synthesis of starch grains in amyloplasts, a decrease in the density of ribosomes, a change in dictyosome morphology, and an increase in the number of small vacuoles in young cap cells followed by a decrease in older cap cells.

Clowes [52], using colchicine-induced chimeras, questions the work of Bell and McCully [13] and Karas and McCully [9] as to the origin of the permanent root cap initials in *Zea mays*. His observations suggest that there is a cap of endodermal origin that is sloughed from the lateral root and that permanent initials are subsequently derived from pericyclic derivatives.

Root caps of most lateral roots are not extensive structures. For example in *Vicia faba*, MacLeod [53] found that newly-emerged lateral roots have a cap of approximately 975 cells while roots 40 mm long have a cap of 3753 cells. This same author estimates that in a single, 11-day-old *Vicia faba* seedling, between 56000 and 85000 root cap cells are added to the rhizosphere each day if one considers all the laterals which develop by this time. MacLeod and McLachlan [45, 46] found that in *V. faba* both the labelling index and the mitotic index are very low in the root caps of newly-emerged laterals and laterals up to 40 mm long. The highest labelling index and mitotic index were found approximately 200 to 300 μm from the root cap tip, the site of root cap initials, although even here the indices are low. Both of these indices show marked increases at later stages of lateral root development.

3.3. Epidermis

From a histogenic point of view, the epidermis of lateral root primordia of most species originates from pericyclic derivatives [6, 12, 15, 16, 22]. In *Malva sylvestris* [22] the protoderm as well as root cap initials is derived from the outermost layer of a 4-layered group of pericyclic derivatives, a situation which is similar in *Ipomoea purpurea* [15]. Cytological changes which occur as epidermal cells differentiate from pericyclic derivatives have not been documented. In *Zea mays* the lateral root epidermis is apparently derived from the parent root endodermis [3, 9, 13]. Here, the endodermis divides, forming a single layer of cells. Periclinal divisions in part of this layer initiate root cap initials. Those cells not involved in root cap formation undergo further anticlinal divisions, become columnar in shape, and pass through a sequence of cytological and physiological changes. The cells have large nuclei, strongly basophilic cytoplasm, numerous small vacuoles, plastids with an electron dense matrix and usually no starch, and dictyosomes which produce vesicles containing electron dense material. Lateral root primordia still within the parent root have this epidermal layer almost back to the parent endodermis, but in emerged laterals only the distal 0.5 mm retains an epidermis with the above cytological characteristics. Bell and McCully [13] point out that columnar epidermal cells are unusual in plants but common in animals where they are invariably associated with secretion or absorption. In *Z. mays,* these cells have a secretory function since a periodic acid Schiff-positive, non-metachromatic material, low in carboxyl and methyl ester residues, coats the outer tangential cell walls and becomes strikingly thick along the epidermis occupying the flanks of the lateral root. This material may prevent self-hydrolysis of primordial cells [13] although it is certainly present around epidermal cells of primary roots as well [54]. In a later study of *Z. mays,* Clowes [52] has reinterpreted the origin of the epidermis in lateral roots based on observations of chimera production with colchicine. He concludes that the epidermis is of pericyclic origin, a concept that is consistent with the observation that in all grass roots the epidermis and cortex are part of the same complex of cells [55].

Changes in a number of parameters of epidermal cell growth have been documented in newly-emerged and elongating lateral roots of *Vicia faba*. Although there is a general overall increase in labelling index in epidermal cells of newly-emerged laterals up to and including laterals of 40 mm in length [45], the mitotic index falls during this period with a significant decrease by the time the root primordium is 2 mm in length [46]. In newly emerged laterals, the mitotic index increases between 200 and 500 μm back from the distal tip of the root cap and then declines, reaching a minimum at 800 μm from the root tip [46]. The number of epidermal cells increases primarily by divisions in which the spindle is oriented along the longitudinal axis of the root [48]. The length of epidermal cells of all sizes of lateral roots up to 40 mm measured at a distance of 400 to 1000 μm from the distal tip of the root did not change [48]. Cell breadth and width, however, increased between 400 and 700 μm and between 700 and 1000 μm in just-emerged laterals and laterals of 2 mm length while cell breadth decreased along the apical 1000 μm of longer roots. Cell width increased between 700 and 1000 μm in laterals between 5 mm and 40 mm in length. The initiation of the epidermis from cortical cells of the parent root in *Pistia stratiotes* [5] needs confirmation.

3.4. Cortex

Cortical initials of lateral root primordia are derived from pericyclic derivatives [1, 5, 6, 7, 15, 21, 22]; an earlier claim that endodermal cells are involved in initiation of the cortex in the Leguminosae and Cucurbitaceae [5] was refuted by Van Tieghem and Douliot [1]. The most detailed account of cortex development is for *Ipomoea purpurea* [15]. Cortical initials are at first uniseriate but become biseriate except at the apex of the primordium. Usually by the time the lateral root is approximately 10 mm long, cortical initials and cell files of the peripheral part of the root cap columella become aligned resulting in an unlayered appearance of the cortical initials. The hypodermis of the lateral root cortex appears to have a separate origin from the inner cortex.

The cortex of *Vicia faba* lateral roots, like the epidermis, has been analyzed in considerable detail. Although the labelling index of cortical cells increases as the roots elongate beyond the parent root, it is only significantly increased when roots are 40 mm long. The labelling index along the length of newly-emerged laterals increases slowly over the apical 200 to 600 μm, remaining fairly constant between 600 and 900 μm and then decreases at 1000 μm. In longer roots there is a general increase in labelling index over the apical 1000 μm [45]. The mitotic index in newly-emerged laterals is maximal in the cortex at 300 μm from the root apex and decreases basipetally [46]. In longer lateral roots there is considerable fluctuation in mitotic index along the length of the root. Since the mitotic index of the cortex and stele is not significantly different in newly-emerged laterals,

MacLeod and McLachlan [46] suggest that the initials for these two tissues may not be differentiated by this time. In contrast, since the mitotic index of epidermis and cortex are significantly different, these initials are differentiated by this stage of development. In further studies of *V. faba* [40, 48], it was found that lateral roots of various lengths from just-emerged to 40 mm have similar numbers of cortical cells in corresponding distances from the root apex. Also, increase in cortical cell size is due primarily to increases in cell breadth and width rather than length.

3.5. Endodermis

In the majority of roots, the endodermis of the main root divides during the early stages of lateral root initiation (see Section 2.2.). Most endodermal divisions are anticlinal and in all roots studied, lead to the formation of a discontinuity in the Casparian band (Fig. 1a), a hydrophobic region in a portion of the radial and transverse walls of the main root endodermis. The discontinuity occurs because the newly-formed radial walls do not contain Casparian bands [6, 9, 27, 28, 56]. The absence of Casparian band development in new walls of the endodermis associated with lateral root formation contrasts with the rapid formation of Casparian bands in the radial walls of the endodermis following anticlinal divisions during secondary (thickening) growth of the main root [57]. During the continued growth and development of the lateral root, Casparian bands, continuous with those of the main root endodermis at the edges of the lateral root, form in the parenchyma cells at the base of the lateral root [1, 6]. After a short period of centripetal development, the Casparian bands begin to form in the endodermal cells of the lateral root and continue to differentiate acropetally [6, 56]. The final result of this sequence of differentiation, seen in Fig. 1b, is that the Casparian band of the lateral root is continuous with the Casparian band of the main root [6, 15, 56, 58]. The possible physiological consequences of the temporary discontinuity in the Casparian band during the development of the lateral root were recognized by Dumbroff and Peirson [56]. They considered that during the very early stages in development, the cell walls of the lateral were probably too thin to provide an apoplastic pathway into the stele of the main root. Later work with fluorescent dyes [59, 60] and lead [61] has indicated that the young walls of lateral roots are impermeable to these tracers. Dumbroff and Peirson [56] point out that the formation of the lateral root Casparian band in *Avena sativa, Lycopersicon esculentum* and *Convolvulus arvensis* lags behind lateral root growth so that at lateral root emergence, probable apoplastic pathways from the main root cortex to its stele occur via mature cell walls at the base of the lateral. Peterson *et al.* [60] found that two fluorescent, apoplastic dye tracers could move from the main root cortex into the stele at the base of the laterals in *Vicia faba* and *Zea mays* when the laterals had just emerged from the parent root cortex. The timing of Casparian

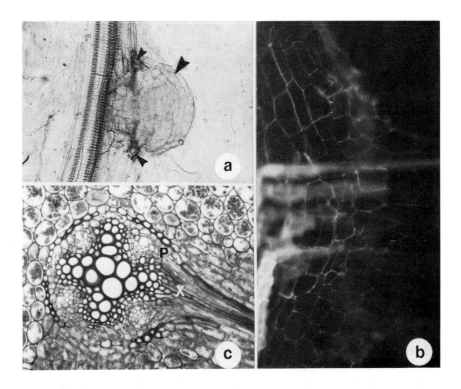

Fig. 1. (a) Cleared leek (*Allium porrum* L.) root stained with chlorazol black E showing a lateral root primordium (*) and a discontinuity in the endodermis of the parent root (arrowheads) due to primordium growth. × 50. (b) Broad bean (*Vicia faba*) clearing of a junction between a lateral root and the main root stained with *Chelidonium majus* root extract and viewed with epifluorescence optics using violet light. The Casparian band is a net-like structure continuous from the main root into the lateral. Brightly fluorescing tracheary elements appear below the plane of focus. × 80. (c) Base of lateral root of *Ranuculus* sp. showing continuity of xylem (X) and phloem (P) with the corresponding tissues of the main root. × 140.

band development in the lateral may vary in different species. In *Daucus carota* [6], the Casparian band is present in the lateral before it emerges from the parent root, whereas in *Humulus lupulus* [62], *A. sativa, L. esculentum* and *C. arvensis* [56] the Casparian band is not evident until after laterals emerge, and in maize it has not yet formed when the lateral has emerged 4.5 mm beyond the main root [9]. The relationship between the length of the lateral root and the development of its Casparian band in *Vicia Faba* depends on the growth rate of the lateral. Some rapidly-growing roots emerge before their Casparian band develops while slowly growing roots have a well-developed Casparian band before emergence [63].

12

3.6. Vascular tissues

In an analysis of lateral root initiation in a number of angiosperm species, Janczewski [5] concluded that the vascular cylinder always arises from the pericycle. The development of primary xylem and phloem in the lateral root primordium proceeds so that these tissues are connected with their counterparts in the main root (Fig. 1c).

3.6.1. Phloem. Very few investigations of lateral root development include a discussion of phloem initiation and maturation. Esau [6], in a paper on *Daucus carota* root, states '*The connection between the vascular tissues of the main and the lateral root is formed through cells of pericyclic origin lying outside the protophloem and protoxylem of the main root. These are parenchyma cells of short diameters arranged like the derivatives of a cambium. [Near the phloem they differentiate] into sieve tubes . . . with which companion cells are associated.*' The most detailed studies of lateral root phloem development are those by Byrne *et al.* for *Glycine max* [29] and *Lycopersicon esculentum* [30]. The following description of phloem development has been extracted from the paper of Byrne *et al.* [29]. The typical main root of soybean is tetrarch with a uniseriate pericycle separated from the vascular tissue by a layer of outer stelar parenchyma (Fig. 2a). At the locus of the main root where the laterals originate, the metaphloem and metaxylem are not yet mature. During inception of a lateral root, a localized region of the pericycle around a protoxylem pole becomes biseriate by means of periclinal divisions. The formation of phloem connector cells (i.e. phloem tissue which connects the phloem of the main root to the phloem of the lateral root) begins with additional divisions in the pericyclic derivatives opposite the two protophloem strands which flank the protoxylem strand over which the lateral root will be centered (Fig. 2b). Cell divisions commence in the order indicated by the numerals in Figure 2b, then spread both acropetally and basipetally in the pericyclic derivatives along the protophloem strands of the main root for 150 to 175 μm in each direction (Fig. 2c). These derivatives eventually form the axial connector phloem. Following the onset of cell divisions in the axial system, divisions begin in the outer stelar parenchyma cells located adjacent to the protophloem near the ends of the axial system, and proceed toward the protoxylem pole, eventually forming two continuous bands bridging the axial connector phloem (Fig. 2d). These horizontal bands form the transaxial connector phloem. Histogenesis of the transaxial connectors is completed in the second quarter of lateral root emergence. The axial and transaxial connectors form a continuous wreath of phloem at the base of the lateral root (Fig. 2e). The axial phloem connector system continues to develop during the second quarter of lateral root emergence, its distal face in contact with the lateral root procambium and its proximal face in contact with the main root proto- and metaphloem. Maturation of the connector phloem begins when the lateral is about three quarters through the parent root cortex. The first connectors to mature are at the junctions of the axial and transaxial connector phloem (Fig. 2d). Maturation continues from both ends of the transaxial connector phloem as indicated in Figure 2d. Maturation of the axial phloem connectors proceeds acropetally from the connectors situated adjacent to the main root phloem into the remainder of the connectors, so that the axial connector system matures concomitant with the emergence of the lateral root. Maturation of the phloem from the procambial cells of the lateral continues acropetally. There appears to be an important correlation between the maturation of the parent root metaphloem and the maturation of the axial connector phloem. Metaphloem maturation in the main root commences late in the second quarter of lateral root pre-emergent growth and continues during the third quarter. The temporal correlation of parent metaphloem and axial connector phloem maturation makes possible the formation of sieve areas and sieve plates between the two phloem systems. Thus, the most important functional connection between the main and lateral root is through the main root metaphloem. The function of the transaxial connector phloem is unknown.

In a subsequent study of lateral root vascularization in *Lycopersicon esculentum*, Byrne *et al.* [30] compared vascularization of this diarch root with the same process in the tetrarch root of *Glycine max*. In the diarch root, the centre of the lateral is positioned between one xylem arm and an adjacent phloem strand. Histogenesis of connector phloem first occurs adjacent to the nearest main root phloem strand but eventually both strands become connected to the phloem of the lateral root. Both axial and transaxial phloem connectors are formed, with the transaxial connector bands lying closer to the centre of the lateral (50–60 μm). The connector and lateral root phloem maturation is acropetal and again the initial events are correlated with main root metaphloem maturation so that the main root metaphloem develops sieve areas with the adjacent connector phloem. Sieve plates are present between adjacent sieve-tube members within the axial connector phloem and between the connector and lateral root phloem. Although studies of phloem development in other species are required to build up a general picture of the process, it is noteworthy that in the two species examined in detail, transaxial as well as axial phloem connectors are formed, and that the connectors form a functional association with the parent root metaphloem.

3.6.2. Xylem. Rywosch's early report [64] on lateral root formation in 20 species of monocotyledons includes descriptions of the maturation of the connector xylem which typically consists of tracheids with reticulate thickenings. With the exception of *Acorus calamus*, connector xylem is always associated with more than one xylem strand of the main root. The number of such strands varies from three in the hexarch root of *Allium porrum* to 13 in *Monstera deliciosa*. Since the total number of xylem strands in *M. deliciosa* numbers 16, connector xylem develops on strands not only on the side of the stele adjacent to the lateral root but also almost totally around the circumference of the main root. The number of

14

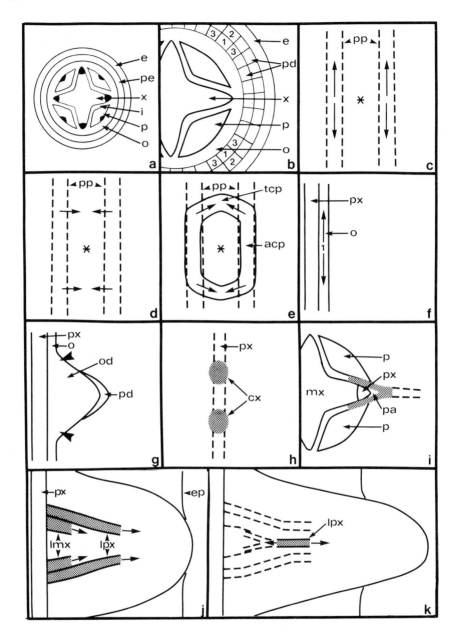

Fig. 2. Lateral root development in *Glycine max*. (a) Cross section of primary root showing endodermis and stele. Shaded areas represent protophloem and protoxylem. (b) Cross section of primary root after the pericycle has become biseriate. The order of cell divisions in the pericyclic deriviates is indicated by numbers; the first divisions occur in the innermost pericyclic derivatives opposite the protophloem poles. The primordium centre is opposite the xylem pole between the two phloem strands. Diagrams c, d and e show tangential longitudinal sections passing through the pericyclic derivatives. Strands of protophloem below the pericyclic derivatives are outlined by dotted lines. The centre of the developing primordium is indicated by an asterisk. (c) Arrows indicate the direction of histogenesis in the axial connector phloem. (d) Arrows indicate the direction of histogenesis in the transaxial connector phloem. (e) Histogenesis of connector phloem is complete. Arrows indicate the direction of transaxial connector phloem maturation. Diagrams (f) and (g) show radial longitudinal sections passing through a protoxylem pole and centre of primordium. (f) Outer stelar parenchyma cells in position 1 begin to divide in many planes. Divisions spread in this layer in the directions indicated by the arrows. (g) A primordium forms from the outer stelar parenchyma derivatives and pericyclic derivatives. Arrowheads indicate the positions of outer stelar parenchyma cells which will divide periclinally to form xylem connector cells. In (h) a tangential longitudinal section is depicted through patches of connector xylem (stippled). The underlying protoxylem is outlined by dotted lines. Diagram (i) shows a cross section of the main root through a patch of connector xylem (stippled). Derivatives of the inner and outer stelar parenchyma contribute to the connector xylem, the former derivatives mature beside the main root metaxylem. The cell(s) directly opposite the main root protoxylem pole remain as parenchyma. The connector xylem strand passes out of the plane of the section (dotted lines). Diagrams (j) and (k) show radial longitudinal sections through a protoxylem pole and developing primordium. In (j) arrows indicate acropetal maturation of the lateral root protoxylem and metaxylem. The area between the protoxylem strands eventually differentiates as metaxylem. A later stage of development is given in (k). At a higher focal plane, another strand of lateral root protoxylem (stippled) differentiates and matures, both acropetally (toward the apex of the lateral) and basipetally (large arrows). It appears to unite with the metaxylem formed earlier (dotted lines). The fourth protoxylem strand matures in a lower focal plane.

Abbreviations

e = endodermis, pe = pericycle, x = xylem, i = inner stelar parenchyma, p = phloem, o = outer stelar parenchyma, pd = pericyclic derivatives, pp = protophloem, tcp = transaxial connector phloem, acp = axial connector phloem, px = protoxylem, od = outer stelar parenchyma derivatives, cx = connector xylem, mx = metaxylem, pa = parenchyma, lmx = lateral root metaxylem, lpx = lateral root protoxylem, ep = epidermis, * = centre of primordium, dashed lines = structure below the plane of the diagram, stippling = lateral root xylem and connector xylem.

main root xylem strands associated with connector xylem is not a simple function of the total number of strands since *Agapanthus umbellatus* has 20–22, only seven of which are linked to the lateral root by connector xylem. The connector xylem matures first adjacent to xylem strands nearest the central point of origin of the lateral, and later near xylem strands increasingly distant from this point. Maturation is completed in the same pattern. In each group of connector tracheids, the direction of maturation is always acropetal from the main root xylem strand into the lateral root. Depending on the widths of the connector xylem patches and the angle they make as they mature, in some plants they become confluent laterally in all but the patches nearest to the lateral root (*M. deliciosa, Syngonium affine*) while in others they do not become confluent (*Clivia minata, A. porrum, Tradescantia albiflora*). In the former group, it is not clear from Rywosch's observations of cross sections where the phloem connectors are formed. Usually the connector xylem matures from derivatives of the pericycle and derivatives of cells between the xylem and phloem strands which also become meristematic. Some xylem connector cells abut the metaxylem in *Monstera* (64, figs. 1 and 3, later reproduced as figs. 91 and 92 respectively by Von Guttenberg, 2) and presumably in others with a similar developmental pattern i.e. *S. affine, Vanda tricolor, Cypripedium insigne, Clivia minata* and *A. umbellatus*. In *Pandanus veitchii* cell divisions also occur in the pith of those lateral roots which replace the mother root. The pith meristem derivatives next to the lateral root mature into a mass of tracheids. The connector xylem tracheids in *Zea mays, Panicum plicatum, Elymus arenaris, Phragmites communis* and *Carex sp.* mature from derivatives of a meristem arising between the main root proto- and metaxylem on the side of the stele where the lateral root is produced. Enlargement of the derivatives of this meristem prior to maturation actually displaces the overlying tissues, including the protoxylem and the adjacent phloem, outward (64, fig. 5). Maturation of the connector tracheids is centrifugal, i.e. from the metaxylem outwards. Here, there is a widespread connection between the connector xylem and the main root metaxylem.

More recently, the development of lateral root xylem and its connection with the xylem of the main root has been reviewed by McCully [3]. The most detailed studies of this process in dicotyledonous plants were made later however, by Byrne *et al.* [29, 30], who provide the following information on xylem development in lateral roots of *Glycine max* [29]. The structure of the main root is shown in Figure 2a. The formation of the connector xylem eventually occurs in derivatives of the outer and inner stelar parenchyma. During the initiation of a lateral root, periclinal divisions occur first in the pericycle, endodermis and inner cortex, then in the outer stelar parenchyma. At this time, the inner stelar parenchyma cells neighbouring the xylem dedifferentiate. In all the layers, cell divisions begin at the centre of the future primordium, then spread both acropetally and basipetally in the main root to the edges of the future primordium (Fig. 2f). A multiseriate layer of outer stelar parenchyma derivatives is formed. Divisions in other planes begin in this layer in the position marked with number 1 in Figure 2f. The innermost derivatives continue to divide periclinally and contribute to the centre of the primordium. Each outer stelar parenchyma cell derivative at the edge of the primordium (arrowheads, Fig. 2g) divides periclinally, forming cells which will eventually mature into xylem connectors. Each group of future connector cells is about 8–11 cells high, 7–8 cells wide and 2–4 cells deep, i.e. extending into the lateral (Fig. 2h). At this stage of development the lateral root primordium is $1/4$ to $1/2$-way through the main root cortex. Cell division begins in the inner stelar parenchyma adjoining the main root metaxylem and continues until emergence of the lateral root. The future xylem connector cells continue to divide and elongate. Derivatives of the outer and inner stelar parenchyma in the connector xylem region (Fig. 2i) mature into vessels at the time of lateral root emergence. When the lateral is 5–10 mm long, acropetal maturation of two strands of xylem, contiguous with the two patches of connector xylem, commences (Fig. 2j). Later, two patches of lateral root metaxylem connectors which differentiate from the derivatives of the inner and outer stelar parenchyma, mature acropetally (Fig. 2j). The origin of tetrarchy in the lateral root comes about by the differentiation of two new files of vessels at a distance of 425 μm from the main root protoxylem (Fig. 2k). Maturation from these new xylem files proceeds acropetally into the lateral root and basipetally toward the parent root, and the xylem strands apparently merge with the metaxylem connectors (Fig. 2k). The cells externally adjacent to the protoxylem usually remain as parenchyma and the functional connection from lateral root xylem to main root xylem is effected through the association of lateral root proto- and metaxylem connectors with the metaxylem of the main root (Fig. 2i).

In a later paper Byrne et al. [30] report that stelar parenchyma cells of Lycopersicon esculentum roots also participate in the formation of connector xylem. When the lateral root nears the surface of the main root, some of the metaxylem of the main root and the adjacent connector xylem begin to mature. After the lateral root has emerged 10–15 mm from the surface of the main root, the connectors are mature and xylem differentiation proceeds acropetally into the lateral. The simultaneous maturation of part of the main root metaxylem and the adjacent connector xylem allows the formation of lateral perforation plates between vessels from the two regions, resulting in the construction of an efficient conduit. This arrangement parallels the association of the main root metaphloem and the axial phloem connectors.

As pointed out by McCully [3], there are a few reports that only the derivatives of the pericyclic cells and not the derivatives of the stelar parenchyma cells are involved in the formation of the xylem connector system [4, 6, 65]. However, in Lupinus albus [66], Glycine max [29], Lycopersicon esculentum [30], Ipomoea purpurea [15], Zea mays [13] and many other monocotyledonous species [64] derivatives of the stelar parenchyma are known to contribute to the connector xylem. These cells must necessarily contribute if connection is made between the lateral root xylem and the metaxylem of the main root. Torrey [67] reports that the diarch or triarch xylem of the lateral root in Pisum sativum originates by the differentiation of a few xylem elements on either side of the protoxylem arm as well as at the outermost point of the radial xylem arm. These elements form xylem strands which converge laterally. Apparently only acropetal differentiation is involved. The early report of basipetal xylem differentiation near the base of the lateral in Daucus carota [23] needs to be confirmed as does the report of Fourcroy [66] which describes degenerative changes in the xylem of the main root at the site of connector xylem association in a number of species. The differing descriptions of some aspects of lateral root xylem development may be in part due to the use of different species for these investigations. In all roots studied, the xylem of the lateral tends to emerge at right angles from the main root, a feature used by Bonnett and Torrey [12] to distinguish between root primordia and endogenous buds in *Convolvulus arvensis* roots.

3.6.3. Transfer cells. Transfer cells, parenchyma cells which elaborate plasmalemma-lined wall ingrowths, are less frequently found in roots than other plant organs [68]. Xylem and phloem parenchyma cells in the main root at the base of lateral root connections and in the lateral root itself are modified as transfer cells in *Hieracium florentinum* plants grown hydroponically [69, 70]. Similar cells were lacking in main root xylem and phloem parenchyma not associated with lateral roots. In another study in which roots of several genera of leguminous plants were examined for vascular transfer cells associated with either nodules or lateral roots, only roots of *Pisum sativum* had xylem and phloem transfer cells at the base of emergent lateral roots [71]. The frequency of transfer cells, thought to be involved in short distance transport of various substances, in lateral roots or in main roots associated with lateral roots appears to be very low; this may be due to the fact that few ultrastructural studies of lateral root connections have been published.

4. Lateral root outgrowth

Lateral roots have an endogenous origin within the main root, necessitating their outgrowth through the cortex and epidermis of the latter. As the young root primordium begins to enlarge, the cortical cells divide in some species [33] but are more commonly stretched (i.e. tangentially flattened) to accommodate the new primordium. The cortical cells in the path of the growing lateral eventually become markedly deformed and die (Fig. 3a). Cell contents of these cortical cells disappear, with the possible exception of starch grains [72] while some or all of the walls remain. Frequently, a cavity is formed in the cortex adjacent to elongating primordia (Fig. 3a). Details of the process of primordium growth through the cortex are not well understood. There are three main ideas on how this is accomplished. 1. Digestive enzymes are secreted by the outer cells of the primordium [73]. These enzymes kill and at least partially digest the cells of the cortex, leaving a cavity into which the primordium grows [1]. 2. Mechanical pressure on the cortical cells by the growing primordium causes the cortical cells to produce digestive enzymes, leading to their autodigestion [74]. 3. Mechanical pressure on the cortical cells by the growing primordium crushes the cells and their contents are absorbed by the surrounding tissues [72]. As Bonnett [27] has pointed out, some combination of mechanical and enzymatic action could be operative. In the early stages of primordium growth, cortical cells are only deformed and not killed e.g. in *Cichorium intybus* [75], *Vicia faba* and *Lupinus alba* [72]. The outermost cells of the root primordium may still be found closely appressed to the cortical cells at later stages e.g. in *Malva sylvestris* [22]. Thibault [23] describes the emergence of *Daucus carota* laterals by mechanical means and Berthon [28] claims that mechanical force alone is sufficient to explain the emergence of laterals in *Pisum sativum, Phaseolus vulgaris, L. alba, Raphanus sativus, Ricinus communis, Cucurbita pepo* and *Zea mays* but does not give any details of his evidence. Pond [72] also suggests that mechanical force can explain the emergence of laterals of *V. faba* and *L. alba*.

Many recent investigations provide indirect evidence of enzyme involvement in lateral root outgrowth. It is well documented that a cavity forms in front of each primordium in *Vicia faba* [38, 76]. Analysis of the contents of these cavities revealed the presence of sugars which may provide nourishment for the primordium which lacks mature phloem [77]. A similar suggestion had been proposed much earlier [1]. In *Zea mays*, Bell and McCully [13] found that the middle lamella between cortical cells adjacent to a lateral root primordium was lysed and, on the basis of toluidine blue 0 staining, polyuronides were removed from the remainder of the wall. The change from uniform to nonuniform birefringence also suggested some structural change in cortical cell walls. Total cortical wall material remaining after primordium growth was less than would have been expected without breakdown, an observation also made by Crooks [21] for *Linum usitatissium.* Weerdenburg and Peterson [78] found that lignin disappeared from the phi thickenings of *Pyrus malus* during lateral root outgrowth. Bell and McCully [13] further documented the removal of cytoplasmic contents from cortical cells and suggested that enzymes were involved. In *Calystegia soldanella*, the appearance of a cavity in front of the lateral root is correlated with the development of the cap which may be the source of the digestive enzymes [16]. Bonnett [27] also followed the disappearance of cortical cell protoplasts in *Convolvulus arvensis* using electron microscopy. In the outermost layer of cells of the primordium, coated vesicles were present, some of which were fused with the plasmalemma contiguous to the outermost wall. He postulated that these vesicles contain hydrolases transported from dictyosomes to the wall and released so that the enzymes diffuse to adjacent cortical cells. The disappearance of complex substances such as proteins, lignin and nucleic acids in the absence of bacteria suggests plant enzyme involvement.

According to Sutcliffe and Sexton [74] in *Pisum sativum*, acid phosphatases are located in main root cortical cells adjacent to the primordium which may be induced by mechanical pressure from the growing primordium. However, Ashford and McCully [31, 32] using *Zea mays*, found much more naphthol AS-B1 phosphatase and β-glucosidase activity in the lateral root (and apex of the main root) than in the cortex of the main root. They concluded that these enzymes are probably not concerned with the breakdown of cortical cells around the lateral root primordium. Bonnett and Torrey [12] and Bonnett [27] argue that the digestive enzymes are secreted by the primordium and not the cortical cells because cortical cell death is associated with lateral root outgrowth but not with endogenous bud outgrowth in roots of *Convolvulus arvensis*.

Bell and McCully [13] noted that, while the central cortex of *Zea mays* roots appeared to offer no resistance to the growth of the lateral roots, the walls of the hypodermis and epidermis were not easily penetrated. These walls have been described as partially lignified [13] and suberized [79, 80]. Bell and McCully [13] observed that hypodermal and epidermal walls may be pushed outward for a considerable distance by the growing primordium before they eventually rupture. Van Tieghem and Douliot [1] also found that the outermost 'cutinisees' root layers resisted the growth of the lateral. Since roots of many species have lignified or suberized hypodermal walls, and similarly-modified epidermal walls are even more common [80, 81], the final emergence of lateral roots may be achieved by mechanical force. It would be instructive to repeat the work of Pond [72] in which he simulated lateral root growth through parenchyma tissue using a glass rod and compared its effect on the parenchyma with that of a root. It may be significant that it was primary roots which were forced to grow through tissues of other plants [72]. According to Van Tieghem and Douliot [1], the lateral root secretes enzymes from a 'poche digestive', a group of cells which is shed after lateral emergence. Since the primary root tip would lack a 'poche digestive', the roots in Pond's experiments may well have grown through tissues by mechanical force alone. There is clearly much scope for work of an experimental nature to resolve the mechanism whereby lateral root primordia penetrate the main root cortex and epidermis.

5. Secondary growth in lateral roots

In plants with perennial root systems, the continuity of vascular tissues between the main root and laterals subsequent to the onset of vascular cambium initiation must be maintained. Esau [4] makes the general statement that in roots with secondary growth, secondary tissues of the main root and laterals differentiate in continuity with each other and that the xylem at the base of the lateral becomes embedded in the xylem of the main root. The continuity between the vascular cambia in the parent root and the base of a lateral root is evident in Figure 3b. Tanaka [82], studying three species of legume roots, showed that there is a correlation between the amount of secondary growth in laterals and the main root. The most complete analysis of secondary growth in laterals has been published for *Glycine max* [29] in which the initiation of vascular cambium in the region of the vascular connection between the lateral and main root was followed. The vascular cambium at the base of a lateral is initiated from derivatives of outer stelar parenchyma cells between the concentric phloem connection and the xylem connection (see Sections 3.6 and 3.7). Divisions in the first cambial initials, which are adjacent to axial phloem connectors, begin just before maturation of late metaxylem connections. The vascular cambium of stelar origin is connected to the pericyclic-derived cambium of the main root for a distance of 700–800 μm vertically above and below the base of the lateral. The former vascular cambium consists of fusiform initials only. The secondary xylem which differentiates from these initials consists of fibres, vessel elements, and abundant axial parenchyma. The vessel elements have simple perforation plates with the terminal end plate joining with a lateral end plate of a vessel element in the main root. The secondary phloem connection occurs later than secondary xylem and in its development crushes some of the axial and transaxial connections. Analysis of vascular cambium initiation is required for more species.

6. Unusual lateral roots

6.1. Proteoid roots

Purnell [83] first described highly-modified tertiary roots, which she called proteoid roots, in several genera of the family Proteaceae. They are present in all 80 species of *Hakea* so far examined [84]. These roots which arise closely spaced in longitudinal rows along lateral roots, all emerge about the same time. In some species, the proteoid roots are produced in such close proximity to each other that they emerge as double files rather than single files due to crowding [85]. The number of files produced reflects the number of protoxylem poles in the parent lateral root, since the proteoid roots arise opposite the xylem poles. Lamont [85] estimates that a 10 mm length of a lateral with seven files of proteoid roots can bear over 1000 individual roots! They elongate for a short time, then form copious, long hairs which are frequently distorted and sometimes forked at the tips. Unlike the normal roots of the plants, the proteoid roots are ephemeral. They form 2 to 6 months after germination and live for an additional 12 to 18 months. Frequently, the proteoid roots form a mat at or near the surface of the soil. Plants with proteoid roots can live in sandy or other extremely nutrient-deficient soils and this root modification is evidently an adaptation for the efficient trapping of nutrients in localized areas. Malajczuk and Bowen [86] found that proteoid roots absorbed ^{32}P supplied as KH_2PO_4 about four times faster than nonproteoid roots. This phosphorus may accumulate in an insoluble form, but not as polyphosphate [87]. The close spacing of proteoid roots may maintain a moist rhizosphere favouring maximal and prolonged nutrient release [87]. In addition, it has been suggested recently [88] that exudates from these roots may solubilize unavailable forms of nutrients. Although Lamont [89] found some evidence for a relationship between proteoid root growth and the concentrations of various ions, later work proved that proteid roots are induced to form by some as yet unidentified soil microbe [86, 90] which is probably located in nutrient-rich

22

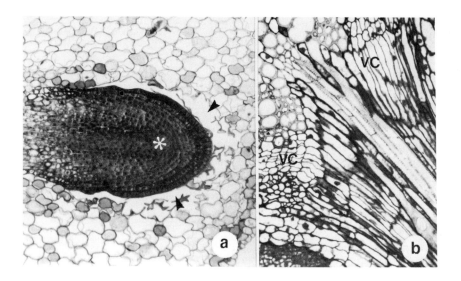

Fig. 3. (a) Root primordium (*) of *Salix* sp. and cavity formation (arrowheads) in adjacent cortex of main root. × 130. (b) *Hieracium florentinum* root showing continuity between vascular cambia (VC) of lateral and main root. × 320.

soil areas. Thus in nature, proteoid roots are concentrated in areas of the soil containing nutrients and micro-organisms; the discrete life-span of the proteoid roots allows the plant flexibility should the location of the nutrient-rich area change, and the advantages of such a root system is consistent with the 'rapid root-turnover' concept [87].

6.2. Dauciform roots

Lamont [84] has described dauciform roots in a sedge, *Cyathochaete avenaceae.* Like proteoid roots, they probably have evolved for efficient nutrient uptake. Dauciform roots are small carrot-shaped roots attached to the parent root by a short stalk-like region. The increased diameter of the swollen region attached to the stalk is due to a striking radial enlargement of the epidermal cells. The roots do not arise in close proximity to each other, nor are they in rows, but they are covered with many, long root hairs. Mature dauciform roots may be invaded by actinomycete hyphae. Somewhat similar swollen lateral roots are found in several genera in the family Cyperaceae [91].

6.3. Hairy roots

Sometimes carrot roots bear numerous very fine laterals known as hairy roots.

They may be induced by nematodes or a virus, or by drought conditions early in the growing season [92].

6.4. Dormant lateral roots

Ginzburg [93] described the formation of dormant lateral roots in the xerophyte, *Zygophyllum dumosum*. These roots cease elongating before emerging from the mother root, but can produce their own laterals and may even proceed to form some secondary vascular tissue. The dormant root and its laterals are visible as a protuberance on the mother root. After short, heavy, seasonal rains, a number of the dormant roots begin to grow. Similar structures have been found in *Tribulus cistoides*, another member of the Zygophyllaceae [94]. These roots undoubtedly represent an adaptation which allows the sudden emergence of a well-elaborated absorptive surface area in the event of moisture availability.

6.5. Short tuberized roots

The formation of unusual roots apparently adapted to periods of drought occurs in the mesophyte, *Sinapis alba* [95, 96]. As Vartanian *et al.* [96] point out *'Even in temperate regions, root systems are often subjected to severe and prolonged periods of drought, followed by re-wetting, particularly in the upper layers of the soil.'* Short, tuberized roots were induced to form on *Sinapis alba* by allowing the soil to dry so that the water potential of the shoot reached -10 bars, a stress which slows the shoot growth rate and causes stomatal closure. The modified roots originate in the pericycle of tap or lateral roots in the upper layers of the soil. Unlike normal roots in this species, they lack hairs and are rich in starch. Their epidermal cells remain turgid whereas epidermal and sometimes even cortical cells of the normal roots to which they are attached desiccate and die. The modified roots can withstand a 3-month drought. When the soil is rewetted, root hairs are rapidly produced on the short tuberized roots and they presumably begin their absorptive activities more rapidly than the normal roots which become suberized during the drought.

6.6. Ectomycorrhizal roots

One of the most striking and important root modifications, usually involving lateral roots, is the establishment of symbiotic relationships between fungi in the Basidiomycotina and Ascomycotina and roots of many gymnosperms and a few angiosperm genera. In this symbiotic association known as ectomycorrhiza, the fungus forms a mantle over the surface of the root (Fig. 4a) with hyphae

24

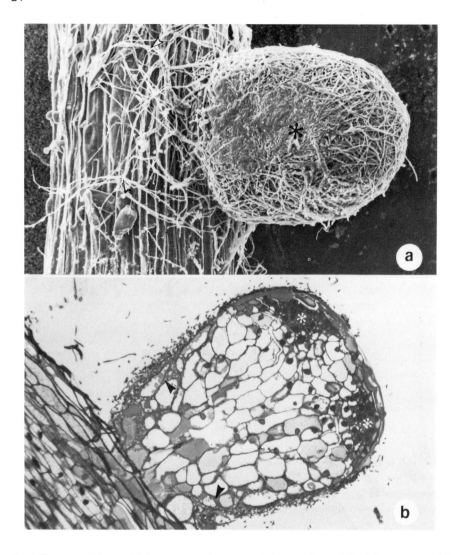

Fig. 4. Ectomycorrhiza established between short roots of *Pinus strobus* and *Pisolithus tinctorius.* (a) SEM of mantle formation (*). Few hyphae (arrowheads) are present on the main root. × 180. (b) Longitudinal section of root in early stage of dichotomy. Two apical meristems (*) are evident. A Hartig net (arrowheads) is forming around cortical cells. × 120.

extending into the soil beyond the root surface, and a Hartig net of intercellular hyphae in the cortex of the root in gymnosperms (Fig. 4b) or between epidermal cells only in many angiosperms.

The gymnosperm *Pinus* probably shows the most pronounced structural changes in roots which become mycorrhizal. The root system of this genus is heterorhizic, i.e. bearing a population of long and short roots [58, 97]. These root types are distinguishable structurally in the primordium stage but subsequent conversion of one type to the other may occur [98, 99]. In *Pinus resinosa*, primordia which will become long roots have a larger diameter (over 50% of that of the mother root) and show more rapid early elongation than primordia which will become short roots [58]. Long root primordia have a rather large, paraboloid-shaped apical meristem while short root primordia have a much smaller hemispherical apical meristem [58]. According to Noelle [98], short roots lack a root cap and secondary growth, have occasional root hairs, branch dichotomously, and are ephemeral. Later studies [58, 100, 101] have shown that a small root cap is present in short roots of *P. resinosa*, *Pinus strobus* and *Fagus sylvatica*.

Short root primordia of *Pinus* are the usual sites of mycorrhizal development [102] although long root primordia may be involved occasionally [58]. The most marked manifestation of mycorrhizal establishment in *Pinus* is dichotomous branching of short root primordia. The origin of the dichotomy involves the suppression of cell division and the onset of vacuolation in central apical meristem cells and the organization of two new meristems lateral to the vacuolated central cells (59, 100 and see Fig. 4b). In *Pinus pinaster*, these events can occur in short roots in the absence of fungal colonization [103], implying that mycosymbionts are not necessary for the expression of these characteristics. In many cases, repeated dichotomies may occur resulting in the formation of coralloid structures.

6.7. Aerial roots

A variety of interesting aerial lateral roots have been discussed by Gill and Tomlinson [104]. These include aerating roots of *Rhizophora mangel*, negatively-geotropic aerating roots of *Avicennia sp.*, dormant lateral roots of *Pandanus*, spinous roots of *Cryosophila warscewiczii* and *Pandanus sp.*, roots which replace the main root in *Rhizophora* and *Pandanus*, and root thorns of *Pandanus spp.*, *Socratea* and *Irartea*.

7. Acknowledgments

We thank SEM Incorporated Ltd., Chicago, Illinois, U.S.A. for permission to reproduce Fig. 4a which was published in Piché *et al.* Scanning Electron Microscopy 1983/III, pp. 1467–1474, Carol Pratt for typing the manuscript and Melanie Chapple for proofreading.

26

8. References

1. Van Tieghem, Ph. & Douliot, H. (1888) Recherches comparatives sur l'origine des membres endogènes dans les plantes vasculaires. Annales des Sciences Naturelles Botanique, 8, 1–660.
2. Guttenberg, H. Von (1968) Der Primäre Bau der Angiospermenwurzel. In Handbuch der Pflanzenanatomie (ed K. Linsbauer), pp. 1–472. Gebrüder Borntraeger, Berlin, Stuttgart.
3. McCully, M.E. (1975) The development of lateral roots. In The Development and Function of Roots (eds J.G. Torrey & D.T. Clarkson), pp. 105–124. Academic Press, London, New York, San Francisco.
4. Esau K. (1965) Plant Anatomy. 2nd ed. John Wiley & Sons, New York, London, Sydney.
5. Janczewski, E. de (1874) Recherches sur le développement des radicelles dans les Phanérogames. Annales des Sciences Naturelles, 20, 208–233.
6. Esau, K. (1940) Developmental anatomy of the fleshy storage organ of Daucus carota. Hilgardia, 13, 175–226.
7. Popham, R.A. (1955) Levels of tissue differentiation in primary roots of Pisum sativum. American Journal of Botany, 42, 529–540.
8. Blakely, L.M., Durham, M., Evans, T.A. & Blakely, R.M. (1982) Experimental studies on lateral root formation in radish seedling roots. 1. General methods, developmental stages, and spontaneous formation of laterals. Botanical Gazette, 143, 341–352.
9. Karas, I. & McCully, M.E. (1973) Further studies of the histology of lateral root development in Zea mays. Protoplasma, 77, 243–269.
10. Kadej, F. & Rodkiewicz, B. (1974) Initial stages of the lateral root development in Raphanus sativus. Electron microscope observations. In Structure and Function of Primary Root Tissues (ed J. Kolek), pp. 113–120. Veda Publishing House, Slovak Academy of Sciences, Bratislava.
11. Blakely, L.M. & Evans, T.A. (1979) Cell dynamics studies on the pericycle of radish seedling roots. Plant Science Letters, 14, 79–83.
12. Bonnett, H.T. & Torrey, J.G. (1966) Comparative anatomy of endogenous bud and lateral root formation in Convolvulus arvensis roots cultured in vitro. American Journal of Botany, 53, 496–507.
13. Bell, J.K. & McCully, M.E. (1970) A histological study of lateral root initiation and development in Zea mays. Protoplasma, 70, 179–205.
14. Charlton, W.A. (1975) Distribution of lateral roots and pattern of lateral initiation in Pontederia cordata L. Botanical Gazette, 136, 225–235.
15. Seago, J.L. (1973) Developmental anatomy in roots of Ipomoea purpurea. II. Initiation and development of secondary roots. American Journal of Botany, 60, 607–618.
16. Bruni, A., Dall'olio, G. & Mares, D. (1973) Anatomia della radice di Calystegia soldanella R. Br. II. Organizzazione e meccanismo di emergenza dell'apice della radice lateral. Annali Dell'università di Ferrara (Nuova Serie) Sezione IV Botanica, 4, 109–125.
17. Bayer, D.E., Foy, C.L., Mallory, T.G. & Cutter, E.G. (1967) Morphological and histological effects of trifluralin on root development. American Journal of Botany, 54, 945–952.
18. Foard, D.E. & Haber, A.H. (1961) Anatomic studies of gamma-irradiated wheat growing without cell division. American Journal of Botany, 48, 438–446.
19. Foard, D.E., Haber, A.H. & Fishman, T.N. (1965) Initiation of lateral root primordia without completion of mitosis and without cytokinesis in uniseriate pericycle. American Journal of Botany, 52, 580–590.
20. Sutherland, J. & McCully, M.E. (1965) A note on the structural changes in the walls of pericycle cells initiating lateral root meristems in Zea mays. Canadian Journal of Botany, 54, 2083–2087.
21. Crooks, D.M. (1933) Histological and regenerative studies on the flax seedling. Botanical Gazette, 95, 209–239.
22. Byrne, J.M. (1973) The root apex of Malva sylvestris. III. Lateral root development and the quiescent center. American Journal of Botany, 60, 657–662.

23. Thibault, M. (1946) Contribution a l'étude des radicelles du carotte. Revue Générale de Botanique, 53, 434–460.
24. Mallory, T.E., Chiang, S.H., Cutter, E.G. & Gifford, E.M. Jr. (1970) Sequence and pattern of lateral root formation in five selected species. American Journal of Botany, 57, 800–809.
25. Clowes, F.A.L. (1978) Chimeras and the origin of lateral root primordia in Zea mays. Annals of Botany, 42, 801–807.
26. Nishimura, S. & Maeda, E. (1982) Cytological studies on differentiation and dedifferentiation in pericycle cells of excised rice roots. Japanese Journal of Crop Science, 51, 553–560.
27. Bonnett, H.T. Jr. (1969) Cortical cell death during lateral root formation. The Journal of Cell Biology, 40, 144–159.
28. Berthon, R. (1943) Sur l'origine des radicelles chez les Angiospermes. Académie des Sciences, Comptes Rendus, Paris, 216, 308–309.
29. Byrne, J.M., Pesacreta, T.C. & Fox, J.A. (1977) Development and structure of the vascular connection between the primary and secondary root of Glycine max (L.) Merr. American Journal of Botany, 64, 946–959.
30. Byrne, J.M., Byrne, J.M. & Emmitt, D.P. (1982) Development and structure of the vascular connection between the primary and lateral root of Lycopersicon esculentum. American Journal of Botany, 69, 287–297.
31. Ashford, A.E. & McCully, M.E. (1970) Localization of naphthol AS-B1 phosphatase activity in lateral and main root meristems of pea and corn. Protoplasma, 70, 441–456.
32. Ashford, A.E. & McCully, M.E. (1973) Histochemical localization of β-Glucosidases in roots of Zea mays. III. β-Glucosidase activity in the meristems of lateral roots. Protoplasma, 77, 411–425.
33. Tschermak-Woess, E. & Doležal, R. (1953) Durch Seitenwurzelbildung induzierte und spontane Mitosen in den Dauergeweben der Wurzel. Östreich Botanische Zeitschrift, 100, 358–402.
34. Webster, P.L. & MacLeod, R.D. (1980) Characteristics of root apical meristem cell population kinetics: a review of analyses and concepts. Environmental and Experimental Botany, 20, 335–358.
35. Davidson, D. (1961) Mechanisms of reorganization and cell repopulation in meristems in roots of Vicia faba following irradiation and colchicine. Chromosoma, 12, 484–504.
36. MacLeod, R.D. & Thompson, A. (1981) Cell proliferation during the development of lateral root primordia. In Structure and Function of Plant Roots (eds R. Brouwer, O. Gašparíková, J. Kolek & B.C. Loughman), pp. 35–42. Martinus Nijhoff/Dr. W. Junk, Dordrecht, Boston, London.
37. Davidson, D. & MacLeod, R.D. (1968) Heterogeneity in cell behaviour in primordia of Vicia faba. Chromosoma (Berlin), 25, 470–474.
38. Friedberg, S.H. & Davidson, D. (1971) Cell population studies in developing root primordia. Annals of Botany, 35, 523–533.
39. MacLeod, R.D. (1972) Lateral root formation in Vicia faba L. 1. The development of large primordia. Chromosoma (Berlin), 39, 341–350.
40. MacLeod, R.D. & McLachlan, S.M. (1975) Tritiated-thymidine labelled nuclei in primordia and newly-emerged lateral roots of Vicia faba L. Annals of Botany, 39, 535–545.
41. MacLeod, R.D. & Davidson, D. (1968) Changes in mitotic indices in roots of Vicia faba. III. Effects of colchicine on cell cycle times. Experimental Cell Research, 52, 541–554.
42. MacLeod, R.D. (1971) The response of apical meristems of primary roots of Vicia faba L. to colchicine treatments. Chromosoma (Berlin), 35, 217–232.
43. MacLeod, R.D. (1973) The emergence and early growth of the lateral root in Vicia faba L. Annals of Botany, 37, 69–75.
44. MacLeod, R.D. (1976) Growth of lateral root primordia in Vicia faba, L. The New Phytologist, 76, 143–151.
45. MacLeod, R.D. & MacLachlan, S.M. (1974) Lateral root elongation in Vicia faba L.: changes in labelling index in the cap, epidermis, cortex and stele. The New Phytologist, 73, 719–729.

28

46. MacLeod, R.D. & McLachlan, S.M. (1974) Lateral root elongation in *Vicia faba* L. Changes in mitotic index in the cap, epidermis, cortex, and stele. Protoplasma, 81, 111–124.
47. MacLeod, R.D. (1977) Proliferating and quiescent cells in the apical meristem of elongating lateral roots of *Vicia faba* L. Annals of Botany, 41, 321–329.
48. MacLeod, R.D. & McLachlan, S.M. (1975) Lateral root elongation in *Vicia faba* L.: the planes of cell division in the epidermis, cortex and stele. The New Phytologist, 74, 465–472.
49. Clowes, F.A.L. (1975) The quiescent centre. In The Development and Function of Roots (eds J.G. Torrey & D.T. Clarkson), pp. 3–19. Academic Press, London, New York, San Francisco.
50. Clowes, F.A.L. (1958) Development of quiescent centres in root meristems. The New Phytologist, 57, 85–88.
51. MacLeod, R.D. & McLachlan, S.M. (1974) The development of a quiescent centre in lateral roots of *Vicia faba* L. Annals of Botany, 38, 535–544.
52. Clowes, F.A.L. (1978) Origin of the quiescent centre in *Zea mays*. The New Phytologist, 80, 409–419.
53. MacLeod, R.D. (1976) Cap formation during the elongation of lateral roots of *Vicia faba* L. Annals of Botany, 40, 877–885.
54. Clarke, K.J., McCully, M.E. & Miki, N.K. (1979) A developmental study of the epidermis of young roots of *Zea mays* L. Protoplasma, 98, 283–309.
55. Clowes, F.A.L. (1961) *Apical Meristems*. Blackwell Scientific Publications, Oxford.
56. Dumbroff, E.B. & Peirson, D.R. (1971) Probable sites for passive movement of ions across the endodermis. Canadian Journal of Botany, 49, 35–38.
57. Weerdenburg, C.A. & Peterson, C.A. (1984) Effect of secondary growth on the conformation and permeability of the endodermis in broad bean (*Vicia faba*), sunflower (*Helianthus annuus*) and garden balsam (*Impatiens balsamina*). Canadian Journal of Botany, 62, 907–910.
58. Wilcox, H.E. (1968) Morphological studies of the roots of red pine, *Pinus resinosa*. II. Fungal colonization of roots and the development of mycorrhizae. American Journal of Botany, 55, 686–700.
59. Peterson, C.A. & Edgington, L.V. (1975) Factors influencing apoplastic transport in plants. In Systemic Fungicides (eds H. Lyr & C. Polter), pp. 287–299. Akademie-Verlag, Berlin.
60. Peterson, C.A., Emanuel, M.E. & Humphreys, G.B. (1981) Pathway of movement of apoplastic fluorescent dye tracers through the endodermis at the site of secondary root formation in corn (*Zea mays*) and broad bean (*Vicia faba*). Canadian Journal of Botany, 59, 618–625.
61. Sieghardt, H. von (1981) Ein histochemischer Nachweis zur Bleiverteilung in juvenilen Wurzeln von *Zea mays* L.: Ein lichtmikroskopische Studie. Mikroskopie (Wein), 38, 193–199.
62. Miller, R.H. (1959) Morphology of *Humulus lupulus*. II. Secondary growth in the root and seedling vascularization. American Journal of Botany, 46, 269–277.
63. Peterson, C.A. and Lefcourt, B. unpublished results.
64. Rywosch, S. (1909) Untersuchungen über die Entwicklungsgeschichte der Seitenwurzeln der Monocotylen. Zeitschrift fur Botaniche, 1, 253–283.
65. Fahn, A. (1967) Plant Anatomy. Pergamon Press, Toronto.
66. Fourcroy, M. (1942) Perturbations anatomiques intéressant le faisceau de la racine au voisinage des radicelles. Annales des Sciences Naturelles Botanique, 3, 177–198.
67. Torrey, J.G. (1951) Cambial formation in isolated pea roots following decapitation. American Journal of Botany, 38, 596–604.
68. Gunning, B.E.S. (1977) Transfer cells and their roles in transport of solutes in plants. Scientific Progress Oxford, 64, 539–568.
69. Peterson, R.L. (1975) The initiation and development of root buds. In The Development and Function of Roots (eds J.G. Torrey & D.T. Clarkson), pp. 125–161. Academic Press, London, New York, San Francisco.
70. Letvenuk, L.J. & Peterson, R.L. (1976) Occurrence of transfer cells in vascular parenchyma of *Hieracium florentinum* roots. Canadian Journal of Botany, 54, 1458–1471.

71. Newcomb, W. & Peterson, R.L. (1979) The occurrence and ontogeny of transfer cells associated with lateral roots and root nodules in Leguminosae. Canadian Journal of Botany, 57, 2583–2602.

72. Pond, R.H. (1907) Emergence of lateral roots. Botanical Gazette, 46, 410–421.

73. Vonhöne, H. (1880) Ueber das Hervorbrechen endogener Organe aus dem Mutterorgane. Flora, 63, 268–274.

74. Sutcliffe, J.F. & Sexton, R. (1968) β-Glycerophosphatase and lateral root development. Nature, 217, 1285.

75. Knobloch, I.W. (1954) Developmental anatomy of chicory – the root. Phytomorphology, 4, 47–54.

76. Francis, D. & MacLeod, R.D. (1977) Some cytological changes accompanying the regeneration of meristematic activity at the apex of decapitated roots of Vicia faba L. Annals of Botany, 41, 1149–1162.

77. MacLeod, R.D. & Francis, D. (1976) Cortical cell breakdown and lateral root primordium development in Vicia faba L. Journal of Experimental Botany, 27, 922–932.

78. Weerdenburg, C.A. & Peterson, C.A. (1983) Structural changes in phi thickenings during primary and secondary growth in roots. 1. Apple (Pyrus malus) Rosaceae. Canadian Journal of Botany, 61, 2570–2576.

79. Ferguson, I.B. & Clarkson, D.T. (1975) Ion transport and endodermal suberization in the roots of Zea mays. The New Phytologist, 75, 69–79.

80. Wilson, C.A. & Peterson, C.A. (1983) Chemical composition of the epidermal, hypodermal and intervening cortical cell walls of various plant roots. Annals of Botany, 51, 759–769.

81. Kroemer, K. (1903) Wurzelhaut, Hypodermis und Endodermis der Angiospermenwurzeln. Bibliotheca Botanica, 12, 1–159.

82. Tanaka, T.N. (1971) Studies on the growth of root systems in leguminous crop plants. IX. On the thickening growth of lateral roots. Proceedings of the Crop Science Society of Japan, 40, 306–310.

83. Purnell, H.M. (1960) Studies of the family Proteaceae 1. Anatomy and morphology of the roots of some Victorian species. Australian Journal of Botany, 8, 38–50.

84. Lamont, B. (1974) The biology of dauciform roots in the sedge Cyathochaete avenacea. The New Phytologist, 73, 985–996.

85. Lamont, B. (1972) The morphology and anatomy of proteoid roots in the genus Hakea. Australian Journal of Botany, 20, 155–174.

86. Malajczuk, N. & Bowen, G.D. (1974) Proteoid roots are microbially induced. Nature, 251, 316–317.

87. Lamont, B. (1982) Mechanisms for enhancing nutrient uptake in plants, with particular reference to mediterranean South Africa and Western Australia. The Botanical Review, 48, 597–689.

88. Lamont, B.B. (1981) Specialized roots of non-symbiotic origin in heathlands. In Ecosystems of the World 9b. Heathlands and Related Shrublands. (ed. R.L. Specht), pp. 183–195. Elsevier Scientific Publishing Company, Amsterdam, Oxford, New York.

89. Lamont, B. (1972) The effect of soil nutrients on the production of proteoid roots by Hakea species. Australian Journal of Botany, 20, 27–40.

90. Lamont, B. & McComb, A.J. (1974) Soil micro-organisms and the formation of proteoid roots. Australian Journal of Botany, 22, 681–688.

91. Davies, J., Briarty, L.G. & Rieley, J.O. (1973) Observations on the swollen lateral roots of the Cyperaceae. The New Phytologist, 72, 167–174.

92. Orzolek, M.D. & Carroll, R.B. (1976) Method for evaluating excessive secondary root development in carrots. HortScience, 11, 479–480.

93. Ginzburg, C. (1966) Formation des racines latérales dormantes chez Zygophyllum dumosum. Académie des Sciences, Comptes Rendus, Paris, 263, 909–912.

94. Allen, V.N. & Allen, E. (1949) The anatomy of the nodular growths on the roots of *Tribulus cistoides* L. Proceedings of the Soil Science Society of America, 14, 179–183.

95. Vartanian, N. (1981) Some aspects of structural and functional modifications induced by drought in root systems. Plant and Soil, 63, 83–92.

96. Vartanian, N., Wertheimer, D.S. & Couderc, H. (1983) Scanning electron microscopic aspects of short tuberized roots, with special reference to cell rhizodermis evolution under drought and rehydration. Plant, Cell and Environment, 6, 39–46.

97. Wilcox, H.E. (1968) Morphological studies of the roots of red pine, *Pinus resinosa*. I. Growth characteristics and patterns of branching. American Journal of Botany, 55, 247–254.

98. Noelle, W. (1910) Studien zur vergleichenden Anatomie und Morphologie der Koniferen-wurzeln mit Rucksicht auf die Systematik. Botanische Zeitung, 68, 169–266.

99. Slankis, V. (1958) Mycorrhiza of forest trees. In Proceedings of First North American Forest Soils Conference. Vol. 1, pp. 130–137. Department of Forestry, Michigan State University, East Lansing.

100. Piche, Y., Fortin, J.A., Peterson, R.L. & Posluszny, U. (1982) Ontogeny of dichotomizing apices in mycorrhizal short roots of *Pinus strobus*. Canadian Journal of Botany, 60, 1523–1528.

101. Clowes, F.A.L. (1954) The root cap of ectotrophic mycorrhizas. The New Phytologist, 53, 525–529.

102. Piche, Y., Peterson, R.L. & Ackerley, C.A. (1983) Early development of ectomycorrhizal short roots of pine. Scanning Electron Microscopy, 3, 1467–1474.

103. Faye, M., Rancillac, M. & David, A. (1981) Determination of the mycorrhizogenic root formation in *Pinus pinaster* Sol. The New Phytologist, 87, 557–565.

104. Gill, A.M. & Tomlinson, P.B. (1975) Aerial roots: an array of forms and functions. In The Development and Function of Roots (eds J.G. Torrey & D.T. Clarkson), pp. 237–260. Academic Press, London, New York, San Francisco.

2. Endogenous and exogenous influences on the regulation of lateral root formation

JOHN G. TORREY

Harvard Forest, Harvard University, Petersham, Massachusetts 01366, USA

1. Studies on seedling roots, cultured excised roots and root parts

1.1. Introduction

According to Sutton and Tinus [1] who adopted the definition by Russell [2], a lateral root is a side root 'that arises by cell division in the pericycle of the parent root. The resulting dome of tissue penetrates the cortex. When the lateral root emerges from the parent root, its apical meristem is comparable with that of the apex of the parent root.'

The initiation of a lateral root typically involves the activation of a series of tangential cell divisions in the single-cell-layered root pericycle. In many roots this takes place opposite the protoxylem of a primary xylem arm of the vascular cylinder. Thereafter, the endodermal cells outside the pericycle may divide, either becoming incorporated into the forming meristem or accommodating its enlargement. With continued periclinal divisions and some anticlinal divisions a root meristem forms, pushing outward on the cortical cells of the root and causing their collapse either by mechanical means or by enzymatic digestion of the cell walls or both [3, 4]. Not until after the developing lateral root meristem has cleared the root cortex in its outward push, can one demonstrate the existence of the elements of the root meristem, i.e. a root cap with root cap initials and a quiescent centre that is demonstrable by the lack of incorporation of DNA precursors [5, 6].

Elsewhere in this volume are described the structural aspects of root initiation and organization, especially of lateral roots (Peterson and Peterson, chapter 1). General considerations of lateral root initiation and development have been published earlier [4, 7, 8, 9, 10, 11, 12].

It is the purpose of this chapter to review the evidence bearing on the physiological, biochemical and genetic factors, primarily endogenous and only secondarily exogenous, which influence the processes leading to lateral root formation. Of primary interest are questions centering on the nature of the stimuli which initiate the first cell changes that signify the initiation process and how these are focussed on the site of initiation both in terms of the endogenous locus in the pericycle opposite a protoxylem point and the determination along the longitudinal axis of the root.

According to the above description we are seeking in most situations factors which elicit cell division in a specific cell in an oriented (tangential) plane in special relation to other surrounding tissues, some of which are functional in long-distance transport within the plant. Even this specification of the process has been questioned. Foard *et al.* [13] described a special case of lateral root initiation not involving cytokinesis. This evidence is discussed below.

Almost without exception in normal development, however, lateral root initiation involves the activation of pericycle cells into cell division. The considerable literature on endogenous and exogenous factors controlling cell division in plant tissues elaborated from plant tissue culture research [14, 15] suggests that any of a number of plant hormones might be involved or serve as the limiting or triggering components. These hormones include indole-3-acetic acid (IAA) or related auxins, cytokinins (CK) such as zeatin or isopentenyl adenine, possibly interacting with abscisic acid (ABA) or ethylene, and possibly further interacting with gibberellins (GA). Vitamins, growth factors and complex organic metabolites may also play a fundamental role in limiting cell division activity. In roots, those shown to be especially important include the B vitamins: thiamine, pyridoxine and nicotinic acid. In addition, myoinositol may play a special role in some cases. Other factors may stimulate cell division, perhaps in their role interacting with the compounds already listed-tryptophan related to auxin synthesis, adenine related to cytokinin synthesis, etc. Finally, inorganic nutrients may influence cell division, especially serving under some circumstances to limit the events of cell division. Here, one places special emphasis on the calcium ion, the borate ion and iron in some soluble form probably because of their critical roles in cell wall formation, membrane function and respiratory metabolism, respectively.

Each of these factors will be examined, using experimental evidence from intact and excised roots grown in culture.

1.2. Hormonal interactions in the root apex

Before discussing the factors eliciting lateral root initiation, it is useful to digress briefly to an examination of the evidence concerning factors acting in the root apex which lead to the formation of the very specialized organization by which it is characterized. In the current view of the root apex of the typical dicotyledonous plant [9], several tissue systems can be described which comprise the functional meristem. The terminal conical-shaped root cap is derived from a highly active meristematic layer, the root cap initials, which delimit the proximal face of the root cap. Proximal to the root cap initials is a hemi-sphere of cells which has relatively low cellular activities and hence is designated the quiescent centre. On the proximal surface of the quiescent centre (Q.C.) lies a population of cells active in division which may extend a few to many cell layers more proximal. This meristematic region is described as the proximal meristem and gives rise to the cylinder of cells which forms the major tissues of the root. Still further proximal progressive cytodifferentiation of the different distinctive cell types occurs which characterizes the mature tissues of the root. These relationships are indicated in Table 1.

For our purposes concerned with interactive chemical and physical forces influencing the organization of the root apex and its homeostatic stability, we can focus on four structures: the root cap, the root cap initials, the Q.C. and the proximal meristem. Lateral roots, like primary roots and radicle apices are similar in structure and function once established. What do we know about the factors acting within such root apices?

The effects of root apex removal on subjacent lateral root initiation has been studied in some detail in the past as an approach to understanding root apical functions. In experiments on seedling roots, lateral roots tend to form on the distal most regions of the decapitated root. This response has been interpreted as either due to the accumulation of a lateral-root-stimulating substance at the cut end or the removal of lateral-root-inhibiting substances originating from the tip. Reference to Table 1 shows some of the hormonal interactions which are believed to occur in the elongating root. Some evidence exists in support of each of these ideas although not all for the same seedling root.

Materials that could be expected to accumulate at the distal end of a decapitated seedling root such as *Zea* or *Pisum* include sucrose transported from the photosynthetic shoot to the growing root tip, vitamins including thiamin, pyridox-

Table 1. Major tissue relationships in roots and some of the hormones involved in root cellular activities (see text for interpretation).

Tissue systems[1]	Tissues	Major cellular activity	Hormones involved
	Lateral Root initiation in pericycle	← Cell division	Auxins Cytokinins
Primary tissues	Epidermis Cortex Endodermis Vascular cylinder Pericycle	Cytodifferentiation	
			Direction of flow Auxins[2]
Primary meristematic tissues	Protoderm Ground meristem Procambium		Sugars Vitamins Cytokinins ↑ (via phloem) (via xylem) ↓
		Cytodifferentiation ↑ Cell enlargement ↑	blocked by decapitation
	Proximal meristem	Cell division	Auxins Cytokinins
	Quiescent centre		
Root apex	Root cap initials Root cap	Cell division ↓ Cell enlargement ↓ Cytodifferentiation	Auxins (ethylene) ABA or ABA-like compounds

[1] Terminology modified from Esau [16]
[2] Polar transport (acropetal)

ine, and nicotinic acid transported via the phloem from the shoot, and hormones including auxin, cytokinins and possibly giberellins transported via the phloem from the shoot, cotyledons or older tissues of the root. Of these, acropetal auxin transport has been demonstrated to be predominantly polar, i.e. toward the root apex. Cytokinins move acropetally via the phloem but can be transported basipetally via the xylem as well. Taken altogether, these substances in raised concentrations could be expected to elicit the initiation of new lateral root primordia at the site of accumulation. In cultured root segments provided with sugar and vitamins, lateral root initiation is still markedly polar in *Pisum* [17] and *Convolvulus* [18] and the evidence points to raised levels of auxin as causative in eliciting lateral root primordia. Somewhat the same effect can be achieved by immersing seedling roots in increased levels of auxin in the exogenous solution. Lateral root primordia are induced in high frequency, even close to the root apex.

Lateral root formation can be prevented in cultured roots by witholding sugar, vitamins and essential cofactors for cell division but IAA appears to be the essential hormonal factor for eliciting the response and its polar movement within the root helps explain the localization of root initiation at specific sites in most cases.

Recent studies have used successfully the methods of excised root culture and microdissection in attempts to unravel the complex relationships between and among these tissue systems. Excision of small tissue pieces from the intact system allows one to isolate the factors limiting or controlling the events leading to lateral root initiation. One must provide to the isolated tissue the nutrients and perhaps hormones which one tissue usually provides the other in order to function normally. The method has its limitations and conclusions must always be reached cautiously and related back to evidence on the intact system for confirmation. Yet there may be clues from this approach. Cultured tissue pieces must always be provided with essential mineral elements and a carbon and energy source as well as favourable pH, temperature and physical environment. Other factors may be limiting and these can be provided to allow the tissue to behave as near normal as possible.

Feldman [19, 20, 21, 22] has been especially diligent in his efforts to unravel the intricate hormonal interactions between tissue systems within the root apex. Because of its favourable structure and ease of manipulation, the root apex of *Zea mays* has been selected for these experiments. The results bear not only on the capacity of the primary root apex to regenerate but shed light also on the subtleties between direct meristem regeneration and the closely related alternative of initiating lateral roots. From his experiments on cultured explants of corn root apical pieces, which include root cap, Q.C., proximal meristem (P.M.) and the Q.C. and proximal meristem cultured as an intact piece, he concluded that neither the P.M. nor the Q.C. could synthesize auxin and cytokinin when cultured alone but when cultured as a unit they regenerated a whole new root without requiring either hormone. The indirect evidence suggested that only when Q.C. and proximal meristem were integrally connected did they function in what was their normal role of synthesizing hormones necessary for regenerating a whole new apex. If the whole P.M. and Q.C. were excised together, the distal end of the seedling root attached to the seedling almost invariably initiated lateral roots at right angles to the main root axis. In this case, hormones essential for lateral root initiation as well as other factors were transported to the distal end of the decapitated root and induced new meristems as lateral roots.

The dependence of corn root tissues, either isolated Q.C.'s or isolated P.M.'s, on exogenous supplies of auxin (IAA) and cytokinin (CK) for regenerative capacity for the reorganization of a whole new meristem is resoundingly demonstrated by these experiments. It is not too imaginative a step to conclude that lateral root initiation also depends on endogenous supplies of auxin and cytokinin in the presence of permissive supplies of sucrose, vitamins, essential inorganic elements and perhaps other probably organic components not yet fully defined. How these all work together and in what sequence remain to be fathomed.

1.3. Defining the site of lateral root initiation

Lateral roots typically originate endogenously in the root pericycle, a uniseriate cylindrical layer of cells early blocked out proximal to the root apical meristem and forming the outermost layer of the primary vascular cylinder. In radish, this cell layer represents 20% of the cell population of the seedling root [23]. When stimulated into division, this tissue gives rise to diploid cells even though spectrophotometric measurements [24] show that the nuclear DNA levels equal 4c. According to Blakely and Evans [24] the pericycle cells of radish possess nuclei arrested at the G2 stage in the nuclear cycle – i.e. having completed DNA synthesis and poised to undergo mitosis when stimulated.

In the transectional view of a root, the cells comprising the circle of pericycle cells appear to be alike. Yet their position differs in relation to the existing or differentiating patterns of the primary vascular tissues internal to the pericycle. Only certain cells aligned radially in special configurations undergo the initial divisions leading to lateral roots. There are two general patterns. In roots with simple vascular patterns, i.e. with few primary xylem arms such as triarch, tetrarch or pentarch, lateral root initiation typically occurs in pericycle cells located immediately outside the protoxylem point. Here, one to several cells in the pericycle undergo tangential divisions and the lateral root develops outward from this point. In polyarchous roots, characteristic of monocotyledenous plants, where numerous primary xylem arms develop, lateral root initiation typically occurs in pericycle cells immediately outside a primary phloem element, usually a sieve-tube element, which lies between two adjacent protoxylem points, (e.g. wheat, Foard *et al.* [13]). In diarch roots, laterals may occur immediately outside the protoxylem points, as in radish, or on either side of a protoxylem point at

45° C, outside primary phloem. The precision of the radial origin of lateral roots formed sequentially and acropetally along the length of the root results in ranks of lateral roots along the length of a seedling root axis reflecting the internal primary vascular tissue pattern.

Determination of lateral root intitiation along the length of the root is less precise than radial orientation but nevertheless reflects endogenous control mechanisms. Typically, lateral root initiation occurs at a given distance proximal to the root apex, depending on the growth rate of the primary root, and new initiations occur in acropetal sequence as the seedling root elongates. In the longitudinal direction new lateral roots tend not to be initiated between already formed lateral roots, although new lateral roots do form on roots which have undergone secondary thickening – that is, roots with diameter increases associated with vascular cambium activation. Statistical studies by Riopel [25, 26] of lateral root position in polyarchous roots demonstrated that the longitudinal occurrence of lateral roots was not random, but showed a disperse distribution with the closest previously formed lateral tending to displace the newly initiating lateral root. He visualized some sort of interaction, more reasonably attributed to competition for metabolites or factors than to the activity of an inhibitor formed by the existing primordium. Charlton [27] also found evidence of an interaction between two successive laterals in the polyarchous roots of *Pontederia,* which resulted in co-ordinated initiation between ranks of primordia. He also speculated about the possible basis for such interactions in terms of each primordium acting as a sink for an essential factor. In any case, although it is less obvious, lateral root formation along the length of the root axis is not only acropetal (which is easily observed) but also is co-ordinated in some special way with existing meristems influencing newly initiating ones. Even in mature trees, the regularity and precision of lateral root formation in patterned arrangements are still discernible [28].

Several other facts are clear concerning the patterning of lateral root initiation. Although the observations are derived from special or exceptional cases, their occurrence places further stringencies concerning the interpretation of the endogenous forces acting on lateral root primordia formation. Initiation of lateral root primordia does not depend upon stimulated cell division or cytokinesis *per se.* Foard *et al.* [13] showed that treating seedlings of wheat with low concentrations of colchicine, which served to block cytokinesis, did not interfere with lateral root initiation. Radially enlarged cells in the pericycle causing radial protuberances of adjacent endodermis formed what these authors called 'primordiamorphs'. DNA synthesis and chromosome replication occurred in these pericycle cells in response to endogenous factors not interfered with by colchicine treatment. Upon removal of colchicine, the primordia proceeded to develop into polyploid lateral roots. Thus, the stimuli for lateral root initiation act on cells in someway predetermined to respond but independent directly on their capacity to undergo cell division. A similar behaviour was reported in colchicine-treated roots of *Pontederia* by Charlton [29].

38

The fact that colchicine-treatment does not interfere with the normal sequence of lateral root initiation, only the subsequent development of characteristic root meristem organization, raises the interesting question about the capacity of already initiated lateral roots to act as either sinks or sources of stimuli in lateral roots later initiated. In 'primordiamorphs', root cap differentiation and other characteristic features of a root meristem do not develop – i.e., the characteristic multicellular organization of the root meristem does not form. Yet these arrested primordia still appear to influence the ontogenetic sequence. It seems unlikely that 'primordiamorphs' are special sources of inhibitors or act as active sinks for metabolic factors required for subsequent lateral root initiation as postulated by Riopel [25] and Charlton [27].

Related to these uncertainties are those concerned with the more general question, how organized must the root apex be to serve as a source of hormones or function as a sink active enough to draw upon resources sufficiently to modify responses of surrounding cells. In analyses of root apex size and root elongation rates in relation to lateral root initiation, Kawata and Ishihara [30] (cited by Barlow and Rathfelder [31]) pointed out the positive correlations which relate to the role of root apex and source and/or sink. Rowntree and Morris [32] illustrate lateral root primordia as sinks in their study of the accumulation of radioactively labelled auxin provided to the shoot of pea seedlings in the newly initiated lateral root primordia.

In their study of lateral root initiation in five species of plants including one fern and four angiosperms Mallory et al. [33] found that lateral roots originated very close to the root apex where the primary meristematic tissues were still dividing and in the process of undergoing early cellular differentiation. All showed lateral root primordia formation closer to the root apex than the occurrence of the first mature protoxylem elements. Evidence from Clowes [5] and from Byrne [6] makes it pretty clear that in lateral root initiation the development of a Q.C. does not occur until the lateral root has already emerged through the root cap and cell elongation in the lateral root is well advanced, (e.g. longer than 5 mm in length in laterals of Malva according to Byrne). Presumably, the proximal meristem has already developed in the lateral root to allow for cell division and elongation essential to its emergence through the root cortex. Although the evidence from Feldman's work cited above strongly supports the view that root apices complete with Q.C. and P.M. are sites of hormone synthesis and also serve as sinks for hormone and metabolite use, we know little or nothing about the pre-emergent activities or capacities of a lateral root primordium.

1.4. Effects of exogenous factors on lateral root initiation

1.4.1. Auxins. Let us now turn to the evidence of experiments which show the influence of exogenous factors on lateral root formation so that we may draw on this information in making inferences about endogenous influences or determinants in the process. Many of these and similar experiments have been considered earlier by reviewers cited in the introduction and need not be repeated here. In actual fact, little new information of the nature of exogenous factors, especially hormonal or catalytic, has been added to our knowledge in the nearly ten years since my last review of the subject [8]. Almost without exception, seedling roots, excised roots in culture and roots of mature plants respond to the application of exogenous auxin by forming larger numbers of lateral root primordia (Fig. 1, Section 2.2.). Chadwick and Burg [34] argued that exogenous auxin effects on roots were mediated by or involved ethylene.

In their development of an experimental system for the study of lateral root formation in cultured roots of *Haplopappus*, Blakely *et al.* [35] routinely incorporated $0.1 \, mg \, l^{-1}$ α-naphthaleneacetic acid (NAA) in the culture medium. This auxin level produced the maximum number of lateral roots per segment length and allowed repeated (clonal) subculture (IAA, which is more subject to oxidation by plant tissues, was only one tenth as effective as NAA in inducing lateral root formation). Lateral root formation by *Haplopappus* root segments without added auxin (presumably reflecting endogenous auxin levels) was only approximately two laterals per cm of cultured segment as opposed to up to 60 laterals per cm in segments with optimum auxin levels). In studies by Blakeley *et al.* [23] of lateral root initiation in seedling roots or root segments of radish (*Raphanus*), lateral root formation by exogenous NAA was markedly increased over that induced by natural stimuli (endogenous auxin). Webster and Radin [36] also reported the stimulation of lateral root formation when the auxins IAA or NAA were provided to excised roots of radish applied via basal feeding in sterile culture.

Wightman and Thimann [37] and Wightman *et al.* [38] re-examined lateral root initiation in seedling roots of the garden pea, using young, developing intact seedlings or seedlings with various organs excised. As a standard test for exogenous stimuli they exposed 3-d-old seedling roots with the root tip excised to a variety of growth factors over a range of concentrations. In general, they found that auxins caused marked stimulation of lateral root initiation. At $10^{-4} M$, the decreasing order of effectiveness in lateral root primordium induction was as follows: 4-chloro-3-indoleacetic acid, 3-indole propionic acid, phenylacetic acid, 3-indoleacetic acid, 3-indolebutyric acid (IBA) and 3-indoleacrylic acid (all of these except IBA were described as natural auxins). Only IAA has been shown to occur naturally in pea seedling roots (Note: these high concentrations of auxin which are near toxic levels for many tissues suggest that the assay used was highly insensitive and unlikely to reflect the natural system).

Robbins and Hervey [39] reported that excised roots of *Bryophyllum calycinum* showed best growth (= most number of root branches and total elongation) when cultured in the presence of $1 \, \mu g$ NAA per 50 ml medium (approx. $10^{-7} M$) combined with cytokinin.

1.4.2. Cytokinins. Cell division in cultured plant tissues, both shoot and root, is usually dependent upon exogenous supplies of both auxin and cytokinin. The morphological response to the exogenous stimuli results in bud formation, more callus formation or root initiation depending upon the ratio of the two plant growth regulators supplied [40]. When one provides cytokinins exogenously to seedling roots or to excised roots in culture, one expects an interaction with exogenously supplied auxins or endogenous auxins. For lateral root formation exogenously supplied cytokinins typically produce inhibition depending upon concentration (See reviews by Torrey [8], Wightman *et al.* [38]). Wightman *et al.* [38] reported stimulations to lateral root initiation when seedling roots of peas with root tips excised were treated with cytokinins to $10^{-6} - 10^{-7}$M and inhibitions at $10^{-4} - 10^{-5}$M. By excising the root apex, a source of endogenous cytokinins, these investigators may have produced an assay more sensitive to cytokinins than auxins. In order of decreasing effectiveness as stimulators of lateral root formation at 10^{-7}M were: kinetin, kinetin riboside, zeatin, zeatin riboside, isopentenyladenine and isopentenyladenosine. The first two substances are synthetic compounds.

Webster and Radin [36] reported inhibition of lateral root formation in radish roots grown in culture by cytokinin applied at 10^{-7}M or higher concentrations.

In cultured roots of *Haplopappus*, Blakely *et al.* [35] reported that exogenous concentrations of benzyladenine from 0.1 to 0.001 mg/l^{-1} inhibited primordium formation in the presence of exogenous NAA. Torrey [42] had reported inhibited lateral root formation by kinetin at 2×10^{-6}M in cultured pea root segments treated with IAA (Figure 1). At 5×10^{-8}M, kinetin stimulated lateral root formation.

1.4.3. Gibberellins. There seems to be little new evidence of an effect of exogenous gibberellins on lateral root formation. In a lonely report which exists in the literature, Butcher and Street [43] reported a stimulus to lateral root formation in cultured excised tomato roots by gibberellins. Wightman *et al.* [38] in their detipped pea seedling root assay reported general inhibition of lateral root formation at GA_3 concentrations of $10^{-3} - 10^{-6}$M. Inhibition by gibberellin of lateral root formation was observed also in pea roots by Goodwin and Morris [44] and Böttger [45] reported no effect of gibberellins on lateral root formation in pea seedlings.

1.4.4. Other growth factors. The general insensitivity of the pea seedling assay with detipped root studied by Wightman *et al.* (37) suggests that it is an inappropriate system in which to test growth factors presumed to be provided to the root from the shoot or cotyledons. Therefore it is not surprising that in their studies little or no response was elicited from exogenously supplied adenine, guanine, or vitamins such as thiamine, pyridoxine or nicotinic acid. Only when abscisic acid and xanthoxin were provided exogenously at 10^{-5}M or higher were inhibitions of lateral root initiation noted. Isolated roots, in contrast, show consistent responses both for elongation in culture and in root branching to additions of thiamine, nicotinic acid for pea roots [41, 46]) and thiamine and pyridoxine for tomato roots (cf. Street and Henshaw, [47]).

In their assay for lateral root initiation, using cultured root segments of *Haplopappus*, Blakely *et al.* [35]) provided the clonally cultured roots the following growth factors: NAA, myoinositol, glycine, nicotinamide, pyridoxine, thiamine, and p-aminobenzoic acid. Whether each of these factors was essential to the growth and lateral root initiation response of the root was not demonstrated rigorously.

Goforth and Torrey [48] demonstrated the essentiality of myoinositol to the continued growth and branching of cultured roots of *Comptonia peregrina*, a woody shrub of the Myricaceae. Myoinositol has proved beneficial to the growth in culture of excised roots of other woody plants (Torrey unpublished).

Interest in abscisic acid and xanthoxin as active factors in root development continues because of evidence suggesting their formation by root apices by Rivier *et al.* [49] and Böttger [50]. Exogenously supplied amounts of xanthoxin are inhibitory to lateral root initiation in excised pea root segments [45] but even at high concentration had little effect in detipped seedling roots of pea [38].

Another substance, previously not reported, recently has been suggested as playing a role in lateral root initiation in peas. It is the compound trigonelline (N-methyl nicotinic acid) claimed by Evans [51] to cause root meristem cells in peas to undergo cell arrest in the G-2 phase of cell division. The substance acts at $10^{-5} - 10^{-7}$M on root meristems in peas, originates in the cotyledons and moves into the roots. A specific effect on the pericycle has yet to be demonstrated but if the pericycle were a specific target for the trigonelline effect, it could serve to 'set-up' pericycle cells for special sensitivity to lateral root inducing stimuli.

1.5. Endogenous factors

In his review in 1976 Torrey [8] brought together the evidence for the existence and, if studied, the approximate concentration of the various plant hormones and growth factors which might be involved in normal root elongation and the initiation of lateral roots. Interested readers are referred to that review. More recent information along this line has been summarized by Wightman *et al.* [38]. Table 2 presents some representative data on the reported endogenous hormone levels in roots of a limited number of species examined by relatively direct methods. Such data are difficult to get and therefore scarce; they are fraught with difficulty of interpretation especially relative to the involvement of such endogenous factors in morphogenetic development and expression. In no case do we understand the mode of action of these plant hormones at the cellular level, let alone at the molecular level. It therefore becomes hazardous to attempt to make

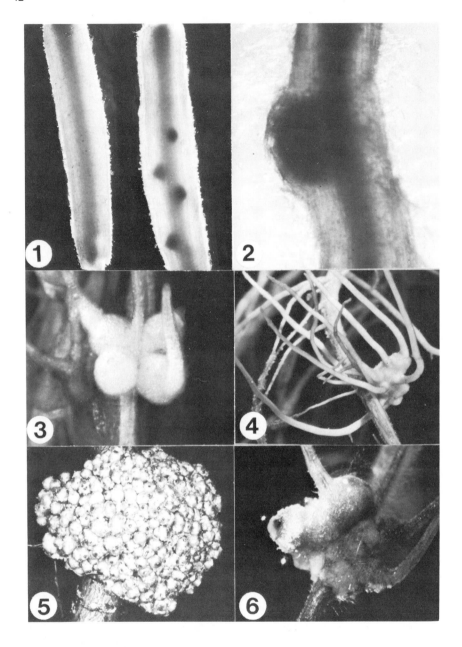

Fig. 1. Cleared segments of excised roots of the garden pea, *Pisum sativum,* stained with Feulgen stain. Left, untreated root showing no lateral root primordia. Right, root primordia induced by 10^{-5} M indoleacetic acid after 5 d. $\times$ 7.5. (Torrey [41]).

Fig. 2. Cleared pea root segment stained with Feulgen stain, showing nodule primordium. Root segment was excised from a seedling root 12 d after infection with *Rhizobium.* $\times$ 16 (Torrey and Barrios [101a]).

Fig. 3. Early nodule formed on seedling root of *Comptonia peregrina* by the soil bacterium *Frankia.* Four modified lateral roots have formed at the same site. The swollen nodule lobes contain the endophyte where dinitrogen fixation occurs. The nodule roots have just begun to extend upward. $\times$ 20. Bowes *et al.* [63].

Fig. 4. Young nodule of *Casuarina cunninghamiana* with multiple lobes, each terminated by an upward-growing nodule root. The endophyte *Frankia* occupies the hypertrophied cortical cells of each nodule lobe where the dinitrogen-fixing enzyme, nitrogenase, develops. $\times$ 4. Torrey [67a].

Fig. 5. Coralloid-type root nodule on root of the red alder, *Alnus rubra,* grown aeroponically. Nodule lobes are modified lateral roots occupied by the infecting soil bacterium *Frankia.* Dinitrogen fixation occurs within the nodule lobes. $\times$ 3.

Fig. 6. Four-lobed young nodule on seedling root of *Alnus rubra* grown in an axenic agar slant in a closed test tube and inoculated with *Frankia* strain ArI3. Note that several of the nodule lobes have developed nodule roots (compare with Fig. 5). Photograph courtesy of A. Berry. $\times$ 15.

more than general estimates of the possible nature of the involvement of such factors in the physiological processes occurring in events such as lateral root initiation.

1.6. Environmental factors influencing lateral root formation

Any external change in the environment, physical or chemical, which in some way modified the internal physiology of the developing root, including the production or supply or endogenous factors, can be expected to influence lateral root initiation and development. Thus a well developed root system interacts with a robust, photosynthetic shoot system upon which the root depends for translocated carbohydrate, vitamins, and other growth factors including shoot-synthesized plant hormones. The practical effects of these environmental parameters influencing the root branching patterns in soil are interestingly discussed by Russell [2] in his book on plant root systems especially with respect to their agricultural implications. Such factors include mineral nutrition, root and shoot temperatures, moisture availability, physical effects of the soil, light quality and intensity impinging on the shoot, and day length. Russell points out, however [p. 87] there exists 'little information on the manner in which the production of endogenous growth hormones is influenced by the quantity of nutrients which enters the roots'.

Table 2. Reported levels of endogenous hormones in roots (expressed on a fresh weight basis unless otherwise stated).

Hormones	Genus	Tissue	Approximate concentration	Method of analysis[1]	Reference
Auxins	*Zea*	Root Cap	$356\,\mu g\ kg^{-1}$ $(5 \times 10^{-7}\,M)$	GC + MS	Rivier and Pilet [73]
	Zea	Root Stele	$53\,\mu g\ kg^{-1}$	GC + MS	Bridges *et al.* [74]
	Zea	Root Stele	$142\,\mu g\ kg^{-1}$	Chromatography and bioassay	Greenwood *et al.* [75]
	Pisum	3 mm-root tip	$130–400\,ng\ g^{-1}$	RIA	Pengelly and Torrey [76]
Cytokinins	*Pisum*	5 mm-root tip	$9.7 \times 10^{-5}\,M$	GLC	Wightman *et al.* [38]
	Pisum	0.1 mm-root tip	$2.5 \times 10^{-5}\,M$	Chromatography and bioassay	Short and Torrey [77]
	Zea	Excised root culture	+	Extraction and bioassay	VanStaden and Smith [78]
	Lycopersicon	Excised root culture	+	Extraction and bioassay	VanStaden and Smith [78]
Gibberellins	*Pisum*	Root	$0.0–0.2\,\mu g\ kg^{-1}$	Dwarf pea assay	Radley [79]
	Lupinus	Root	$24–64\,\mu g\ kg^{-1}$	Lettuce hypocotyl assay	Dullaart and Duba [80]
	Glycine	Root	$0.16\,mol\ g^{-1}$ dry wt	TLC + GC	Williams and Sicardi [81]
	Myrica	Root	$0.6–1.7\,\mu g\ kg^{-1}$	Lettuce hypocotyl assay	Wheeler *et al.* [72]
Abscisic acid	*Pisum*	Root	$70\,ng\ g^{-1}$ $(2.7 \times 10^{-7}\,M)$		Böttger [82]
	Zea	Root	+	GC + MS	Rivier *et al.* [49]
Cis-trans xanthoxin	*Pisum*	Root	$20\,ng\ g^{-1}$		Böttger [80]

[1] Analytical methods: GC + MS – gas chromatography – mass spectrometry;
GLC – gas – liquid chromatography;
TLC – thin layer chromatography;
RIA – radio immunoassay.

1.7. Conclusions

We do not know the endogenous factors controlling lateral root formation, at any of the levels – molecular, biochemical or physiological. We can hazard some guesses from the kinds of evidence reviewed above and by earlier authors. There seem to be two levels of activity about which we may make some tentative conclusions. A root capable of undergoing lateral root formation must be in a metabolically capable state – that is, actively growing and metabolizing. In the intact seedling the factors which determine this state encompass the physiology of the whole plant. Any of a number of factors might block lateral root formation if they become limiting: e.g. sucrose from the photosynthetic shoot in the seedling or provided in the medium in excised aseptically cultured roots or root parts. Similarly as integral parts of the metabolic machinery but particularly vulnerable in excised roots or root parts are the B vitamins – thiamine, nicotinic acid, pyridoxine – all serving known roles in normal cell metabolism, the usual list of inorganic elemental requirements with certain ones especially critical including calcium, boron, phosphate and iron and affected by ambient pH of nutrient solutions – optimally between pH 5.0 and 6.0. Myoinositol belongs in the list of 'permissive' cofactors, utilized in the operating machinery in the root – either for cell walls or cell membranes or both.

Even with all of these endogenous factors provided and operational, root branching need not occur. Lacking are the triggers to directed cell divisions – most probably endogenous plant hormones. The primary trigger which sets off lateral root initiation appears to be auxin. The response seems to be concentration dependent such that endogenous levels of auxin frequently can be augmented by an exogenous supply. Another primary trigger is probably cytokinin, in some unknown way interacting with auxin. Here the evidence is even less direct that for auxin. Both of these hormones are known in other systems to be potentially interactive with other hormones – ethylene with auxin, gibberellins and abscisic acid with cytokinins, supplementing or negating the effects of the primary triggers and thereby modifying the primary response.

Lacking in all of these speculations are ideas or notions about the control of oriented cell divisions, ordered sequences leading to organised root primordia. Seeking control elements for these events in plant hormones *per se* seems fruitless.

We need wholly new concepts of regulation to cope with these phenomena – the regulation of spindle orientation for mitoses, the programmed cessation of cell division activities coupled with cell enlargement leading to what we recognize – the phenotypic expression of lateral root formation.

2. The influence of micro-organisms on lateral root formation

2.1. Introduction

In recent years a considerable amount of attention has been paid to the intimate relationships among soil micro-organisms and the ramifying root systems of plants which invade and occupy the soil. The rhizosphere has a terminology of its own to give precision to some of the complex interactions known to occur. The subject of these interactions has been reviewed repeatedly and from many points of view – the plant coping with low soil nutrients [52, 53, 54] plants dealing with dessication [55, 56] plant roots responding to soil micro-organisms [57, 58] plant roots establishing mutualisms with soil micro-organisms [59, 60, 61] and plant roots forming symbiotic relationships with soil micro-organisms. This is a vast and complex field of study and one still in its infancy.

In a surprisingly large number of instances, soil micro-organisms interact with plant roots modifying lateral root initiation, even, it seems, influencing endogenous factors initiating, inhibiting or, in some other way, affecting or effecting lateral root forma-tion. In this root-rhizosphere-soil micro-organism complex I plan to focus directly on those aspects concerned with lateral root formation, in the expectation that close examination of these interactions will help us understand better the natural events of root formation in normal development in the soil.

Much of the experimental material discussed in the first part of this review dealt with roots or root systems studied under laboratory conditions – e.g. plants grown with their roots in nutrient solutions in water or sand culture, or in artificial, usually sterile, soils or substrates. In many cases in order to make analysis possible root physiologists have used aseptic organ culture or axenic systems save for the single micro-organisms introduced into the root environment for study. These experimental simplifications of the real-world situation are justifiable as demonstrated by the considerable information which has come from them. But that they are far removed from what goes on in plants grown in the soil must be recognized and efforts made to relate the facts established in the laboratory to the situation as it occurs in the field. By focussing on the very special root response of lateral root initiation our subject may be less intractable.

Soil micro-organisms are extremely active in their metabolism. They grow in the soil using whatever substrates are available. Many of the substrates are dead, dying or living plant parts. Many of the metabolic by-products of bacterial or fungal biochemical activities are substances the plant roots will respond to – soluble organic molecules of wide variety including simple sugars, amino acids, purines and pyrimidines, vitamins, plant growth hormones, plant growth regula-tors and many other factors. Most of these substances are in turn metabolised by other micro-organisms and are short lived. In some cases, the presence of these exogenous substances profoundly influences root development. In the extreme cases, the micro-organisms find their way into the roots of receptive hosts,

establish themselves and live in symbiotic relationship, often markedly influencing root morphology, structure, physiology and biochemistry.

One simple way to address the complexities of relating micro-organisms to root
systems is to treat them from the point of view of microbial systematics. For each
taxonomic group of micro-organisms there exists a separate subject and therefore
literature. Let us examine lateral root formation as it occurs or is modified in each
of the following higher plant microbial associations: actinorhizal plants, *Rhizobium*-legumes, *Rhizobium* and non-legumes, proteoid root formation, free-
living rhizosphere interactions influencing roots, and special pathogenic situations.

2.2. Actinorhizal plants

Nodules induced on the roots of a wide range of woody dicotyledonous plants,
collectively referred to as actinorhizal plants, represent the most striking case of
lateral root formation under the direct stimulus of a microbial symbiont. The
filamentous soil bacterium *Frankia* of the Actinomycetales infects the root hairs
of susceptible host plants [62] or invades intercellular spaces via the epidermis and
cortex (Miller and Baker, unpublished), penetrates the root cortical cells, there
stimulates the initiation of multiple lateral roots [63, 64] and then progressively
invades and occupies the cortical tissues of the basal swollen lobes of the nodule
made up of these closely packed lateral roots. The analogy can be made to a
'witch's broom' involving in this case many lateral roots compacted together
rather than shoot buds. The micro-organism within the host plant proliferates and
successively induces new sites for its occupancy in the form of swollen, modified
lateral roots.

Actinorhizal nodules are of two morphological types: the coralloid or *Alnus*-
type and the nodule-root forming or *Myrica*-type [65, 66]. Although the resulting
morphological structures appear quite dissimilar (Figs. 3, 4, 5, 6), they are
basically alike in their origins and anatomy. The more obvious structure is the
Myrica-type nodule whose developmental ontogeny has been described in considerable morphological and anatomical detail in *Comptonia peregrina* [63, 64], in
Myrica gale [66, 67] and in *Casuarina* [67a]. These papers also review the earlier
descriptions of this nodule type. In *Comptonia* as many as 10–12 lateral root
primordia are initiated within a millimeter or two of the initial infection site in a
very short time course (Fig. 3). These primordia elongate to form short, truncated
nodule lobes within which the endophyte occupies the cortical cells, forming the
symbiotic relationship resulting in dinitrogen fixation. From the terminal meristem at the apex each nodule lobe develops a determinate nodule root which
typically grows vertically upward to about 2–3 cm (Figs. 4, 8). Nodule roots
remain free of the endophyte which is restricted to the swollen basal tissue lobes
(Fig. 5).

The coralloid root nodule typical of *Alnus* and the majority of actinorhizal plants is formed in fundamentally the same developmental sequence [68, 69]. However, the meristematic cell population at the terminal part of the nodule lobe is permanently arrested, forming not an elongate nodule root but a terminal papilla on each lobe (Figs. 5, 6).

Each of these two morphologically distinct actinorhizal nodule types is perennial. Promptly after the proliferation of lateral root primordia has occurred, setting off early nodule development, the lateral root on which the new nodule has formed undergoes secondary thickening, assuring the retention of the nodule on the plant root [64]. Depending on seasonal fluctuations, such as changes in temperature or moisture availability, new nodule lobes are formed in successive developmental waves (cf. [67]) and a progressively larger and more highly branched structure results, with nodules becoming quite woody at the base or centre and, in some cases, reaching many centimeters in diameter.

What are the endogenous stimuli, presumably originating from the invading micro-organism, which induce this remarkably specialized response in lateral root initiation? That *Frankia* produces the inciting stimulus or stimuli seems evident. Very little information is available on the question of what are the metabolic products of *Frankia* which might cause this response. The reason for this is that only recently has *Frankia* been isolated and grown in pure culture [70] and therefore little direct study of metabolic activities in culture has been possible. Wheeler *et al.* [71] reported that *Frankia* in culture produces indoleacetic acid. This observation, while expected, is an encouraging first step and it could well be that localized increases of IAA produced by *Frankia* could be involved in nodule initiation.

The next best approach, lacking evidence of the production of inciting stimuli by pure cultures of the micro-organism, is to analyze root nodules for endogenous hormones, seeking evidence of unusual levels of substances originating either from the infecting organism or from the symbiotic state which could help explain the anomalous plant structures represented by the nodule. This approach has been taken, notably by Wheeler and his associates, for actinorhizal nodules. A review of their work [72] summarizes much of this literature. From the point of view of lateral root forming stimuli, such analyses of mature structures unfortunately miss the events concerned with the early incitement.

As might be expected of a dense, highly meristematic mass of plant tissue, young actinorhizal nodules are filled with relatively high levels of plant hormones, especially if they are compared to neighbouring, uninfected root tissues. In Table 3A are listed some of the hormones which have been reported for some of the actinorhizal species. The plant source, the hormone, the method of analysis and the authors are cited. This list is not exhaustive and there exist all of the usual qualifications and problems presented by such determinations but the table does give an impression of the situation one finds.

Actinorhizal root nodules contain substantially higher levels of a number of

Table 3. Reported levels of hormones in root nodules and in uninfected roots (values in $\mu g\ kg^{-1}$ fresh weight unless otherwise stated).

Hormone	Host plant species	Hormone concentration		Method of analysis*	Reference
		in nodules	in roots		
A. *Frankia* – actinorhizal plant symbiosis.					
Auxins	*Alnus glutinosa*	385–1430	90–205	Spectro-fluorescence	Dullaart [83]
	Alnus glutinosa	250	50–100	HPLC	Wheeler *et al.* [72]
	Alnus glutinosa	20 mg kg^{-1}	–	Extraction and bioassay	Silver *et al.* [84]
	Myrica cerifera	10 mg kg^{-1}	0	Extraction and bioassay	Silver *et al.* [84]
Gibberellins	*Alnus glutinosa*	4.6–4.9	0.4–1.6	Lettuce hypocotyl bioassay	Henson and Wheeler [85, 86]
	Myrica gale	11.1–11.9	0.6–1.7	Lettuce hypocotyl bioassay	Wheeler *et al.* [72]
Cytokinins	*Alnus glutinosa*	97–617	16–38	Soybean bioassay	Henson and Wheeler [87]
	Myrica gale	62–95	16–24	Soybean bioassay	Henson and Wheeler [87]
Abscisic acid	*Alnus glutinosa*	250–700 ng g^{-1}	100–200 ng g^{-1}		Bano *et al.* [88]
	Alnus glutinosa	+	+		Watts *et al.* [89]
B. *Rhizobium* – legume symbiosis.					
Auxins	*Vicia faba*	400	150	HPLC	Wheeler *et al.* [72]
	Lupinus luteus	250–500	30–160	Spectro-fluorescence	Dullaart [90]
Gibberellins	*Pisum sativum*	240	0–0.2	Dwarf pea bioassay	Radley [79]
	Lupinus luteus	420–3680	24–64	Lettuce hypoc. bioassay	Dullaart and Duba [80]
	Glycine max	1.34 nmol g^{-1} dry wt	0.16 nmol g^{-1} dry wt	TLC + GC	Williams and Sicardi- de Mallorca [81]
Cytokinins	*Vicia faba*	139–163	11–12	Soybean bioassay	Henson and Wheeler [85]
	Pisum sativum	20–27	–	Soybean bioassay	Syōno *et al.* [90, 91]
Abscisic acid	*Glycine max*	2.21 nmol g^{-1} dry wt	0.18 nmol g^{-1} dry wt	TLC + GC	William and Sicardi- de Mallorca [81]

[1] Analytical methods: HPLC – high pressure liquid chromatography;
TLC + GC – thin layer chromatography and gas chromatography.

known plant hormones including indole-3-acetic acid (IAA), gibberellins (GA), cytokinin (CK), and abscisic acid (ABA), than occur in uninfected roots. Differences between nodule and root tissue in different actinorhizal genera vary considerably. For example, in *Alnus glutinosa* nearly 40 times more cytokinin occurs in nodules than in roots and 5–15 times more auxin in the nodules than in the roots. Such crude comparisons serve to provide little understanding as to cause and effect. One can say that the actinorhizal nodules are rich in a number of known plant hormones and that their perpetuated meristematic state may be related to these active substances. Wheeler *et al.* [72, 93] have attempted to document the seasonal changes in some of these parameters. Aside from showing that changes in carbohydrates and nitrogenous compounds do occur over the season in the temperate zone and some of the nodular plant hormones change as well, no clear correlation could be made between these changes and nodule ontogeny. The two genera serving as the major subjects for these studies (Table 3A), viz., *Alnus glutinosa* and *Myrica gale,* are, respectively, of the coralloid and nodule-root types. They show no consistent differences between types.

Some interesting observations have been made (Berry and Torrey, unpublished) on nodulated seedlings of *Alnus rubra* grown in inorganic nutrient agar in closed sterile test tubes. Inoculation with pure cultures of *Frankia* sp. HFPArI3, which caused root nodule formation, occasionally led to the formation of nodule roots on the terminal lobes of root nodules (Fig. 6). This anomalous behaviour for *Alnus* was observed also by Rodriguez-Barrueco [94] in *Alnus jorullensis* and by Becking [95] in *Alnus* species following cross-inoculation with nodule-suspensions prepared from different hosts. Such expression suggests that in both types of nodule, the terminal meristematic mass of cells in each lobe is an arrested lateral root meristem which has the potential to be expressed as an elongate root, depending upon the internal stimuli or inhibitors provided to it by the endophyte. This difference in expression might serve as a useful subject for further study in the unravelling of the hormonal stimuli involved.

The upward growth of nodule roots in nodules of the *Myrica*-type has provided a fascination of its own. Tjepkema [96] has provided evidence that under conditions of low oxygen, nodule roots, which are filled with extensive intercellular air space, could serve to channel air, and therefore, oxygen and nitrogen, to the tissues actively fixing dinitrogen at the base of the nodule lobe. Excision of nodule roots in plants grown with their roots subjected to low O_2 tensions led to a decrease in acetylene reduction activity. Whether environmental conditions modify nodule root elongation or structure has not been studied. Silver *et al.* [84] were interested in the question whether the hormone status within the root nodule and nodule roots of *Casuarina* and *Myrica* could explain the anomalous upward growth of nodule roots which contrasts strikingly with the downward growth typical of laterals formed along the rest of the root. They showed that the enzyme complex, indoleacetic acid oxidase, which leads to IAA destruction in tissues, was substantially higher in the root nodules of both species forming nodule roots than in *Alnus*. They suggested that nodule roots lacked the normal positive geotropic response, due to lowered levels of auxin. Unfortunately, their comparisons of auxin levels were not made within the same plant species, but across species. They did not examine ethylene production which is known to interact with auxin production, especially in situations of restricted gas exchange.

It seems clear from the above brief review of hormonal relations in actinorhizal root nodules that these structures are ususually rich sites of plant hormones resulting almost certainly from the presence of microbial endophytes. Although the interaction between host and microbial symbiont is complex, and can involve a range of expressions depending on the genera concerned and the environment in which they grow, actinorhizal nodule development may offer a useful system to explore further the nature of the endogenous factors influencing lateral root initiation and development.

2.3. Interactions between Rhizobium and the roots of legumes

Having explored in some detail the hormonal relationships between the ac-tinomycete *Frankia* and the various hosts in which root nodules are induced, it is appropriate that we examine briefly from this same perspective the situation which exists between the soil bacterium *Rhizobium* and the legumes which serve as host and symbiont. This subject is an older and larger one, initiated by such observations as Thimann [97] and Kefford *et al.* [98] that rhizobial root nodules of *Pisum* contained unusually high concentrations of auxin. The possibly reciprocal relationship between occurrence of nodules and lateral roots especially with respect to sites of initiation was explored by Nutman [99] in clover, (*Trifolium*) using experimental and genetic approaches. Considerable effort has been de-voted to analyses of hormone production by *Rhizobium* in pure culture and to hormone levels in root nodules. These studies, while somewhat removed from the main line of our discussion, provide as we shall see, further insight into the endogenous factors involved in lateral root formation in legumes.

Root nodules induced by *Rhizobium* on roots of legumes, unlike actinorhizal root nodules, are not modified lateral roots. In many legumes root hair infection by *Rhizobium* is followed by bacterial thread invasion into inner cortical cells of the root which are stimulated to proliferate, producing *de novo* a localized tissue proliferation, typically polyploid with the central tissue usually occupied by a population of bacteroid-containing cells. These events have been carefully and completely described and the literature reviewed elsewhere by Dart [100, 101] and by Newcomb [101a]. Frequently nodules originate at sites normally occupied by lateral roots, e.g., opposite protoxylem points of the primary vascular system. Also they are endogenous, but originate from inner cortex rather than pericycle (Fig. 2). The bulk of the nodule tissue is polyploid whereas the meristem of the lateral root is typically diploid.

Legume root nodules may be organized in several different tissue arrangements [102]. Nodules of members of the Phaseoleae, such as *Glycine* (soybean) and *Phaseolus* (snap bean) have determinate growth with a closed vascular structure at maturity and a globose to oblate shape. In the Vicieae and Trifolieae, including such genera as *Pisum* (pea) *Vicia* (broad bean) and *Trifolium* (clover), nodules show indeterminate growth with an apical meristem, an open-ended vascular structure and cylindrical shape with continued elongation and occasional nodule branching. In some special circumstances [103] nodules of the indeterminate type may undergo physiological changes which result in lateral root development from the terminal meristem, giving the appearance not dissimilar from nodule roots in actinorhizal plants.

Plant hormone production by *Rhizobium* grown in pure culture is well documented (see Table 4). Different strains of *Rhizobium* grown on a range of media may produce IAA, GA, CK, or ABA in culture and might be expected to elicit plant tissue responses such as stimulation of meristematic activity, stimulation of root hair formation, cell enlargement, polyploidization and other familiar hor-

Table 4. Reported occurrence and level of plant growth regulating substances produced by bacteria in culture.

Bacterial genus	Plant growth substance[1]	Concentration in medium	Reference
Rhizobium	IAA	+	Badenoch-Jones *et al.* [106]
	IAA	+	Wheeler *et al.* [71]
	IAA	0.1 ng cm^{-3}	Wang *et al.* [101]
	CK	1.5 μg KE g^{-1} dry wt	Phillips and Torrey [108, 109]
	CK	0.1 μg ml^{-1}	Azcon *et al.* [110]
	ABA	0	Williams and Sicardi de Mallorca [81]
	GA	1.0 nmol l^{-1}	Williams and Sicardi de Mallorca [81]
	GA	+	Katznelson and Cole [111]
	GA	0.1 μg ml^{-1}	Azcon *et al.* [110]
Azotobacter	IAA	0.3 μg ml^{-1}	Azcon *et al.* [110]
	IAA	+	Brown [112]
	CK	0.05 μg ml^{-1}	Azcon *et al.* [110]
	CK	+	Brown [112]
	GA	0.06 μg ml^{-1}	Azcon *et al.* [110]
	GA	+	Brown [112]
Frankia	IAA	+	Wheeler *et al.* [71]

[1] Plant growth substances: IAA – indole-3-acetic acid;
CK – cytokinins;
KE – kinetin equivalents;
ABA – abscisic acid;
GA – gibberellins.

mone-related tissue responses. Explicit experimental approaches designed to test these ideas were described by Libbenga *et al.* [104, 105].

Root nodules of legumes also show notably increased hormone levels in the nodule tissues compared to normal root tissues. Examples of these differences are summarized in Table 3B. Legume nodules, as specialized centres of hormone production, are similar to actinorhizal nodules both in the hormones produced and the degree of amplification of hormone production over uninfected non-meristematic root tissues. In making comparisons between the two nodule types it is clear that the basal nodule lobe in actinorhizal nodules containing the polysaccharide-encapsulated *Frankia* filaments and the swollen dinitrogen-fixing vesicles of this endophyte is most analogous to the bacteroid-containing dinitrogen-fixing tissue mass of the rhizobial nodule. Both are stimulated by the presence of the bacteria, probably by the metabolic products of the microorganisms, at least some of which fall into several classes of higher plant hormones.

2.4. Rhizobium-induced root nodules in the Ulmaceae

In 1976 Trinick (112a) reported root nodules in a tropical tree in the family Ulmaceae which he showed were produced by *Rhizobium*. The genus was *Parasponia* (originally described erroneously as *Trema*) and there are several species in islands of the South Pacific which form dinitrogen-fixing root nodules following the invasion and infection by *Rhizobium* strains belonging to the cowpea miscellany. These nodules are particularly interesting as they show features half-way between typical rhizobial nodules and actinorhizal nodules [113]. The ultrastructural details of the infection process and early development of these root nodules have been described recently [114, 115]. The following events occurred in axenically grown seedlings of *Parasponia rigida* exposed on agar slants to a strain of *Rhizobium* which readily nodulates host members of the cowpea miscellany such as *Macroptillium*. Seedling roots first show a response to the presence of the *Rhizobium* by forming multicellular root hairs, either uni- or biseriate. Cortical cells at the base of these root hairs show hypertrophy and cell division accompanied by the proliferation of bacterial cells within the intercellular spaces with initial entry presumed to be at loosened intercellular sites in the root epidermis. No root hair infection occurs. The bacteria invade deep into the cortical tissues of the *Parasponia* root by proliferation within intercellular spaces, modifying middle lamellae and primary cell walls. This aspect of the infection process is reminiscent of some aspects of infection in *Arachis* (peanut) and in *Stylosanthes* [116, 117]. Finally, the bacteria dissolve the primary cell wall of one or more inner cortical cells and form typical infection threads with a polysaccharide thread wall surrounding and accompanying the continually dividing bacterial cells. Once established within threads in the infected cells the bacteria never leave

Fig. 7. Young root nodules of the tropical tree *Parasponia rigida* of the elm family Ulmaceae produced by infection with a cowpea strain of *Rhizobium*. Each lobe of the coralloid nodule is a modified lateral root containing proliferated cortical cells filled with bacterial threads containing rhizobia. The surfaces of the nodules are covered with small swollen masses of lenticel-like aerenchyma. × 5.5. Lancelle and Torrey [114].

Fig. 8. A well developed root nodule cluster on a seedling root of *Casuarina cunninghamiana* grown aeroponically. Inoculation of the seedling was with a suspension of a pure culture of *Frankia* strain HFPCcI3. Swollen nodule lobes form the central dinitrogen-fixing tissues of the nodule and elongate determinate nodule roots extend outward from them. × 2.5.

Fig. 9. Excised cultured root of *Convolvulus sepium* which has been derived from root transformed after infection by the soil bacterium *Agrobacterium rhizogenes*. Note the abundant lateral root formation. Culture courtesy of J. Mugnier. × 0.9.

the threads which repeatedly branch and ramify, filling the whole centre of the infected cortical cells.

Concurrent with the bacterial invasion of the root, lateral root primordia are initiated in the immediate vicinity of the original bacterial invasion site [115]. These lateral root primordia are almost certainly initiated in response to the presence of the bacterial endophyte. Their normal development is arrested and the bacterial threads from cortical cells pass from cell to cell within the lateral root, forming swollen, lobed multi-branched nodules (Fig. 7) structurally analogous to *Frankia*-occupied coralloid nodules in actinorhizal plants. The ramifying, bacterial-filled threads fill cortical cells which typically occupy a horse-shoe-shaped tissue arrangement around the central vascular cylinder of the modified lateral root. Repeated endogenous branching of lateral roots and suppressed elongation under the control of the endophyte in the root cortex results in a coralloid-type nodule filled with *Rhizobium* held within threads within which nitrogenase activity develops. Thus, nodules in *Parasponia*, while unique, represent the combined features of certain *Rhizobium* nodules of the cowpea miscellany and *Frankia* nodules of the coralloid type. We have essentially no information about the stimuli produced by the bacteria which elicit these host responses in this symbiosis. Presumably, the rhizobial strains infective and effective in *Parasponia* combine physiological features of both *Rhizobium* and *Frankia*, features which may have in common the production of higher-plant hormones and related growth regulating substances.

Before leaving consideration of root nodules and their relation to lateral root formation, we should mention that growth regulating substances applied exogenously to roots may elicit the development of nodule-like structures called 'pseudonodules'. Often these nodule-like protuberances of the root are modified or inhibited lateral roots. Such structures were reported by Arora *et al.* [118] following treatment of tobacco roots with the synthetic cytokinin, kinetin. Rodriguez-Barrueco and De Castro [119] treated roots of *Alnus glutinosa* with kinetin and with the cytokinin, 2-isopentenyl adenine, and induced similar nod-

56

ule-like structures. Torrey (unpublished) found that externally applied substances known to block auxin transport, such as naphthyl phthalamic acid and related compounds, caused multiple, nodule-like structures to form on white clover roots in the absence of *Rhizobium*. Such pseudo-nodules contained no bacteria, did not fix dinitrogen and were in fact highly modified lateral roots. Doubtless other growth-stimulating substances or substances which interfere with normal hormonal physiology within the root can be expected to elicit abnormal structures often centering around the modification of normal root branching.

2.5. Cluster–root formation (proteoid roots)

A remarkable phenomenon involving specialized lateral root formation, found especially in the Australian plant family Proteaceae, is claimed to be the result of lateral root stimulation by unidentified soil micro-organisms. Many hair-like lateral roots form in dense clusters along the root, especially in poor soils, where they facilitate uptake of minerals and enhance the nutrient uptake of the plants. The subject has been reviewed comprehensively by Lamont [53]. In his review Lamont refers to several different types of 'cluster-rooted plants', including the proteoid type which are well documented in over 900 species of 30 genera, especially endemics of Western Australia and South Africa. Because of the proliferation of groups of longitudinal rows of densely clustered determinate lateral roots, the root surface area of these plants may be increased by up to five times or greater [120] with demonstrated increases in ion uptake, especially phosphate, of from two to ten times or more. These rootlets contain no endosymbionts yet there are reports [121, 122, 123, 124] that proteoid roots are microbially induced. The micro-organisms active in their initiation and the mechanism of induction remain for further study. Initiation is stimulated by wet soil conditions [125] and occurs most frequently in nutrient-poor soils, showing inhibition or total suppression in the presence of rich nutrients. Proteoid roots tend to be short-lived and serve a short-term function in increased nutrient uptake. Cluster-root formation occurs in both rhizobial nodulated plants [126, 127, 128] and in actinorhizal plants [129].

Other cluster-rooted plants described by Lamont [53] include dauciform, capillaroid and stalagmiform roots, each type involving modified lateral root initiation and deserving careful study from the perspective of exogenous and endogenous stimuli for lateral root initiation. Their structures are described elsewhere in this volume by Peterson and Peterson (chapter 1).

2.6. Effects on root development produced by free-living bacteria

In recent years considerable interest has developed over the report that free-living bacteria capable of fixing atmospheric nitrogen are associated intimately with root surfaces and stimulate plant growth by providing fixed nitrogen [130, 131]. This subject has been reviewed repeatedly from various points of view and continues to be controversial [59, 61, 132]. To the extent that combined nitrogen is available to the roots of plants, such as the tropical grass *Paspalum*, in association with soil bacteria, such as *Azospirillum* or *Azotobacter*, the generalized benefit to the plant no doubt involves increased plant development including root branching.

One alternate explanation for the beneficial effects of root-associated soil bacteria aside from nitrogen fixation might be their production of plant growth regulating substances including known plant growth hormones [57, 133]. If produced in sufficient concentrations in the immediate vicinity of the root, these substances could be expected to influence root development, including lateral root formation, causing either stimulation or inhibition. Most of the evidence suggests that bacterially produced plant growth regulating substances either inhibit root development [57] or are absorbed by the root and transported to the shoot. The positive evidence that soil bacteria benefit plant growth or root development by plant hormone production is difficult to find. Another explanation for beneficial effects of specific soil bacteria introduced experimentally to enhance plant growth is attributed [134, 135] to siderophores produced by 'plant growth-promoting rhizobacteria' which complex the iron in the soil, reducing the development in the microflora of potentially deleterious fungi and bacteria. Increases in plant yield of up to 144 per cent have been reported which must inevitably be a reflection, among other responses, of improved root growth. No detailed documentation of effects on root development *per se* have been given.

Doubtless other, as yet unexplained, interactions between free-living soil bacteria and root development remain to be disinterred.

2.7. Other microflora–root interactions involving lateral root formation

Having attempted a cursory exploration of the literature on the interactions between fungi, both beneficial such as ecto and endomycorrhizal fungi, and detrimental such as parasitic and pathogenic fungi, I have opted not to attempt to review the interactions which may occur in these situations influencing lateral root formation. In the same way, I have elected not to discuss bacterial pathogens which affect root form – with a single exception with which I have chosen to conclude this review. It is the especially interesting and appropriate disease condition caused by the soil bacterial pathogen *Agrobacterium rhizogenes* which has been known for a long time but whose mode of operation has only recently begun to be unravelled at the molecular level.

The experiments are those of Tepfer and Tempé [136], Tepfer [137] and Chilton *et al.* [138]. Carrot root discs (*Daucus carota*) were placed in axenic culture and inoculated with *Agrobacterium rhizogenes,* a pathogen causing 'hairy root' disease in dicotyledonous plants. The infected tissues formed large numbers of lateral roots over the surface of the tissue in 2–3 weeks. These roots could be subcultured as excised roots which outgrow the bacteria of the original inoculum and then can be subcultured aseptically at regular intervals free of the inciting bacteria. These cultured roots could be shown to contain the opines, agropine or mannopine, characteristic of the transformed state. In a system analogous to the much studied *Ti*-plasmid from *A. tumefaciens* [139, 140] which causes the tumorigenic crowngall disease in plants, Chilton *et al.* [140] have shown that T-DNA which occurs in a large plasmid in *A. rhizogenes* designated the *Ri* (root inducing) plasmid is transferred from the bacterium to the host cells and is there stably integrated into the plant nuclear DNA and is thereafter reproduced within the host cells and tissues. In the transformed state the roots bearing the *Ri* plasmid T-DNA show the following characteristics: they elongate and branch more than non-transformed roots in culture, they are plagiotropic, and they produce high levels of opines – a diagnostic feature of *Agrobacterium* transformation. In place of the tumourigenesis which characterizes *A. tumefaciens* transformation, these plants show enhanced lateral root formation. Similar transformations have been reported for root tissues of *Nicotiana tabacum* and species of *Convolvulus* (Fig. 9).

From the cultured, transformed roots of carrot it was possible to induce callus tissue by treatment with plant growth substances (2, 4-D plus kinetin), produce cell suspensions in liquid culture and, upon transfer to medium lacking plant growth substances, regenerate from embryoids whole plants which looked normal. Such transformed whole plants were modified by the presence of the T-DNA from the *Ri* plasmid. They contained opines, the shoots showed reduced apical dominance, the leaves were wrinkled and their biennial behaviour had been changed to annual with somewhat reduced seed production [137]. All of these characters segregate together acting as a dominant Mendelian gene. The potential of this *Ri* plasmid as a vector for introducing new genes into plants has been explored by Chilton *et al.* [138] and by Tepfer [137].

For our discussion the significant feature is that in the *Ri* plasmid T-DNA transferred to the plant are the genes coding for lateral root initiation (*l.r.i.*) a discrete assemblage of DNA which conveys the complex instructions for the initiation of the events leading to lateral roots. This piece of DNA codes for lateral root initiation, controlling the 'endogenous factors influencing lateral root formation'. Clearly this genetic information is the basis for the events we have been trying to understand. Characterization of the *l.r.i.* genes from the *Ri* plasmid will provide explicit information for some or all of the events leading to the phenotypic expression of root initiation. Liu and Kado [141] and Liu *et al.* [142]

have already demonstrated that indoleacetic acid production is a plasmid function of *Agrobacterium tumefaciens* and part of the sequence leading to tumor formation in the crown gall disease.

This same sequence of events appears to be encoded in the T-DNA of the *Ri* plasmid [143] and, as one might expect, is an essential step for 'rootiness'. Other genes which lead to the root initiation phenotype await identification from the T-DNA of the *Ri* plasmid. Genes controlling cytokinin biosynthesis have begun to be explored in the *Agrobacterium tumefaciens* tumourigenesis system and analogous genes are very likely involved, perhaps under different control in the *Ri* plasmid system. A comparable sequence of *l.r.i.* genes must occur in the DNA of the plant genome of all normal plants. To understand their activities is to really understand the endogenous influences affecting the singular event of lateral root initiation.

One cannot help but extrapolate the ideas generated by the *Ri* plasmid evidence for a *l.r.i.* gene to the other microbially induced events discussed earlier. Are proteoid roots the expression of a comparable transformation produced by not-yet-identified vectors? Does *Frankia* carry a plasmid bearing the genes for modified lateral root formation leading to actinorhizal root nodules or is that genetic information borne directly in the DNA of the bacterium? Already the evidence has begun to accumulate concerning the 'nodulin' genes in *Rhizobium* controlling nodule initiation [144] – is this another auxin-cytokinin synthetic sequence?

2.8. Conclusions

Clearly, new tools are available to begin to unravel the puzzles of the events of plant and specifically root morphogenesis. The route now available using techniques of gene cloning, DNA hybridization and genetic engineering may provide, not only information on the sequential metabolic events leading to morphological phenotypic expression, but also to the understanding of the molecular mode of action of the plant hormones involved.

From the evidence already available, it is clear that the genetic analysis of the lateral root initiating genes will bring us face to face with the sequential events leading to the first cell divisions, the subsequent events of primordial organization and the elongation and differentiation of the discrete structures of the phenotype 'root'. Evidence summarized in the first half of this review will bear on the analysis of these events. Auxins and cytokinins already appear to be primary reactants in the early events and the control of their biosynthesis may be key regulatory events in lateral root primordia initiation. Use of the *l.r.i.* gene cluster may allow one to determine the importance of other reactants in this complex morphogenetic event and begin to elucidate with greater clarity the nature of the endogenous factors controlling lateral root initiation. Ten years hence we may hope to have a better story to tell.

3. References

1. Sutton, R.F. & Tinus, R.W. (1983) Root and Root System Terminology. Forest Science Monograph 24, Society of American Foresters, Washington, D.C.
2. Russell, R.S. (1977) Plant Root Systems, McGraw Hill, London.
3. Bonnett, H.T. Jr. (1969) Cortical cell death during lateral root formation. Journal of Cell Biology 40, 144–159.
4. McCully, M. 1975) The development of lateral roots. In The Development and Function of Roots (eds J.G. Torrey, and D.T. Clarkson), pp. 105–124. Academic Press, London, New York, San Francisco.
5. Clowes, F.A.L. (1958) Development of quiescent centres in root meristems. The New Phytologist, 57, 85–88.
6. Byrne, J.M. (1973) The root apex of *Malva sylvestris*. III. Lateral root development and the quiescent center. American Journal of Botany 60, 657–662.
7. Torrey, J.G. (1965) Physiological bases of organization and development in the root. Handbook of Plant Physiology XV/I. (ed A. Lang), pp. 1256–1327. Springer-Verlag, Berlin.
8. Torrey, J.G. (1967) Root hormones and plant growth. Annual Review of Plant Physiology, 27, 437–459.
9. Torrey, J.G. & Feldman, L.J. (1977) The organization and function of the root apex. American Scientist, 65, 334–344.
10. Ellmore, G.S. (1982) The organization and plasticity of plant roots. Scanning Electron Microscopy III, 1083–1102.
11. Feldman, L.J. (1984) Regulation of root development. Annual Review of Plant Physiology, 35, 223–242.
12. Charlton, W.A. (1983) Patterns and control of lateral root initiation. In Growth Regulators and Root Development, Monograph 10 (eds. M.B. Jackson and A.D. Stead), pp. 1–14. British Plant Growth Regulator Group, Wantage, U.K.
13. Foard, D.E., Haber, A.H. & Fishman, T.N. (1965) Initiation of lateral root primordia without completion of mitosis and without cytokinesis in uniserate pericycle. American Journal of Botany, 52, 580–590.
14. Yeoman, M.M. (1970) Early development in callus cultures. International Review of Cytology, 29, 483–509.
15. Bhojovani, S.S. & Razdan, M.K. (1983) Plant Tissue Culture: Theory and Practise. Elsevier, Amsterdam.
16. Esau, K. (1965) Plant Anatomy, 2nd edition. John Wiley, New York.
17. Torrey, J.G. (1950) The induction of lateral roots by indoleacetic acid and root decapitation. American Journal of Botany, 37, 257–264.
18. Bonnett, H.T. Jr. & Torrey, J.G. (1965) Chemical control of organ formation in root segments of *Convolvulus* cultured *in vitro*. Plant Physiology, 40, 1228–1236.
19. Feldman, L.J. (1979) Cytokinin biosynthesis in roots of corn. Planta, 145, 315–321.
20. Feldman, L.J. (1979) The proximal meristem in the root apex of *Zea mays* L. Annals of Botany, 43, 1–9.
21. Feldman, L.J. (1980) Auxin biosynthesis and metabolism in isolated roots of *Zea mays*. Physiologia Plantarum, 49, 145–150.
22. Feldman, L.J. & Torrey J.G. (1976) The isolation and culture *in vitro* of the quiescent centre in *Zea mays*. American Journal of Botany, 63, 345–355.
23. Blakely, L.M., Durham, M., Evans, T.A. & Blakely, R.M. (1982) Experimental studies on lateral root formation in radish seedling roots. I. General methods, developmental stages, and spontaneous formation of laterals. Botanical Gazette, 143, 341–352.
24. Blakely, L.M. & Evans, T.A. (1979) Cell dynamics studies on the pericycle of radish seedling roots. Plant Science Letters, 14, 79–83.

25. Riopel, J.L. (1966) The distribution of lateral roots in *Musa acuminata* 'Gros Michel'. American Journal of Botany, 53, 403–407.
26. Riopel, J.L. (1969) Regulation of lateral root positions. Botanical Gazette, 130, 80–83.
27. Charlton, W.A. (1975) Distribution of lateral roots and pattern of lateral initiation in *Pontederia cordata* L. Botanical Gazette, 136, 225–235.
28. Henderson, R., Ford, E.D., Renshaw, E. & Deans, J.D. (1983) Morphology of the structural root system of Sitka Spruce. 1. Analysis and quantitative description. Forestry, 56, 121–135.
29. Charlton, W.A. (1977) Evaluation of sequence and rate of lateral-root initiation in *Pontederia cordata* L. by means of colchicine inhibition of cell division. Botanical Gazette, 138, 71–79.
30. Kawata, S. & Ishihara, I. (1977) Correlations among size of root apex, root apex – lateral distance, and elongation rate in rice roots. Japanese Journal of Crop Science, 46, 228–238 (in Japanese with English summary).
31. Barlow, P.W. & Rathfelder, E.L. (1984) Correlations between the dimensions of different zones of grass root apices, and their implications for morphogenesis and differentiation in roots. Annals of Botany, 53, 249–260.
32. Rowntree, R.A. & Morris, D.A. (1979) Accumulation of ^{14}C from exogenous labelled auxin in lateral root primordia of intact pea seedlings (*Pisum sativum* L.) Planta, 144, 463–466.
33. Mallory, T.E., Chiang, S.-H., Cutter, E.G. & Gifford, E. Jr. (1970) Sequence and pattern of lateral root formation in five selected species. *American Journal of Botany*, 57, 800–809.
34. Chadwick, A.V. & Burg, S.P. (1970) Regulation of root growth by auxin-ethylene interaction. Plant Physiology, 45, 192–200.
35. Blakely, L.M., Rodaway, S.J. Hollen, L.B. & Croker, S.G. (1972) Control and kinetics of branch root formation in cultured root segments of *Haplopappus ravenii*. Plant Physiology, 50, 35–42.
36. Webster, B.D. & Radin, J.W. (1972) Growth and development of cultured radish roots. American Journal of Botany, 59, 744–751.
37. Wightman, F. & Thimann, K.V. (1980) Hormonal factors controlling the initiation and development of lateral roots. 1. Source of primordia-inducing substances in the primary root of pea seedlings. Physiologia Plantarum, 49, 13–20.
38. Wightman, F., Schneider, E.A. & Thimann, K.V. (1980) Hormonal factors controlling the initiation and development of lateral roots. II. Effects of exogenous growth factors on lateral root formation in pea roots. Physiologia Plantarum, 49, 304–314.
39. Robbins, W.J. & Hervey, A. (1978) Auxin, cytokinin and growth of excised roots of *Bryophyllum calycinum*. American Journal of Botany, 65, 1132–1134.
40. Skoog, F. & Miller, C.O. (1957) Chemical regulation of growth and organ formation in plant tissues cultured *in vitro*. Symposium of the Society for Experimental Biology, 11, 118–130.
41. Torrey, J.G. (1956) Chemical factors limiting lateral root formation in isolated pea roots. Physiologia Plantarum, 9, 370–388.
42. Torrey, J.G. (1962) Auxin and purine interactions in lateral root initiation in isolated pea root segments. Physiologia Plantarum, 15, 177–185.
43. Butcher, D.N. & Street, H.E. (1960) The effect of gibberellins on the growth of excised tomato roots. Journal of Experimental Botany, 11, 206–216.
44. Goodwin, P.B. & Morris, S.C. (1979) Application of phytohormones to pea roots after removal of the apex: effect on lateral root production. Australian Journal of Plant Physiology, 6, 195–200.
45. Böttger, M. (1974) Apical dominance in roots of *Pisum sativum* L. Planta, 121, 253–261.
46. Torrey, J.G. (1954) The role of vitamins and micronutrient elements in the nutrition of the apical meristem of pea roots. Plant Physiology, 29, 279–287.
47. Street, H.E. & Henshaw, G.G. (1966) Growth, differentiation and organogenesis in plant tissue and organ culture. In Cells and Tissues in Culture Vol. 3. (ed E.N. Willmer) pp. 459–689 Academic Press, London, New York, San Francisco.

62

48. Goforth, P.L. & Torrey, J.G. (1977) The development of isolated roots of *Comptonia peregrina* (Myricaceae) in culture. American Journal of Botany, 64, 476–482.
49. Rivier, L., Milon, H. & Pilet, P-E. (1977) Gas chromatography-mass spectrometric determinations of abscisic acid levels in the cap and the apex of maize roots. Planta, 134, 23–27.
50. Böttger, M. (1978) The occurrence of cis-trans- and trans-trans-xanthoxin in pea roots. Zeitschrift für Pflanzenphysiologie, 86, 265–268.
51. Evans, L.S. (1979) Developmental aspects of roots of *Pisum sativum* influenced by the G_2 factor from cotyledons. American Journal of Botany, 66, 880–886.
52. Bowen, G.D. (1981) Coping with low nutrients. In The Biology of Australian Plants (eds J.S. Pate and A.J. McComb) pp. 33–64. University of Western Australia Press, Nedlands.
53. Lamont, B. (1982) Mechanisms for enhancing nutrient uptake in plants, with particular reference to Mediterranean South Africa and Western Australia. Botanical Review, 48, 597–689.
54. Harley, J.L. & Smith, S.E. (1983) *Mycorrhizal Symbiosis*. Academic Press, London.
55. Paleg, L.G. & Aspinall, D. (eds) (1979) Physiology and Biochemistry of Drought Resistance in Plants. Academic Press, Sydney.
56. Vartanian, N. (1981) Some aspects of structural and functional modifications induced by drought in root systems. Plant and Soil, 63, 83–92.
57. Brown, M.E. (1972) Plant growth substances produced by micro-organisms of soil and rhizosphere. Journal of Applied Bacteriology, 35, 443–451.
58. Lynch, J.M. (1982) Interactions between bacteria and plants in the root environment. In Plants and Bacteria. (eds M. Rhodes-Roberts and F.A. Skinner), pp. 1–23. Academic Press, London, New York, San Francisco.
59. Dobereiner, J. (1974) Nitrogen-fixing bacteria in the rhizosphere. In *The Biology of Nitrogen Fixation* (ed A. Quispel), pp. 86–120. North Holland, Amsterdam, The Netherlands.
60. Gaskins, M.H. & Hubbell, D.H. (1979) Response of non-leguminous plants to root inoculation with free-living diazotrophic bacteria. In The Soil-Root Interface (eds J.L. Harley and R.S. Russell), pp. 176–182. Academic Press, London.
61. Brown, M.E. (1982) Nitrogen Fixation by free-living bacteria associated with plants – fact or fiction? In Bacteria and Plants (eds M.E. Rhodes-Roberts and F.A. Skinner), pp. 25–41. Academic Press, London, New York, San Francisco.
62. Callaham, D., Newcomb, W., Torrey, J.G. & Peterson, R.L. (1979) Root hair infection in actinomycete-induced root nodule initiation *in Casuarina, Myrica* and *Comptonia*. Botanical Gazette 140, (Suppl.) S1–S9.
63. Bowes, B., Callaham, D. & Torrey, J.G. (1977) Time-lapse photographic observations and morphogenesis in root nodules of *Comptonia peregrina* (Myricaceae). American Journal of Botany, 64, 516–525.
64. Callaham, D. & Torrey, J.G. (1977) Prenodule formation and primary nodule development in roots of *Comptonia* (Myricaceae). Canadian Journal of Botany, 55, 2306–2318.
65. Torrey, J.G. (1978) Nitrogen fixation by actinomycete-nodulated angiosperms. BioScience 28, 586–591.
66. Torrey, J.G. & Callaham, D. (1978) Determinate development of nodule roots in actinomycete-induced root nodules of *Myrica gale*. Canadian Journal of Botany, 56, 1357–1364.
67. Schwintzer, C.R., Berry, A.M. & Disney, L.D. (1982) Seasonal patterns of root nodule growth, endophyte morphology, nitrogenase activity, and shoot development in *Myrica gale*. Canadian Journal of Botany, 60, 746–757.
67a. Torrey, J.G. (1976) Initiation and development of root nodules of *Casuarina* (Casuarinaceae). American Journal of Botany, 63, 335–344.
68. Angulo Carmona, A.F. (1974) La formation des nodules fixateurs d'azote chez *Alnus glutinosa* (L.) Vill. Acta Botanica Neerlandica, 23, 257–303.
69. Becking, J.H. (1975) In *The Development and Function of Roots*. (eds J.G. Torrey and D.T. Clarkson) pp. 507–566. Academic Press, London, New York, San Francisco.

70. Callaham, D., DelTredici, P. & Torrey, J.G. (1978) Isolation and cultivation *in vitro* of the actinomycete causing root nodulation in *Comptonia*. Science, 199, 899–902.

71. Wheeler, C.T., Crozier, A. & Sandberg, G. (1984) The biosynthesis of indole-3-acetic acid by *Frankia*. Plant and Soil, 78, 99–104.

72. Wheeler, C.T., Henson, I.E. & McLaughlin, M.E. (1979) Hormones in plants bearing actinomycete nodules. Botanical Gazette, 140 (Suppl.), S52–S57.

73. Rivier, L. & Pilet, P.-E. (1974) Indolyl-3-acetic acid in cap and apex of maize roots: identification and quantification by mass fragmentography. Planta, 120, 107–112.

74. Bridges, I.G., Hillman, J.R. & Wilkins, M.B. (1973) Identification and localization of IAA in primary roots of *Zea mays* by mass spectrometry. Planta, 115, 189–192.

75. Greenwood, M.S., Hillman J.R., Shaw, S. & Wilkins, M.B. (1973) Localization and identification of auxin in roots of *Zea mays*. Planta, 109, 369–374.

76. Pengelly, W.L. & Torrey, J.G. (1982) The relationships between growth and indole-3-acetic acid content of roots of *Pisum sativum* L. Botanical Gazette, 143, 195–200.

77. Short, K.C. & Torrey, J.G. (1972) Cytokinins in seedling roots of pea. Plant Physiology, 49, 155–160.

78. Van Staden, J. & Smith, A.R. (1978) The synthesis of cytokinins in excised roots of maize and tomato under aseptic conditions. Annals of Botany, 42, 751–753.

79. Radley, M. (1961) Gibberellin-like substances in plants. Nature, 191, 684–685.

80. Dullaart, J. & Duba, L.I. (1970) Presence of gibberellin-like substances and their possible role in auxin bioproduction in root-nodules and roots of *Lupinus luteus* L. Acta Botanica Neerlandica, 19, 877–883.

81. Williams, P.M. & Sicardi de Mallorca, M. (1982) Abscisic acid and gibberellin-like substances in roots and root nodules of *Glycine max*. Plant and Soil, 65, 19–26.

82. Böttger, M. (1978) Levels of endogenous indole-3-acetic acid and abscisic acid during the course of the formation of lateral roots. Zeitschrift für Pflanzenphysiologie, 86, 283–286.

83. Dullaart, J. (1970) The auxin content of root nodules and roots of *Alnus glutinosa* (L.) Vill. Journal of Experimental Botany, 21, 975–984.

84. Silver, W.S., Bendana, F.E. & Powell, R.D. (1966) Root nodule symbiosis. The relation of auxin to root geotropism in roots and nodules of non-legumes. Physiologia Plantarum, 19, 207–218.

85. Henson, I.E. & Wheeler, C.T. (1976) Hormones in plants bearing nitrogen-fixing root nodules: the distribution of cytokinins in *Vicia faba* L. The New Phytologist, 76, 433–439.

86. Henson, I.E. & Wheeler, C.T. (1977) Hormones in plants bearing nitrogen-fixing root nodules: gibberellin-like substances in *Alnus glutinosa* (L.). The New Phytologist, 78, 373–381.

87. Henson, I.E. & Wheeler, C.T (1977) Hormones in plants bearing nitrogen-fixing root nodules: cytokinin levels in roots and root nodules of some non-leguminous plants. Zeitschrift für Pflanzenphysiologie, 84, 179–182.

88. Bano, A., Watts, S.H., Hillman, J.S. & Wheeler, C.T. (1983) Abscisic acid and nitrogen fixation in *Faba vulgaris* (*Vicia faba*) and *Alnus glutinosa*. In Interactions between Nitrogen and Growth Regulators in the Control of Plant Development, Monograph 9 (ed M.B. Jackson), pp. 5–31. British Plant Growth Regulator Group, Wantage, U.K.

89. Watts, S.H., Wheeler, C.T. Hillman, J.R., Berrie, A.M.M., Crozier, A. & Math, V.B. (1983) Abscisic acid in the nodulated root system of *Alnus glutinosa*. The New Phytologist, 95, 203–208.

90. Dullaart, J. (1967) Quantitative estimation of indole-acetic acid and indole carboxylic acid in root nodules and roots of *Lupinus luteus* L. Acta Botanica Neerlandica, 16, 222–230.

91. Syōno, K., Newcomb, W. & Torrey, J.G. (1976) Cytokinin production in relation to the development of pea root nodules. Canadian Journal of Botany, 54, 2155–2162.

92. Syōno, K. & Torrey, J.G. (1976) Identification of cytokinins in root nodules of the garden pea, *Pisum sativum* L. Plant Physiology, 57, 602–606.

93. Wheeler, C.T., Watts, S.H. & Hillman, J.R. (1983) Changes in carbohydrates and nitrogenous

64

compounds in the root nodules of *Alnus glutinosa* in relation to dormancy. The New Phytologist, 95, 209–218.

94. Rodriguez-Barrueco, C. (1966) Fixation of nitrogen in nodules of *Alnus jorullensis*. H.B. & K. Phyton (Buenos Aires) 23, 103–110.

95. Becking, J.H. (1968) Nitrogen fixation by non-leguminous plants. In Nitrogen in Soil. pp. 47–74. Dutch Nitrogenous Fertilizer Review No. 12.

96. Tjepkema, J. (1978) The role of oxygen diffusion from the shoots and nodule roots in nitrogen fixation by root nodules of *Myrica gale*. Canadian Journal of Botany, 56, 1365–1371.

97. Thimann, K.V. (1936) On the physiology of the formation of nodules on legume roots. Proceedings of the National Academy of Science (U.S.A.), 22, 511–514.

98. Kefford, N.P., Brockwell, J. & Zwar, J.A. (1960) The symbiotic synthesis of auxin by legumes and nodule bacteria and its role in nodule development. Australian Journal of Biological Sciences, 13, 456–467.

99. Nutman, P.S. (1951) Studies on the physiology of nodule formation. III. Experiments on the excision of root tips and nodules. Annals of Botany, 16, 79–101.

100. Dart, P.J. (1975) Legume root nodule initiation and development. In The Development and Function of Roots. (eds J.G. Torrey and D.T. Clarkson), pp. 467–506. Academic Press, London, New York, San Francisco.

101. Dart, P.J. (1977) Infection and development of leguminous nodules. In A Treatise on Dinitrogen Fixation. Section III. Biology. (eds R.W.F. Hardy and W.S. Silver), pp. 367–472. John Wiley & Son, New York.

101a. Newcomb, W. (1981) Nodule morphogenesis and differentiation. International Review of Cytology. Supplement 13, 247–296.

101b. Torrey, J.G. and S. Barrios (1969) Cytological studies on rhizobial nodule initiation in *Pisum*. *Caryologia* 22, 47–62.

102. Sprent, J.I. (1980) Root nodule anatomy, type of export product and evolutionary origin of some Leguminosae. Plant, Cell and Environment, 3, 35–43.

103. Roughley, R.J., Dart, P.J. & Day, J.M. (1976) The structure and development of *Trifolium subterraneum* L. root nodules. I: in plants grown in optimal root temperatures. Journal of Experimental Botany, 27, 431–440.

104. Libbenga, K.R. & Harkes, P.A.A. (1973) Initial proliferation of cortical cells in the formation of root nodules in *Pisum sativum* L. Planta, 144, 17–28.

105. Libbenga, K.R., VanIren, F., Boyers, R.J. & Shraag-Lamers, M.F. (1973) The role of hormones and gradients in the initiation of cortex proliferation and nodule formation in *Pisum sativum* L. Planta, 144, 29–39.

106. Badenoch-Jones, J., Summons, R.E., Djordjevic, M.A., Shine, J., Letham, D.S. & Rolfe, B.G. (1982) Mass spectrometric quantification of indole-3-acetic acid in *Rhizobium* culture supernatants: relation to root hair curling and nodule initiation. Applied Environmental Microbiology, 44, 275–280.

107. Wang, T.L., Wood, E.A. & Brewin, N.J. (1982) Growth regulators, *Rhizobium* and nodulation in peas. Planta, 155, 345–349.

108. Phillips, D.A. & Torrey, J.G. (1970) Cytokinin production by *Rhizobium japonicum*. Physiologia Plantarum, 23, 1057–1063.

109. Phillips, D.A. & Torrey, J.G. (1972) Studies on cytokinin production by *Rhizobium*. Plant Physiology, 49, 11–15.

110. Azcon, R., Azcon, G.-De Quilar, C. & Barea, J.M. (1978) Effects of plant hormones present in bacterial cultures on the formation and responses of VA endomycorrhiza. The New Phytologist, 80, 359–364.

111. Katznelson, H. & Cole, S.E. (1965) Production of gibberellin-like substances by bacteria and actinomycetes. Canadian Journal of Microbiology, 11, 733–741.

112. Brown, M.E. (1976) Role of *Azotobacter paspali* in association with *Paspalum notatum*. Journal

Applied Bacteriology, 40, 341–348.

112a. Trinick, M.J. (1976) *Rhizobium* symbiosis with a non-legume. In Proceedings 1st International Symposium on Nitrogen Fixation. (eds W.E. Newton and C.J. Nyman), pp. 507–517. Washington State University Press, Pullman.

113. Trinick, M.J. & Galbraith, J (1976) Structure of root nodules formed by *Rhizobium* on the non-legume *Trema cannabina var. scabra*. Archives of Microbiology, 108, 159–166.

114. Lancelle, S.A. & Torrey, J.G. (1984) Early development of *Rhizobium*-induced root nodules of *Parasponia rigida*. I. Infection and early nodule initiation. Protoplasma, 123, 26–37.

115. Lancelle, S.A. & Torrey, J.G. (1985) Early development of *Rhizobium*- induced root nodules of *Parasponia rigida*. II. Nodule morphogenesis and symbiotic development. Canadian Journal of Botany, 63, 25–35.

116. Chandler, M.R. (1978) Some observations on infection of *Arachis hypogaea* L. by *Rhizobium*. Journal of Experimental Botany, 29, 749–755.

117. Chandler, M.R., Date, R.A. & Roughley, R.J. (1982) Infection and root-nodule development in *Stylosanthes* species by *Rhizobium*. Journal of Experimental Botany, 33, 47–57.

118. Arora, N., Skoog, F. & Allen, O.N. (1959) Kinetin-induced pseudonodules on tobacco roots. American Journal of Botany, 46, 610–614.

119. Rodriguez-Barrueco, C. & De Castro, F (1974) Cytokinin-induced pseudonodules on *Alnus glutinosa*. Physiologia Plantarum, 29, 277–280.

120. Dell, B., Kuo, J. & Thomson, G.J. (1980) Development of proteoid roots in *Hakea obliqua* R. Br. (Porteaceae) grown in water culture. Australian Journal of Botany, 28, 27–37.

121. Purnell, H.M. (1960) Studies of the family Proteaceae. I. Anatomy and morphology of the roots of some Victorian species. Australian Journal of Botany, 8, 38–50.

122. Malajczuk, N. & Bowen, G.D. (1974) Proteoid roots are microbially induced. Nature, 251, 316–317.

123. Lamont, B.B. & McComb, A.J. (1974) Soil micro-organisms and the formation of proteoid roots. Australian Journal of Botany, 22, 681–688.

124. Lamont, B. (1981) Specialized roots of non-symbiotic origin in heathlands. In Heathlands and Related Shrublands of the World. B. Analytical Studies. (ed R.L. Specht), pp. 183–195. Elsevier, Amsterdam.

125. Lamont, B. (1976) The effects of seasonality and waterlogging on the root systems of a number of *Hakea* species. Australian Journal of Botany, 24, 691–702.

126. Lamont, B. (1972) 'Proteoid' roots in the legume *Viminaria juncea*. Search, 3, 90–91.

127. Trinick, M.J. (1977) Vesicular-arbuscular infection and soil phosphorus utilization in *Lupinus*. The New Phytologist, 78, 297–304.

128. Walker, B.A., Pate, J.S. & Kuo, J. (1983) Nitrogen fixation by nodulated roots of *Viminaria juncea* (Schrad. & Wendl.) Hoffmans (Fabaceae) when submerged in water. Australian Journal of Plant Physiology, 10, 409–421.

129. Diem, J.G. Gueye, I., Gianinazzi-Pearson, V., Fortin, J.A. and Dommergues, F.R. (1981) Ecology of VA mycorrhizae in the tropics: the semi-arid zone of Senegal. Acta Oecologia/ Oecologia Plantarum 2, 53–62.

130. Dobereiner, J. & Day, J (1976) Associative symbiosis in tropical grasses: characterization of micro-organisms and dinitrogen fixing sites. In *Proceedings of the First International Symposium on Nitrogen Fixation*. (eds W.E. Newton and C.J. Nyman), pp. 518–538. Washington State University Press, Pullman.

131. Dobereiner, J., Burris, R.H. & Hollaender, H. (1978) *Limitations and Potentials for Biological Nitrogen Fixation in the Tropics*. Plenum Press, New York.

132. Stewart, W.D.P., Rowell, P. & Lockhart, C.M. (1979) Associations of nitrogen-fixing pro-karyotes with higher and lower plants. In Nitrogen Assimilation of Plants (eds E.J. Hewitt and C.V. Cutting), pp. 45–66. Academic Press, London, New York, San Francisco.

133. Barea, J.M. & Brown, M.E. (1974) Effects on plant growth produced by *Azotobacter paspali*

related to synthesis of plant growth regulating substances. Journal of Applied Bacteriology, 37, 583–593.

134. Kloepper, J.W., Leong, J., Teintze, H. & Schroth, M.N. (1980) Enhanced plant growth by siderophores produced by plant growth-promoting rhizobacteria. Nature, 286, 885–886.

135. Schroth, M.N. & Hancock, J.G. (1982) Disease-suppressive soil and root colonizing bacteria. Science, 216, 1376–1381.

136. Tepfer, D.A. & Tempé, J. (1981) Production d'agropine par des racines formées sous l'action d'*Agrobacterium rhizogenes* souche A4. Comptes Rendus des séances de l'Académie des Sciences, Paris, 292, 153–156.

137. Tepfer, D. (1983) The potential uses of *Agrobacterium rhizogenes* in the genetic engineering of higher plants: Nature got there first. In Genetic Engineering in Eurkaryotes (eds P.F. Lurquin and A. Kleinhofs), pp. 153–164. Plenum Press, New York.

137a. Tepfer, D. (1984) Transformation of several species of higher plants by *Agrobacterium rhizogenes:* sexual transmission of the transformed genotype and phenotype. Cell, 37, 959–967.

138. Chilton, M.-D., Tepfer, D.A., Petit, A., David, C., Casse-Delbart, F. & Tempé, J. (1982) *Agrobacterium rhizogenes* inserts T-DNA into the genomes of the host plant root cells. Nature, 295, 432–434.

139. Schell, J. (1975) The role of plasmids in crown gall formation by *A. tumefaciens.* In Genetic Manipulations with Plant Materials. (ed L. Ledoux), pp. 163–181. Plenum Press, New York.

140. Chilton, M.D., Drummond, H.J., Merlo, D.J., Sciaky, D., Montoya, A.L., Gordon, M.P. & Nester, E.W. (1977) Stable incorporation of plasmid DNA into higher plant cells : the molecular basis of crows gall tumorigenesis. Cell, 11, 263–271.

141. Liu, S.T., & Kado, C.I. (1977) Indoleacetic acid production: a plasmid function of *Agrobacterium tumefaciens* C 58. Biochemical Biophysical Research Communications. 90, 171–178.

142. Liu, S.-T., Perry, K.L., Schardl, C.L. & Kado, C.I. (1982) *Agrobacterium* indoleacetic acid gene is required for crown gall oncogenesis. Proceedings of the National Academy of Sciences (U.S.A.). 79, 2812–2816.

143. Nester, E.W. Gordon, M.P., Amasino, R.M. & Yanofsky, M.F. (1984) Crown gall: A molecular and physiological analysis. Annual Review of Plant Physiology 35, 387–413.

144. Long, S.R., Buikema, W.J. & Ausubel, F.M. (1982) Cloning of *Rhizobium meliloti* nodulation genes by direct complementation of Nod⁻ mutants. Nature, 298, 485–488.

3. Adventitious roots of whole plants: their forms, functions, and evolution

PETER W. BARLOW[1]

Agricultural and Food Research Council Letcombe Laboratory, Wantage, Oxfordshire, OX12 9JT, UK

[1] Present Address: Long Ashton Research Station, University of Bristol, Long Ashton, Bristol, BS18 9AF, UK

1. Introduction

Many plants have two root systems that differ in origin. One is the *primary root system* whose origin can be traced back to the radicle developed during embryogenesis. The other is an *adventitious root system* which arises on parts of the plant not originating from the embryonic root – that is, the roots arise on parts of the shoot. Adventitious roots usually initiate endogenously from tissue within the parent plant (see Chapter 4) though a few cases of exogenous origin are known [1]. Roots which arise on the primary root out of the usual acropetal sequence that characterises lateral roots, either as a normal part of development or after experimental treatment, are sometimes also called adventitious. The term *adventive* can perhaps be applied to such roots to distinguish them from roots of shoot origin.

Because their origin is on the shoot system, and shoots are often (but not invariably) above ground, the adventitious roots of some species have had opportunity to develop a structure and function somewhat different from those of the underground primary root system. Nevertheless, a principal physiological function of both root systems is extracting water and nutrients from their surroundings. An aerial habit, however, enables adventitious roots to exploit a part of the plant's environment inaccessible to the primary root system. For instance, adventitious roots arising on the branches of the maple, *Acer macrophyllum*, (Aceraceae) penetrate detritus accumulating among epiphytes growing in the canopy of this tree [2] and so tap mineral resources in a locality other than the soil. A second property of a shoot-borne root system is that it may function as an additional mechanical support for the shoot system and thereby permit further shoot enlargement: adventitious roots can be pillars and buttresses, in addition to serving as devices that enable the shoot to climb or scramble over neighbouring plants and the terrain [3, 4]. These features may also be an important adjunct in the vegetative propagation of the individual. Aerial roots may also have a 'glandular' function, synthesising certain specialized metabolites such as cytokinins and amino acids needed for the growth of nearby shoot tissues [5]. In a few species, adventitious roots are modified as tubers and serve as stores of protein and carbohydrate.

Adventitious roots are common throughout the angiosperms; the species that bear them range from xerophytes to hydrophytes. The form and origin of the roots are correspondingly wide-ranging and probably help members of a given species adapt to their characteristic environment. Moreover, whether a species forms an adventitious root system or not can depend on the environment. For example, adventitious roots are unknown on *Eucalyptus robusta* (Myrtaceae) growing in Australia, but they form freely in the moister climate of Hawaii [6].

The aim of this chapter is to draw attention to the variety of forms of adventitious roots. An exhaustive treatment is not attempted; the reader interested to learn more about the various forms and origins of adventitious roots is referred to the scholarly treatises by Troll [7] and Goebel [8, 9] and the useful book devoted to roots by Weber [9a]. For our purpose the most convenient way of illustrating this variety of forms is to select species whose roots show some particular feature related to the functions mentioned above. The examples also illustrate that the adventitious root system can rival the more familiar shoot system in diversity of structure, colour and form. In addition, study of the adventitious system suggests ideas about how root initiation and development are controlled, and the contribution of that system to the growth of the individual plant and the perpetuation of the species.

2. Adventitious roots of terrestrial plants

Terrestrial plants fall broadly into species that are rooted in the ground, rooted on the surface of other plants (epiphytes), or rooted at each of these sites during some stage of their life (hemi-epiphytes). Adventitious roots are present in all three groups and may help the species adapt to each of these modes of life. One way in which this is achieved is through the development of alternative forms of adventitious roots. For example, two anatomically and functionally distinct types of root are found in many hemi-epiphytes and lianes [10, 11]. Root dimorphism is also found in water plants; their adventitious root systems will be discussed in Section 3.

2.1. Aerial roots of hemi-epiphytes and epiphytes, and the problem of root dimorphism

Epiphytes and hemi-epiphytes often have a well developed aerial adventitious root system. Went [10] made many interesting observations on the adventitious roots of such plants, and his illustration of the aerial roots of *Philodendron melanochrysum* (Araceae) (a liane that is also a hemi-epiphyte) is reproduced in Fig. 1. This plant (and some other species in Araceae) has two types of root, clasping roots and feeding roots, which develop at the same node on the stem. The root system is, therefore, dimorphic. As a rule, the clasping roots are more numerous and grow horizontally to girdle the stem of the support plant, while the feeding roots descend to the ground. The two roots also differ in their internal anatomy. The feeding roots have well developed vascular tissues but poorly developed sclerenchyma; the converse holds for the clasping roots.

 How such alternative root forms are established and maintained with only rare intermediate forms has not been investigated experimentally, though it is obviously a fundamental developmental issue. Presumably, the future development of the primordium of each root type is determined at its inception or at a very early stage of growth. Differences in the positioning of the primordium on the

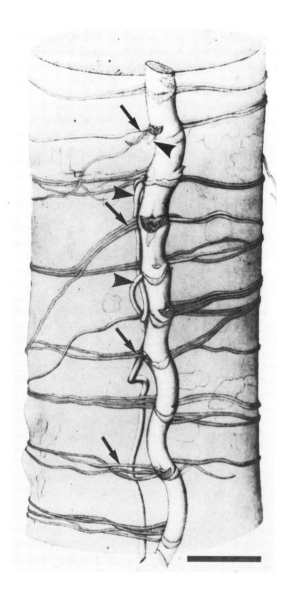

Fig. 1. Clasping and feeding roots of the liane *Philodendron melanochrysum.* A number of clasping roots (small arrows) arise at each node on the shaded side of the liane's stem and twine around the trunk of the support tree as horizontal thongs. A single feeding root (arrowheads) arises below each leaf scar on the opposite (outward-facing) side of the stem to the clasping root and grows downwards. Note also the difference in thickness of the two types of root. Scale bar is 5 mm. From [10].

stem, or the time of formation with respect to one another or in relation to some change in the physiology of the parent tissue, may be partly responsible for the particular course of determination. It is known, for example, that small spatial differences across the meristematic plaque in the angle of petiole and stem of water-cress (*Rorippa nasturtium-aquaticum* (Cruciferae) seem to be associated with the development of either adventitious shoots or roots from this site [12]. Shoots arise at the outside of the plaque, roots at the inside. The apices of the two types of roots, once they have been formed and determined, continue to produce tissues of the appropriate type. However, changes in the external environment of the root can modify its internal structure; anatomical changes that occur when aerial roots of *Ficus benghalensis* (Moraceae) and *Monstera deliciosa* (Araceae) enter the soil are listed in Table 1. These changes are presumably the result of a redetermination of the meristem causing it to produce a new pattern of tissue differentiation.

2.1.1. Stranglers. One of the most striking forms of adventitious roots amongst the hemi-epiphytes is associated with the 'strangling' habit. Examples of 'stranglers' are *Clusia rosea* (Guttiferae), *Ficus* spp. (Moraceae) and *Metrosideros* spp. (Myrtaceae). All these species are trees when adult, and all form a trunk from roots of adventitious origin which add to and fuse with the primary root. It must be admitted, however, that in most cases the exact details of trunk formation are unknown since these trees start life as epiphytes high in the canopy of the support tree.

Root development of *Clusia rosea* was fully described by Schimper [15]. The seed of *C. rosea* germinates in humus that has collected in the bark of the support

Table 1. Characters of aerial roots before and after entry into the soil and their consequent conversion into terrestrial roots.

Species	Character	Root	
		Aerial	Terrestrial
Ficus benghalensis	Hairs	Absent	Present
	Lenticels	Many	Few
	Endodermis	Poorly developed	Well developed
	Pith	Distinct	Absent
	Secondary vascular	Many fibres	Few fibres
	tissue	Few small vessels	Numerous large vessels
		Little parenchyma	More parenchyma
Monstera deliciosa	Hairs	Absent	Present
	Trichosclereids	Many	Few
	Raphides	Some	Many

Data from [13] for *Ficus,* and [14] for *Monstera.*

72

tree, or in the persistent leaf bases if the support tree is a palm. The primary root and its laterals remain small and are sufficient only to establish the young seedling. Later, adventitious roots arise from the stem and grow in all directions over the surface of the support tree. They clasp the tree by means of their root hairs and extract nutrients from the surface of the bark. They also prop up the growing shoot system. Later, one root out of this mass of adventitious roots becomes positively geotropic and grows vigorously downwards and enters the soil. This first-formed adventitious root system is thus differentiated into horizontal clasping roots and a vertical feeding root. The adventitious root system continues to develop producing the occasional feeding root along with many clasping roots. The clasping roots are slow growing and are said to show a strong negative phototropism; they die if after reaching a certain length they do not find a support. The feeding root, on the other hand, elongates quickly and shows no phototropism.

The same general features hold for strangling figs, 'the thugs of the vegetable kingdom' [16]. In *Ficus religiosa* it is the primary root and its laterals that make the major contribution to the strangling 'root-trunk' system [17], though adventitious aerial roots may later add to it. The roots which descend the stem have a great propensity to fuse [17, 18] (Fig. 2) by a process of natural grafting [19]. Kerner von Marilaun [20] writes of these roots, which: '. . . *nestling to the substratum, flatten and spread like a doughy plastic mass; the adjacent roots fuse together, and in this way irregular lattice works or incrusting mantles, only interrupted here and there by gaps, are formed, which lie on the supporting trunks and are firmly fastened and cemented to them without fusing with it or deriving nourishment from it.'* Fusion of the roots continues underground [21] enhancing the anchorage of the plants in the soil and also re-arranging the paths of nutrient and water flow from the soil into the aerial parts.

The rata of New Zealand, *Metrosideros robusta*, owes its habit to a root development similar to that of the strangling figs. The rata in its epiphytic form (occasionally plants grow directly in the ground) sends a strong root, probably the tap root, down the trunk of the support tree. Aerial roots, which may be adventitious though this is not properly documented, fuse with this main root. From this root-trunk lateral, or adventive, roots arise which bind the root-trunk to the trunk of the support tree [22, 23, 24].

The 'strangling' of the support tree by the roots of the fig and rata has perhaps been exaggerated. Kirk [22] writes of the roots of *Metrosideros robusta* that the stem of the support tree 'is literally strangled by their iron embrace'. Probably the reason for the death of the support tree is that it is already fairly old when the epiphyte lodges in it, and that its demise is hastened by its canopy being shaded by the canopy of the more vigorous epiphyte. A support tree such as the pururi (*Vitex littoralis* (Verbenaceae)) can survive the 'iron embrace' of the rata, probably because its canopy is more tolerant of shade [22]. The encircling lattice of roots, particularly one as extensive as that of the fig, may also cause an unfavoura-

Fig. 2. A species of *Ficus* with the banyan habit. Aerial roots descend from the shoots and branches and often fuse when they touch each other. In the centre of the photograph the roots can be seen to have fused to form a massive root-trunk. Photograph kindly provided by Dr Caroline M. Pannell, Botany School, Oxford University, England.

ble exchange of gases (O_2, CO_2, ethylene) in the underlying living cells in the supporting trunk and so perhaps impair their vitality.

Whatever the cause of death of the support tree after its encounter with the strangler, the result is that its site is occupied by the former epiphyte which now exists as a tall tree in its own right. In the case of strangling figs, the crumbling remains of the support tree occupy the hollow centre of the fig's root-trunk [18].

The successful establishment of *Clusia* and *Ficus* depends on the vigorous growth of their adventitious clasping roots, and in particular the orientation of this growth. The process that orientates clasping roots is totally unexplored. However, many authors have written that these roots are negatively phototropic (see Schimper [15] on *Clusia*), or light-avoiding (see Kerner von Marilaun [20] on *Ficus*); or, as Corner [25] puts it, the aerial roots of *Ficus* may grow towards the trunk 'as though they were able to see it'. The roots may indeed 'see' the trunk and branches around which they will later grow; they may possess some mechanism for sensing and growing towards dark areas rather than simply growing away from light (cf. observations on stems of seedlings of *Monstera gigantea* which bend and grow towards dark objects from up to 2 m way [26]). If a dark-sensing mechanism exists in aerial roots it would explain their ability to grow towards a support. Maybe such roots are also insensitive to gravity since a geotropic response of any sort (positive, negative or diageotropic) would be of no advantage and could even be disadvantageous.

Another related question is how the aerial roots adhere to a surface. Writing of a specimen of *Ficus repens*, Tennet [16] recalls: *'one which has fixed itself on the walls of a ruined edifice, . . . its roots streaming downwards over walls as if they had once been fluid, follow every sinuosity of the building and terraces till they reach the earth.'* At much the same time, Charles Darwin was observing the climbing behaviour of these roots and supplied the answer to this question. He noticed that the young root exuded a clear liquid containing caoutchouc. He presumed the root re-absorbed part of the liquid, but not the caoutchouc which dried and cemented the root to the surface [3].

2.1.2. Banyan trees. Fig trees with the banyan habit owe their remarkable structure to their adventitious aerial roots which grow down from, and ultimately support, the branches (Fig. 2). These prop, or pillar, roots allow the banyan, *Ficus benghalensis*, to reach a spectacular size: specimens have been found with a canopy 500–600 m in circumference supported by many hundreds of prop roots, some up to 3.7 m in girth [27]. Some other species of *Ficus* (e.g. *F. aurea*, *F. microcarpa*) also display the banyan habit. However, the prop roots of both *F. benghalensis* and *F. microcarpa* only develop in shady, wet habitats, such as the jungle; if these trees grow in more open or in arid habitats the aerial roots either do not develop or do not reach the ground [28].

A banyan starts life as an epiphytic strangling fig in the manner described earlier. J.D. Hooker [29] in his 'Himalayan Journals' of 1854, writing of the banyan tree (*F. benghalensis*) in Calcutta Botanical Gardens – 'the pride and ornament of the garden' – recalls that in 1782 its site was occupied *'by a Kujoor (Date-palm), out of whose crown the banyan sprouted, and beneath which a Fakir sat'*. In India the banyan is sacred and the young aerial roots are encouraged to form by applying mud cakes to the branches; the new roots are then provided with bamboo tubes in which to grow to a ground specially prepared for them [29]. Hooker continues with a description of the further development of the prop roots: *'They are mere slender whip-cords before reaching the earth, where they root, remaining very lax for several months; but gradually, as they grow and swell to the size of cables, they tighten, and eventually become very tense. This is a curious phenomenon, and so rapid, that it appears to be due to the rooting part mechanically dragging down the aerial.'* A similar tightening of the aerial roots of tropical lianes occurs when they enter the ground [30]. We know now that the aerial part of the *Ficus* prop root is not tightened by 'the rooting part mechanically dragging down the aerial', but is actually due to the contraction of the aerial part as the result of the formation of tension wood [31]. Entry into the soil is the stimulus for tension wood development; entry into water is not sufficient. In their final phase of development these prop roots are under compression rather than tension owing to the weight of the branch they support. They develop secondary thickening and so are better able to support this weight. Thus, the roots pass through three phases during their life: a freehanging phase, a contraction phase and a consolidation phase [31].

The original trunk of the banyan eventually dies together with many of its horizontal branches leaving a grove of separate trees. The trunk of each of these trees is derived from a prop root of the original tree. In 1905 Trelease [18] recounted that this disintegrated state was now the condition of the Calcutta Banyan described by Hooker [29] 50 years earlier. The integrity of this massive tree had therefore lasted for some 120 years.

2.1.3. Epiphytes. Epiphytes are a prominent feature of the tropical rain forest [11, 15, 30]. Many of them belong to the families Araceae, Bromeliaceae and Orchidaceae. Their root systems are mainly of adventitious origin and vary considerably in quantity and form. In fact, roots of orchids are solely adventitious and range in form from the usual type of root to one that is a mere pad conjoint with the stem, as in *Cheirostylis phillipinensis* [7, 32]. Orchids of the genus *Microcoelia* are practically all root (Fig. 7A) and have a greatly reduced shoot, while *Tillandsia usneoides* (Bromeliaceae) has practically no root system in the adult state.

The root systems of epiphytes have been classified by Goebel [8] into two broad types: 'nest roots' and 'assimilation roots'. In the former, which is found in aroids (e.g. *Anthurium* spp.) and orchids (e.g. *Cymbidium* spp.), the root system develops as a tangled mass resembling a bird's nest. Humus collects among the roots which, in the case of certain orchids (e.g. *Grammatophyllum, Ansellia*), are thin, stiff and upward-growing, and are well adapted for this purpose [32]. Assimilation roots are not particularly adapted for humus collection, but like the nest roots they absorb water which they also store in dry periods when the plant has lost its leaves [8].

The adventitious root system of certain orchids and other species have developed an important association with ants and are used by them to form part of their residential 'ant gardens' [33, 34]. The 'garden' gives shelter to the associated ant colony, and the ants, besides protecting the plant against predators and dispersing its seeds, also provide additional minerals to the plant. Ant gardens encourage the production of adventitious roots: for example, clasping roots extending from the ant garden on *Codonanthe uleane* (Gesneriaceae) permit this plant to live as a epiphyte rather than as a terrestrial plant, which is its form in the absence of the ant association [33].

Among the epiphytes that benefit from an association with ants is the curious plant *Dischidia rafflesiana* (Asclepiadaceae). This plant produces two types of leaf; one of these is an 'ant-leaf' and is formed like a pitcher with a lid [35, 36]. Then, from the petiole one or two adventitious roots grow into the pitcher (Fig. 3). Water, humus and the detritus brought in by colonizing ants collect in the pitcher and cause the roots to branch [38]. These roots extract water and nutrients from the contents of the ant-leaf and probably play an important role in the nutrition of the plant. A functionally somewhat similar structure exists on the fern *Solanopteris brunei*. Here, the rhizome forms tubers the interior of which breaks down when the tuber is mature. The inner surface of the resulting cavity then sprouts roots. These roots initially may recycle the minerals released by cell lysis, but more importantly in the long term, they absorb minerals brought into the hollow tubers by ant colonies [39].

One other aspect of the biology of roots of epiphytes and hemi-epiphytes that deserves mention concerns their interaction with the support plant. Although intimate contact exists between the clasping roots of an epiphyte and its support, this is usually no more than the result of root hairs securing them firmly to the substrate [10]. Such contact is by and large harmless. But observations on the contact between certain epiphytic orchids and their support plant raise the question of whether these epiphytes parasitize the support plant. The suspicion of parasitism first arose from the belief that plantations of coffee and citrus trees carrying epiphytes were less healthy than plants cleared of them [40]. Later work by Johansson [41] on the effect of the orchid *Microcoelia exilis* on trees of *Terminalia mollis* tends to confirm an unhealthy interaction between the two plants: branches that support the epiphyte become deformed and the leaves wilt and eventually die. These effects seem to result from a complex interaction (epiparasitism) between the support, the epiphyte and a fungal endophyte associated with the latter. Roots of *Microcoelia* spp., for example, in contact with a support branch possess a fungal endophyte [42, 43]. At the contact site, the endophyte penetrates the bark of the branch which subsequently disintegrates. The orchid itself cannot be considered a parasite, however, since none of its cells penetrate those of the host and it has no specialized adaptation, such as a haustorium, for doing so.

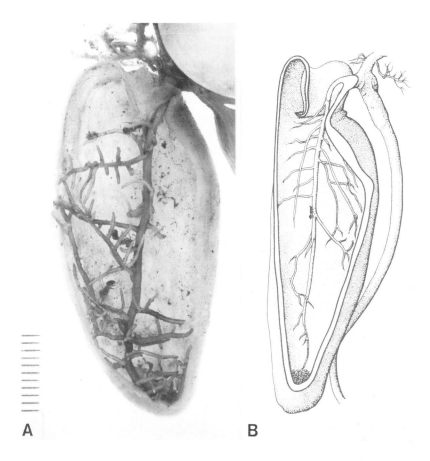

A   B

Fig. 3. The ant-leaf of *Dischidia rafflesiana* containing adventitious roots. A. Adventitious roots arise on the petiole and grow into the pitcher-like ant-leaf, one side of which has been cut away; a cut-away drawing is shown in B. Each division of the scale in A is 1 mm. The plant material was kindly provided by Mr. M. Jebb, Botany School, Oxford University, England; the drawing in B is reproduced, with permission, from [37].

2.2. Stilt roots, root spines and spine roots

Stilt roots are thick adventitious roots that emerge from the trunk or stem, sometimes up to 1 m or more above the ground, bend downwards, perhaps under their own weight, and enter the soil (Fig. 4). They are characteristic of small or medium sized-trees of the tropical rain forest, though some of the ground herbs also possess them (Fig. 4B and Table 2). Stilt roots may be round or flattened in cross-section depending on the species and on the symmetry of their secondary thickening. Roots that are flattened perhaps serve as stays, while those that are cylindrical serve as props. Well known examples of stilt roots are found in mangroves (*Rhizophora* spp.) [45] (Fig. 4A), species of *Clusia*, screw-pines

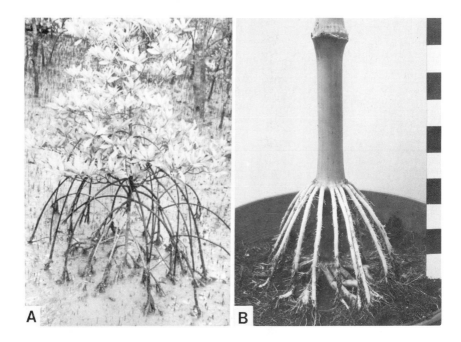

Fig. 4. Silt roots growing out from stems of (A) a small tree of *Rhizophora apiculata* and (B) *Zea mays*. Each division of the scale in B is 2 cm. The photograph of *Rhizophora* was kindly provided by Dr D.J. Mabberley, Botany School, Oxford University, England.

(*Pandanus* spp.), as well as in *Zea mays* (where they are more usually called nodal roots) (Fig. 4B). In the screw-pines their distribution (e.g., Fig. 12D) differs from species to species and can serve as a diagnostic feature for classification [45]. In some species, *Musanga smithii* (Moraceae), for example, a new shoot may arise where the stilt enters the ground [46]; the stilt root system can therefore be an important means of rapid vegetative propagation, particularly in a species such as *M. smithii* which only flowers when the tree is very old.

Striking examples of stilt roots are found in palms. Those of *Iriartea exorrhiza* have been beautifully described by Wallace [47]: '. . . *what most strikes attention in this tree, and renders it so peculiar, is, that the roots are almost entirely above ground. They spring out from the stem, each one at a higher point than the last, and extend diagonally downwards till they approach the ground, when they often divide into many rootlets, each of which secures itself in the soil. As fresh ones spring out from the stem, those below become rotten and die off; and it is not an uncommon thing to see a lofty tree supported entirely by three or four roots, so that a person may walk beneath them, or stand with a tree seventy feet growing immediately over his head.'*

It is probable that stilt root formation is a specific property and in some circumstances is also encouraged by a damp environment around the base of the stem or trunk [11, 48]. In the lower levels of a forest there is usually enough moisture in the atmosphere to permit stilt root development, but in some species stilts may arise on the trunk only in response to flooding. E.J.H. Corner (personal communication to E.A. Menninger [44]), has stated that in *Elaeocarpus macrocereus* (Tiliaceae), a tree of Malayan forest rivers and creeks, the height of the uppermost stilts corresponds to the height of maximal flooding: this may be up to 10 m above the usual water level. On the other hand, stilts develop in a few species that naturally grow on well drained, unflooded sites: e.g. *Dillenia reticulata* (Dilleniaceae) and *Xylopia ferruginea* (Annonaceae) [11].

Other types of adventitious roots, in addition to stilt roots, may be formed on the trunks of palm trees. For example, on young plants of the Central American palm *Cryosophila guagara*, McArthur and Steeves [49] found three types of adventitious roots that differed not only in location but also in form: there were stilt roots at the base of the trunk, crown roots among the bases of the living fronds, and spiny trunk roots below the crown among the bases of the dead and detaching fronds. However, older specimens (6 m or more) had a different pattern of roots: stilt roots were actively developed at higher levels on the trunk, trunk roots were found at lower levels than previously, and crown roots were absent. Presumably, the crown roots that had formed earlier had become detached. Crown roots are also lacking in other species of *Cryosophila* (*C. nana* and *C. warscewiczii*) [45].

Little work has been done on the factors responsible for regulating the type of root or their positioning on the trunk. There may be a genetic control over the rooting pattern which is strongly linked with environmental signals. The crown

Table 2. Some species of trees and herbs which possess stilt roots

Family	Species	Distribution
Trees		
Annonaceae	*Xylopia ferruginea*[1]	Malaya
Bombacaceae	*Pachira aquatica*	Tropical America
Casuarinaceae	*Casuarina sumatrana*[1]	Malaya
Combretaceae	*Rhizophora mangle*[1]	Tropical America
Dilleniaceae	*Dillenia reticulata*[1]	Malaya
Euphorbiaceae	*Bridelia micrantha*[1]	Africa
Guttiferae	*Tovomita* spp.	South America
Moraceae	*Musanga smithii*[1]	Africa
Palmeae	*Iriartea exorrhiza*[1]	South America
Sapotaceae	*Palaquium xanthochymum*	Malaya
Tiliaceae	*Elaeocarpus macrocereus*[1]	Malaya
Herbs		
Araceae	*Culcasia striolata*	Africa
Cyperaceae	*Mapania* spp.	Malaya
Gramineae	*Zea mays*[1]	North America

[1] Species referred to in this chapter. Data mainly from [11] and [44].

roots mentioned above may have developed in response to water accumulating in the bases of the fronds [45]; their absence in older plants may be a result of a changed pattern of water retention at the crown or an intrinsic change in the type of root that can be formed by the stem.

The trunk roots of *Cryosophila guagara* grow downwards to a length of 6–12 cm, then stop growing and transform into a spine. The anatomy of crown roots also alters during their life. They initially grow upwards and then turn down, perhaps under their own weight (but a change in their georesponse cannot be ruled out) and finally they, too, become spinous. Lateral roots on these two types of roots, as well as those on the stilt roots, also become spinous [49]. De Granville [50] and Kahn [51] believe that some of these short spiny laterals, such as are found on roots of *Iriartea exorrhiza,* have a ventilating function; this function has led such roots to be called 'pneumorhizae'. South American Indians use the spiny surface of the stilt roots of *I. exorrhiza* as a grater for preparing flour from the adventitious root-tubers of cassava (*Manihot esculenta* (Euphorbiacea)) [47].

The transformation of the root into a spine occurs with the cessation of mitotic activity in the meristem and the differentiation of the cells right up to the tip. In *Cryosophila guagara* the lateral root sheds its outer cortex and cap and becomes spiny by virtue of the elongation of the conical apex and the lignification of the remaining cells of stele and inner cortex [49]. Spines (pneumorhizae) of the palm *Euterpe oleracea* develop in much the same way, though here the root cap is lost and the exposed tip comes to be occupied by a loose mass of cells that consitutes an aerenchyma [50].

In many respects the pattern of spine formation is similar to that which occurs in the development of thorns from lateral shoots (e.g. in gorse, *Ulex europaeus*). The development of gorse thorns has been shown to be under hormonal and environmental control [52]; whether similar controls operate in the development of spiny roots remains to be seen. It would be of particular interest, for instance, to know what factor(s) regulate the length of these roots. Chandler [53] states that the spiny roots of *Cryosophila nana* (formerly *Acanthorhiza aculeata*) rarely exceed a length of 15 cm if they are unbranched, but if branched they can reach 40–50 cm. Clearly, the spiny tip does not develop in response to root length alone. The observation suggests that the root meristem is intrinsically determinate in activity and capable of producing only a certain amount of tissue before converting to a spine. However, a lateral primordium on such a root is the source of a new, active meristem; but it, too, is able to achieve only a certain amount of growth before becoming spiny. Some of these potentially spiny roots appear capable of changing to a less determinate mode of growth if provided with moisture as, for example, when they are close to, or enter, the ground [53].

The spiny nature of these inhibited roots has led them to be called 'root spines' or 'spine roots' by Jeník and Harris [54]. These authors use the term 'root spine' if the length of the root is $< \times 10$ the thickness, and the term 'spine root' if the length is $> \times 10$ the thickness.

Adventitious spiny roots are not confined to the palms; they have also been described on the trunks of dicotyledonous trees from tropical Africa (e.g. Euphorbiaceae, Ixonanthaceae, Sterculiaceae). In *Macaranga barteri* (Euphorbiaceae) stilt roots originate from spine roots, the transformation probably occurring when conditions are exceptionally wet [54]. The spines of *Bridelia micrantha* and *B. pubescens* (Euphorbiaceae) form in much the same way as described above for *Cryosophila*; even the pith becomes lignified and the root cap hardens [54, 55]. Sometimes the spines bear lenticels [54] which may be important for gas exchange (cf. the pneumorhizae of palms).

Spinous roots are probably protective, discouraging animals from browsing on the stems of the trees that bear them. They may also protect perennating organs such as tubers and corms (e.g. in *Dioscorea prehensilis* (Dioscoreaceae) and *Moraea* spp. (Iridaceae) respectively) [56]. In *D. prehensilis* the tuber is enclosed in a subterranean cage of spine-bearing roots, while in *Moraea* the spine roots are stiff and wiry and arise from the base of the plant; they radiate in all directions, curve and interlace 'and suggest a vegetable hedgehog more than anything else' [56]. Short root spines cover the tuberous base of the epiphytic ant-plant *Myrmecodia tuberosa* (Rubiaceae) [57]. These probably give protection to ants (which inhabit chambers within the tuber) as they wander over the plant's surface. Roots at the proximal end of the tuber, or in contact with damp material, do not become spinous but elongate and secure the plant to its support.

2.3. Tabular roots and their buttresses

Tabular roots are found in many tropical trees and originate at, or just below, the junction between the trunk and the tap root. The roots extend horizontally and are called 'tabular' because secondary thickening predominates on the upper (dorsal) side of the root causing it to increase in height. The proximal end of such a root is often characterized by a conspicuous buttress, a plate of tissue that forms in the angle between the trunk and the root (Fig. 5). The relative contributions of trunk and root to the buttress probably varies from species to species and determines the angle at which the buttress meets the trunk as well as its extent up or along the trunk or root. Thus, where the buttress meets the trunk at a steep angle (approaching 90°), as in *Trattinickia rhoifolia* (Burseraceae) [51] or some *Cynometra* spp. (Leguminosae) [58], it looks as though most of the tissue is derived from the root; but where the angle of approach is shallower, to give a more fluted buttress, there is a greater contribution to the buttress from the cambium of the trunk (see also [59] for similar views).

Persistent cambial activity in the buttress and trunk elevates the buttress and increases its surface area. Buttresses may become so high that a man may be hidden from view behind one of them (Fig. 5). Richards [11] mentions buttresses extending 9 m up the trunk and an almost equal distance out from the trunk.

82

Fig. 5. Root buttress of *Koompassia excelsa* (Leguminosae). Photograph kindly provided by Dr D.J. Mabberley, Botany School, Oxford University, England.

Dimensions of buttresses of some Australian trees are given by Francis [60].

Occasionally buttresses and stilt roots both form on the same tree, e.g. *Grewia coriacea* (Tiliaceae) [11, 48]; and in some species, e.g. *Tarrietia utilis* (Sterculiaceae), the stilt roots are flattened in the vertical plane just like the tabular root [61]. While much remains to be discovered about the development of tabular roots, flattened stilt roots, and buttresses, all represent different spatial patterns of cambial activity. It may not be too far-fetched to regard buttresses, particularly fluted buttresses, as a type of fasciated adventitious aerial root where a vertical strip of cambium in the trunk has been activated but has not grown out of the trunk as a discrete root. A cylindrical, aerial stilt root represents the more usual result of cambial activity, being the outgrowth of a disc of primordial root cells.

The ability to form buttresses is genetically determined, but environmental factors may, in some species, play a role in their development [11, 59]. The position where they develop on the trunk may also be causally linked with their biological function. Senn [62] examined European populations of *Populus nigra* var. *italica* (Salicaceae) and although this species forms rather small buttresses, in 92 per cent of the trees examined (461 trees were sampled) the most prominent buttress was located on the windward side of the trunk. In both Cuban and West African populations of *Ceiba pentandra* (Bombacaceae) it also seems reasonably clear that buttressing develops most conspicuously in relation to the direction of the prevailing wind [63, 64]. The correlation between wind direction and buttress position has led to the suggestion that buttresses are not so much a support structure resisting the push given the trunk by the wind, but rather a traction structure for pulling the wind-blown trunk back into place [63, 64]. Buttresses may also arise in response to an asymmetric distribution of weight in the canopy [64].

2.4. Internal and hidden roots

A number of species within diverse families have adventitious roots embedded within the body of the plant [65, 66, 67]. These so-called 'internal roots' grow for a time within the cortex of the parent root or shoot before emerging. One curious example of internal adventitious roots is found in *Asphodelus tenuifolius* (Liliaceae) where they arise from the flattened disc of the stem base and then grow down the cortex of the persistent primary root [66]. They branch within the primary root and eventually emerge either singly or in clusters through breaks in its periderm.

Adventitious roots persisting in the stem have been known for a long time in *Lycopodium* (Lycopodiaceae) [68, 69] and is typical of erect species of this genus. The roots arise from the pericycle just behind the shoot apex but emerge only near the base of the stem [65]. After their initiation, the young roots turn and grow downwards through the cortex, dissolving and crushing the cells that lie in their path. In prostrate species, however, the roots grow across the cortex, taking a more direct course to the exterior. The primordia again arise near the shoot apex but many seem to become dormant after having grown half way to the outside. Successful rooting of cuttings can be achieved by using these apical portions of the stem [70].

In other species, notably members of the Velloziaceae and Xanthorreaceae (Liliales) [71, 72, 73], as well as *Navia schultesiana* and perhaps some *Guzmannia*

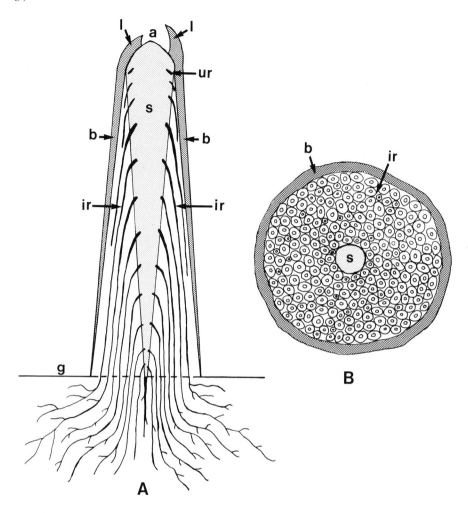

Fig. 6. The hidden internal roots of the *Velloziaceae*. A. A sketch showing the roots arising near the shoot apex and growing down to the ground in the space which they create around the stem by penetrating the thick cover of dead leaf bases. The stem eventually dies at its base leaving the roots as the only support of the plant. B. A drawing of a transection of a stem near the base of a plant of *Vellozia macarenensis*. 226 roots lie between the stem and the cover of leaf bases. Both redrawn from [74].

Key to lettering: a – shoot apex; ur – unemerged root; ir – internal root; l – crown of living leaves; b – cover of leaf bases; s – stem; g – ground level.

spp. (both Bromeliaceae) [71, 74], the roots are hidden from view, not by enclosure within the cortex, but by the bases of the leaves (Fig. 6) (see Troll [7] for other examples). The roots of these plants are initiated a few centimetres from the crown of the shoot; as the shoot grows upwards it widens and the number of roots initiated also increases. Tall specimens (approx. 6 m) of *Kingia australis* (Xanthorreaceae) may contain about 2500 of these aerial roots, plus another 1000 roots that have arisen on them as laterals, at a level 200 cm above the ground; the density of roots, as seen in a cross-section of stem, approaches 15 per cm^2 [72].

In the early stages, the roots of *Kingia* and *Xanthorrea* seedlings probably have a contractile function (as do adventitious roots of many bulbous geophytes) and their stems become established underground [73]. Later, when the plant has grown taller, the densely packed roots create a pseudo-stem, or caudex, which supports the aerial portion of the stem. The caudex eventually becomes the only support for the stem since its basal part eventually dies; this occurs when the plant is 300–400 years old [72]. These roots are thus rather similar to the stilt roots of those palms whose stem base also dies (see p. 12 and [47]).

The roots of *Kingia australis* grow at about 16 cm per year and so in tall specimens they spend many years in the air [72]. The roots are enclosed by the persistent leaf bases which besides protecting the roots from the dry environment, also intercept rainfall and channel water, as well as nutrients dissolved from dust and detritus gathered on the leaf base surface, into the caudex. Here the nutrients and water are absorbed by the roots and translocated into the stem and leaves [71, 75].

Clearly the 'hidden' roots have great adaptive significance for plants growing in an arid environment with sporadic rainfall. Such roots are also interesting from an evolutionary point of view; hidden roots are present in the tree fern *Cyathea* (Cyantheaceae) and have also been found in some of the more ancient ancestors of modern plants [68] of which *Lycopodium* (mentioned above) is a present day relative.

3. Adventitious roots of water plants

3.1. The forms of adventitious roots

One of the first written records of adventitious roots concerns those of a water plant. In 1844 Rev. W.H. Coleman [76] wrote of *Oenanthe phellandrium* (now *Oe. aquatica* (Umbelliferae)) that: '*From the joints proceed numerous whorled pectinated fibres of which the lower ones are as stout as the original fusiform root: these, descending in a conical manner to the bottom of the water, form a beautiful system of shrouds and stays to support the stem like a mast in an erect position...*' We may infer that the adventitious roots ('the pectinated fibres') provide both a nutritional and a supportive function – functions they share with many of the adventitious root systems of terrestrial plants previously described.

Although it is usual for there to be only a single type of adventitious root in water plants, there are many species where the roots are of two types. Each type of root may perform a particular function essential in the life of the plant. Some species of water plant in which two types of root have been identified, and which will be mentioned later, are listed in Table 3.

86

In Table 3 the terms used to designate the two types of root describe either their location in the environment (e.g. water, mud) or have some functional connotation (e.g. absorbing, anchoring). The locational terms seem preferable to the functional terms since the actual function of each type of root is imperfectly known. Ellmore [81], in describing the dimorphic root system of *Ludwigia peploides,* prefers to call the two types of root of this plant 'upward-' and 'downward-growing' to avoid the possibility of ascribing an unwarranted physiological function.

All the species listed in Table 3 possess vertical, downward-growing mud roots. These are long (reaching the muddy substratum of the aqueous habit), rarely branched, and may bear hairs. The second type of root, the water root, is usually shorter, thinner, lacks root hairs, and is often profusely branched. These latter roots tend to grow horizontally, and sometimes may even turn upwards to reach the surface of the water (e.g. *Typha*). This is said to be an aerotropic response [79, 84]. Upward growth is well expressed in *Ludwigia* and *Jussiaea*. In these plants the structure of the upward-growing root is quite distinct from that of the downward-growing root (Fig. 7), though in *Jussiaea* an intermediate form of root also exist [80]. The upward-growing root of *Ludwigia peploides* has a characteristic aerenchyma formed by the radial expansion of the cortical cells [85]. Roots of this type break the surface of the water and atmospheric gases pass into them and thence to the submerged stem [85]. The downward-growing roots have a more compact cortex and are probably better able to absorb nutrients.

It is not known whether water and mud roots differ in function. Conceivably there may be qualitative differences between them – for example, one type of root may selectively import certain ions which the other type cannot – but it is more likely that any difference is of a quantitative nature due to differences in

Table 3. Species of water plants with dimorphic roots which are mentioned in the text.

Family	Species	Terms given by the author to the two types of root		Reference
Elatinaceae	*Bergia capensis*	Water	Soil	[77]
Gramineae	*Phragmites communis*	Water	Mud	[78]
Gramineae	*Typha latifolia*	Water	Soil	[79]
Onagraceae	*Jussiaea repens*	Aérifère (aeriferous)	Rougeâtres (reddish)	[80]
Onagraceae	*Ludwigia peploides*	Upward-growing	Downward-growing	[81]
Pontederiaceae	*Heteranthera zosterifolia*	Saugorgan (feeding root)	Haftorgan (anchoring root)	[82]
Restionaceae	*Sporodanthus traversi*	Absorbing	Anchoring	[83]

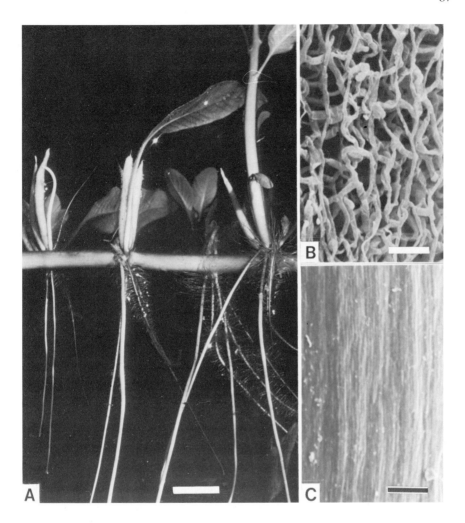

Fig. 7. Upward- and downward-growing roots of *Ludwigia peploides*. A. Part of a horizontal rhizome bearing upward- and downward-growing roots at the nodes. B and C. Scanning electron micrographs of the surface of the two kinds of roots: (B) an upward-growing root showing the open, aerenchymatous meshwork of cells; (C) the downward-growing root showing a smooth, unbroken surface. Scale bar in A is 2 cm, and in B and C is 0.1 mm. Photographs kindly provided by Dr G.E. Ellmore, Department of Biology, Tufts University, Medford, Mass., USA.

surface area/volume ratio and metabolic efficiency determined by the oxygen tension of the surrounding medium. The fine water roots of *Phragmites communis* are known to contribute to stem growth since removing them decreases the mean height of the stem by 37 per cent after a 70 day period [86]. The nature of the roots' input to stem growth, and whether the amount of stem growth is related to the surface area of the roots, was not determined, however. Weaver and Himmel [79] suggest that the finely branched water roots of *Typha* are better placed to absorb oxygen from the water. Oxygen absorption is a property of all roots, but in the case of the water roots the oxygen which they take up may maintain a more efficient absorption of mineral ions than can be achieved by the deeper, potentially oxygen deficient mud roots. It is unlikely that the absorbed oxygen would exist in gaseous form and diffuse from the roots to other parts of the plant. However, the water roots of *Bergia capensis* contain chloroplasts within their cortical cells [77] and may evolve gaseous oxygen which could then diffuse within the extensive cortical aerenchyma.

The two types of roots in the water plants *Sporodanthus traversi* and *Phragmites communis* probably play another, very different role in the life of the plant in addition to their usual physiological role. They are instrumental in building and maintaining the community in which the plants grow. Pallis [78] describes a particular population of *Ph. communis* which grows in the Danube delta as an aggregate of tussocks. The rhizomes and the short water roots form a tangled mass filled with silt and detritus; the longer mud roots anchor the community preventing it from floating away in the current. *S. traversi* is a species present in raised bogs in New Zealand. Its fine water roots constitute a mesh on which other bog plants grow, while the long mud roots are said to support the weight of the bog vegetation [83].

The two types of roots in water plants arise from a common source, such as a node of an underwater stem. This implies that there is some sort of physiological control regulating the type of root which will grow out from the stem. The various combinations of the two root types found in *Heteranthera zosterifolia* [62] illustrate phenomena which physiological explanations of root dimorphism need to encompass. Two roots emerge at each leaf base: one is short and branches freely (water root), the other grows rapidly down to the substratum and develops few laterals (mud root). Occasionally both roots of the pair are of the water type, or both are of the mud type. If only one root develops at a leaf base it is usually the water root, less frequently it is the mud root. Nothing is known of the circumstances leading to the different patterns of development, although environmental factors may play a role since the root dimorphism was noted in *Heteranthera* plants growing naturally in Brazil [82], but was not found in plants raised in England [87] (it was not stated, however, which type of root was developed).

3.2. Factors influencing root form

The two different types of roots described in the preceding section may, in some species, both arise from the same node, or from nearby nodes, of an underwater upright stem or horizontal rhizome. Vertical stems of *Typha latifolia* develop mud roots on the lower part of a node and water roots on the upper part [79]. However, in some other species, the two types develop on distinct parts of the stem. In *Phragmites communis*, for example, the mud roots form on portions of horizontal rhizome which are buried in the mud, while water roots form on the rhizome where it emerges from the mud and enters the water, and also at nodes on the upper parts of vertical stems [78, 79, 88]. Intermediate zones can give rise to roots of both types [78]. Under natural conditions, the distribution and density of the two types of root in *Phragmites* is highly dependent upon environmental factors, and perhaps on the genotype also [88].

Ludwigia and *Jussiaea* have both been used in experiments to determine the controls of root dimorphism. The first root to develop on the dorsal and ventral sides of a horizontal *Ludwigia* rhizome is that nearest the stipule of a leaf scar; it is always downward-growing [81]. Roots which arise in positions slightly more removed from the node are usually upward-growing if on the dorsal side, but downward-growing if on the ventral side. The frequency of upward- or downward-growing roots varies with their distance from the node (Table 4). The primordia on each side are anatomically indistinguishable until shortly before they emerge, but at least those on the dorsal side have a particular physiological status since rotating the node 180° renders more than half of them incapable of further growth and does not convert them into downward-growing roots. Hormonal differences between dorsal and ventral sides, or differential illumination, may possibly influence the growth and course of development of the respective primordia. Such differences probably also explain the extreme dimorphism of roots in *Nymphaea candida* and *Nuphar lutea* where only the adventitious root primordia on the ventral side of the rhizome are capable of full development; primordia on the dorsal side are arrested and do not emerge [89].

The hormonal milieu appears to influence the type of root emerging from rhizomes of *Jussiaea repens* [90]. Cuttings of *J. repens* normally develop aerenchymatous, upward-growing roots and non-aerenchymatous, downward-grow-

Table 4. Frequencies of upward- and downward-growing roots on the rhizome of *Ludwigia peploides* in relation to their site of origin.

Side	Root	Per cent			
		Root position			
		1	2	3	4
Dorsal	UGR	0	39	86	80
	DGR	100	61	14	20
Ventral	UGR	1	2	31	78
	DGR	100	98	69	22

The frequency (percentage) of upward-growing (UGR) or downward-growing (DGR) roots on the dorsal or ventral side of nodes 40–75 cm from the apex was scored with respect to their position from the stipule (root 1 closest to stipule, root 4 furthest away). Only those nodes bearing 4 roots per side were considered. 70 nodes were scored from 10 plants. Previously unpublished data of Dr G.E. Ellmore.

ing roots in the ratio 0.27:1. Naphthylene acetic acid (NAA) and gibberellic acid (separately at 10^{-7}–10^{-5} M) completely suppress aerenchymatous root formation and cause all the primordia to develop as downward-growing roots. The effect of NAA can be overcome by the cytokinin benzylaminopurine (BAP) at 10^{-6} M. BAP alone at 10^{-7}–10^{-6} M causes the primordia to grow out as aerenchymatous roots rather than as downward-growing roots. Internal hormonal changes may also be responsible for environmental and correlative effects on root type. For example, *Jussiaea* cuttings placed vertically, or in darkness never develop aerenchymatous roots but produce only the downward-growing type of root [90]. A great deal more experimental work needs to be done to clarify the factors regulating the development of dimorphic root systems.

4. Extreme forms of adventitious roots

The root and shoot systems can vary tremendously in their contribution to the total mass of the plant. At the two extremes, there are plants which completely lack roots, and others which, in the non-flowering state, are almost all root. The latter are the more interesting in the present context since in the examples cited the roots are adventitious in origin. Also described in this section are the remarkable modifications of adventitious roots found in certain parasitic plants. All these examples illustrate ways in which plants can exploit novel environmental niches through modification of their adventitious roots.

4.1. Podostemaceae

The family Podostemaceae has a wide, though tropical, geographic distribution and consists of plants that live only in rapidly-running streams and cataracts. When the Podostemaceae were first discovered they were understandably mistaken for algae, lichens, and even mosses. The plants consist of a thallus that cements itself to rocks by means of specialized organs called haptera (Fig. 8). The thalli of all Podostemaceae are resistant to exposure during periods when the water level falls and regenerate new plants when re-submerged. They also readily regenerate new tissue when broken. Thalli range in form from near-cylindrical threads up to 60 cm long which grow at their tip (as in *Dicraea elongata*) (Fig. 8A) to near-circular plates that encrust the rocks on which they are found (as in *Hydrobryum olivaceum*) (Fig. 8C). In all cases the thallus is in reality a highly modified root complete with a cap covering its growing tip (Fig. 8A). The flattened thallus of *Lawia* species is particularly curious in that its ventral side seems to have root characteristics, while its dorsal side is shoot-like [93].

Dicraea elongata has two types of thallus – a 'creeping' thallus which is the first to form and which attaches to rocks by means of haptera, and a 'drifting' thallus

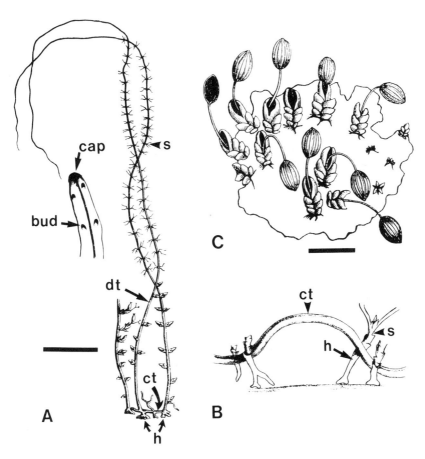

Fig. 8. Examples of the root-thallus of Podostemaceae. A. *Dicraea elongata.* A creeping thallus is attached to the substrate by haptera. From the creeping thallus arises a drifting thallus which bears small shoots that flower when older (i.e. on more basal parts of the thallus). To the left is an enlargement of the growing tip of the drifting thallus to show the terminal root 'cap' and the shoot buds which form very close to the apex. B. A creeping thallus of *Podostemon ceratophyllum* to show haptera and shoots, both of which arise at much the same location on the thallus. Enlarged. C. *Hydrobryum olivaceum.* The thallus, which grows at its rim, forms a crust on rocks; its upper surface bears vegetative and flowering shoots (both shown). Scale bar in A is 2 cm and in C is 5 cm. A and C from [91]; B from [92]. Key to lettering; ct – creeping thallus; dt – drifting thallus; h – hapteron; s – shoot.

which arises as a branch on the creeping thallus and, with other such thalli, form the bulk of a mature plant (Fig. 8A). Both creeping and drifting thalli produce small, floriferous shoots [91, 92, 93].

Members of the Podostemaceae have never been grown successfully in culture and all observations on their growth and life history have had to be made *in situ*. The torrential waters in which they live has limited the number of observations on the early stages of thallus development. Much of what we know of these stages is due to the work of J.C. Willis [93] and the best examples were obtained for *Podostemon subulatus* and *Dicraea elongata* [93]. In both species there is little doubt that the thallus arises adventitiously at the base of the hypocotyl (which in *Dicraea* is swollen into a sort of tuber) following the formation of a number of leaves at one end of the insignificant primary axis. The thallus seems to arise endogenously, though in *D. elongata* Willis could not be sure that it did not sometimes arise exogenously. Haptera arise exogenously from the thallus or shoot very early after germination and attach the seedling to its rock. No species of Podostemaceae investigated has been shown to develop a definite primary root.

Earlier writers such as Warming and Goebel (see [93]) considered the hapteron to be an organ *sui generis*, though Willis, concurring with this view, qualifies this by saying 'there seems no absolute improbability in their being modified adventitious roots'. Their rather irregular occurrence and exogenous origin makes them unlike lateral roots even though they seem to grow at their tip and have some sort of terminal cap. Perhaps the ordinary canons of morphology have to be laid aside in front of such organisms, for certainly in dealing with the Podostemaceae we are confronted with one of the most curious forms in the plant kingdom.

4.2. 'Shootless' orchids

The orchid genera *Microcoelia*, *Taeniophyllum* and *Campylocentrum* within the tribe Vandeae are characterised by an apparently 'shootless' or 'aphyllous' form. In the main, the plants consist of roots (Fig. 9). Leaves are reduced to tiny scales (Fig. 9C) and have been replaced by roots (Fig. 9C) as the principal photosynthetic organs. The plants are epiphytes and their nutritional requirements may also be supplemented by a fungal association between the roots and the support plant (see Section 2.1.3.). The significance and evolution of 'shootlessness' in relation to the epiphytic habit has been discussed by Benzing and Ott [94].

In the aphyllous orchids the shoot, although extremely reduced, is the main propagating axis of the plant from which roots arise adventitiously; in the Podostemaceae mentioned earlier the root is the main axis and buds off shoots. In *Microcoelia* older regions of the shoot disintegrate and the flourishing part usually does not exceed a length of 10 mm (the species, *M. perrieri*, illustrated in Fig. 9B has one of the longest stems in the genus, reaching up to 90 mm); by contrast, the numerous roots which sprout from it, may attain lengths of up to 90 cm [42].

Orchid roots in general have a distinctive anatomy, particularly through the possession of a modified epidermis (velamen), gas-filled cells (pneumathodes), and cells that contain crystals of calcium oxalate. Aerial roots of the aphyllous orchids share these features [42, 95]. The roots are generally green in colour owing to the presence of chloroplasts (the various shades of colour can be used as

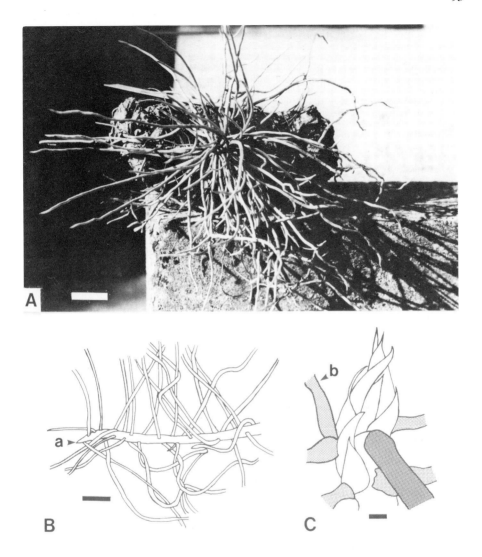

Fig. 9. The 'shootless' orchid *Microcoelia* A. Photograph of *Microcoelia exilis*. B. Drawing of a stem of *M. perrieri* with roots (which are up to 40 cm long) emerging from it. a – shoot apex. C. Drawing of a shoot apex of *M. exilis* with roots arising close to its tip. (Roots are the stippled outgrowths; the outgrowth labelled 'b' is the base of an inflorescence.) Scale bar in A is 5 cm and in B and C is 1 mm. B and C are drawings made from photographs in [42].

a taxonomic feature in *Microcoelia* [42]), since they contain some of the key enzymes of photosynthesis [96], are able to fix carbon [94]. The photosynthetic ability of the roots, as well as the biology of these 'shootless' orchids is under-researched. Of particular interest is the physiological cause of the extensive growth of the roots and the corresponding reduction of the shoot

4.3. Haustoria of parasitic plants

Parasitic plants adopt a wide range of forms and may attack their hosts through either the host's root system or shoot system. In each case a specialised organ, the haustorium, penetrates the host and forms a bridge to bring nutrients from the host to the parasite. There is little doubt that the haustorium is a modified root [97]. Thus, a haustorium developed by a root should be regarded as a modified branch root (e.g. *Pholisma* spp. (Lennoaceae)) but where the shoot bears the haustorium (as in dodder, for example) the latter should be regarded as an adventitious root.

In those species of mistletoes (Loranthaceae) that inhabit the branches of trees, the haustoria are borne on 'epicortical roots' which grow along the host branches [97, 98]. The epicortical roots are themselves adventitious since they arise from the shoot. In *Struthanthus orbicularis* the epicortical roots emerge at the nodes of the shoot, and the haustoria that form on them quickly penetrate the host. The main mode of propagation of this parasite is through the growth of its scrambling shoot. On the other hand, in *Ileostylus micranthus* it is the epicortical roots that grow over the branches of the host, sprouting leafy shoots and sinking haustoria as they go [98]. The creeping, ageotropic nature of epicortical roots resembles the adventitious roots of certain epiphytes mentioned earlier (Section 2.1.1.): they both share a clasping habit that is facilitated by an exudate that helps them stick to the branch, and they may also be negatively phototropic [98].

The dodders (*Cuscuta* spp. (Convolvulaceae)) are some of the most familiar European parasitic plants, and the development and structure of their haustoria are known in some detail [99, 100]. Haustoria arise endogenously [99] and are confined to the inner surface of the twining stem, or haustorial coil (Fig. 10). The emergence of the haustorium is preceded by the production of an ephemeral organ called the 'prehaustorium' [101] which is composed of elongated cells of the contact surface of the haustorial coil [99, 101]. These cells drive into the host tissue like a wedge, seemingly dissolving the host cells to make an entrance into which the haustorium proper will grow. The primary root of the dodder plantlet dies soon after the first haustorial contacts have been made. Thereupon all contact with the soil is lost and the dodder relies solely on its haustoria for nutrition.

4.4. The leaf-root of Drosera

To conclude Section 4, it is of interest to mention the so-called 'leaf-root' (Blattwurzel) described by Goebel [9] on certain Australian rootless species of *Drosera* (Droseraceae). In *D. erythrorhiza* root-like organs which bear hairs and function as roots appear at the base of the scale leaves. Goebel could not decide whether these structures were roots or leaves, and since they have not been studied further their true status remains uncertain.

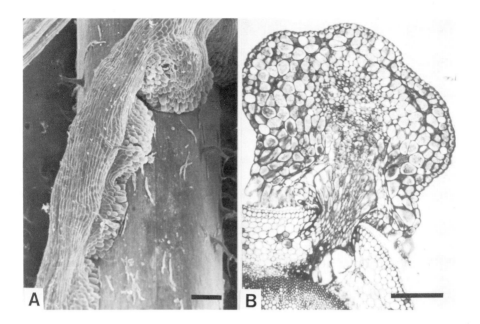

Fig. 10. The haustorium of the dodder, *Cuscuta europaea*. A. Scanning electron micrograph of a haustorial coil twining around a grass stem. B. Transection through a haustorium showing its penetration of the host and contact with host vascular tissue. The surface of the parasite has pulled away from the stem during the preparation of the section and shows the imprint of the epidermal surface of the host. Scale bars in A and B are 0.2 mm. Preparations kindly made by S.F. Young.

5. Pathologically induced adventitious roots

Adventitious roots sometimes develop on shoot tissue as a result of infection by pathogens. For example, stems of tomato develop adventitious roots in response to infection by the bacteria *Pseudomonas solanacearum* and *Corynebacterium michiganense* [102] and the virus (or mycoplasma-like organism) aster yellows [102a]. The cause of the additional roots is not known. Grieve [102] finds that the roots develop in advance of the columns of bacteria invading the xylem vessels. Aster yellows, on the other hand, invades the phloem and disturbs its development; xylem differentiation is unaffected [102a]. Both classes of pathogen may induce the additional roots through some feature of the infection process, as occurs when tomato stems are infected by the fusarium wilt fungus, *Fusarium oxysporum* [103]. In this case the gas ethylene, liberated as the result of the infection, has been proposed as the rhizogenic stimulus [104]. In all these works with infected tomato it is not clear to what extent the additional adventitious roots are the result of the activation of preformed but dormant root primordia, or the initiation of new primordia.

The beet necrotic yellow vein virus has been found to induce additional roots on sugar beet, *Beta vulgaris* [105]. One of the symptoms of infection is the development of a 'beard' of roots on the beet and has caused the disease to be named 'rhizomania'. These additional roots are adventive, rather than adventitious; the physiological changes that lead to these additional roots are poorly known.

An interesting and economically important transformation occurs to the adventitious root primordia of the tropical legume, *Sesbania rostrata,* after infection with the nitrogen-fixing organism *Rhizobium* [105a]. The primordia occur as small 'mammilae' regularly arranged along the stem and will grow and develop normally given appropriate conditions. However, when infected with *Rhizobium* the region behind the tip develops as a nodule whose centre becomes occupied by the N-fixing bacteroids. The nodule is a distinct organ with its own characteristic vasculature and photosynthetic parenchymatous exterior. The original apical meristem is retained and can still organize a normal root if the nodule is placed in water.

Root-producing tumours are familiar on the woody stems of *Forsythia intermedia* [106, 107] and occur rarely on gooseberry, *Ribes grossularia* [108]. In neither case has a pathogen been identified as the causal agent; the genetic constitutions of the plants have been proposed as being responsible for the tumours [106, 108].

Undoubtedly, one of the most interesting agents to cause adventitious roots on stems is the bacterium *Agrobacterium rhizogenes* (see also Chapter 2, Section 2.7). This bacterium, first isolated by Riker and colleagues [109], causes the condition known as 'hairy root' or 'woolly knot' on nursery apple trees. The condition is characterized by the production of large numbers of roots that arise directly from the root or stem, or from local swellings at graft unions. The isolated bacterium can also induce 'hairy root' in a number of other species, e.g. rose, honeysuckle, sugar beet [109, 110], and *Forsythia* [106]. The hairy root disease may have some practical application since infected apple trees are able to withstand drought better than non-infected trees [111], presumably because of their additional roots.

Root formation in response to *Agrobacterium rhizogenes* is used as a marker in laboratory experiments designed to give information about the organization and expression of the genetic material of this bacterium. It is now known that root formation from the bacterium-induced tumour is determined by genes carried on a plasmid in the bacterial cell [111, 112], part of which integrates into the DNA of the host plant nucleus. This plasmid can also be transferred to strains of bacteria that normally do not induce roots but which then become able to do so [112]. Particular mutations in the plasmid DNA can 'switch on' varying amounts of adventitious root formation [111]. Different mutations in the plasmid of the crown gall bacterium, *A. tumefaciens,* control the type of organ (shoots or roots) formed on the tumour [113], as well as the size of the tumour itself, by influencing the endogenous auxin/cytokinin balance [114]; the plasmid of *A. rhizogenes* may act

Fig. 11. Adventitious roots induced on stems of *Poa nemoralis* by the presence of larvae of the midge *Mayetiola poae.* A and B. Root-enveloped galls on stems of *P. nemoralis.* C. Split region of a stem to show the larvae in a chamber above which the roots arise. Photographs kindly provided by Dr Françoise Dreger-Jauffret, Botanical Institute, University of Strasbourg, France.

in the same way. Recently, the DNA of two root-inducing plasmids of *A. rhizogenes* has been mapped using restriction endonucleases [115, 116] and it will surely not be long before the base sequence of the entire plasmid is known and the molecular mechanism for root initiation in infected (and in normal) tissue elucidated.

It is rare that insect-induced galls on shoots give rise to roots, but two cases are known. Beyerinck [117] described a gall induced by the midge *Mayetiola poae* on stems of *Poa nemoralis* (Gramineae). Here, perfectly normal root primordia form endogenously in a callused portion of the stem close to where the midge larvae are housed (Fig. 11). Beyerinck [118] also described adventitious roots growing from a gall on a detached leaf of *Salix purpurea* (Salicaceae). The gall was a hollow sphere caused by the larva of the wasp *Pontania proxima;* the roots had arisen on the inside surface of the gall cavity and emerged through the exit hole made by the larva.

These examples of new root formation are of great interest because the roots often form at locations that do not normally bear them, or do so to only a limited extent. Whether there is any causal similarity between the *Agrobacterium*-induced root-producing tumour and the root-bearing gall on *Poa* remains to be seen; in both situations roots arise as the result of redifferentiation and redetermination of cells in the plant tissue, presumably as a consequence of changes in endogenous growth regulators. Work with *Agrobacterium* hints at the prospect of identifying the genes concerned with root formation and thereby the possibility of understanding the genetic basis of organogenesis.

6. Positions of adventitious roots and the classification of adventitious root patterns

In many of the examples discussed in the preceding sections rather little attention has been given to exactly where on the plant the adventitious roots form. At a gross level, the distribution of these roots obviously follows the distribution of those parts of the stems that are able to form them. If, for example, the stems are underground they may be small discs, as in the bulbs and corms of geophytes, or they may be elongated axes, such as the creeping rhizomes of grasses and sedges; in these cases the adventitious roots are correspondingly clustered or more widespread in the soil.

When root distribution is studied in closer detail certain interesting features emerge that could be useful for establishing the causes underlying the positioning of adventitious roots. Gill [119] studied the distribution of aerial roots on the stems of trees, shrubs and vines in an elfin forest in Puerto Rico. The roots of the trees and shrubs usually emerge behind, rather than within, the terminal leafy zone, and the distance from the leafy zone at which they emerge is particular to the species (see also Chapter 4). The aerial roots of vines, however, always occur within the leafy zone. The precise locations of root placement vary within the vines. In some species, roots form at random along the stem, but in many others they form at or near a node. Additional variation is found in the relation of the roots with the nodes and their associated leaf (or leaves). For example, in *Marcgravia sintenisii* (Marcgraviaceae) aerial roots run acropetally from the leaf base, but in *Gonocalyx portoricensis* (Ericaceae) they run basipetally [119].

The species examined by Gill [119] belong to different families and so different patterns of rooting are not unexpected. Weber [120], however, records species specific differences of root placement within a single family, the Labiatae. The reasons for the differences in root position in the Labiatae and vines have not been determined. There may be something particular about a node, or, in trees, a branch junction [2], that favours adventitious roots to form there. It may be worthwhile to try to understand positioning of adventitious roots in terms of

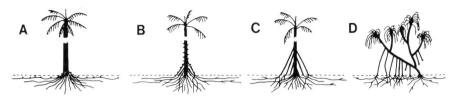

Fig. 12. Major types of root systems in palms. The species serving as the model for each type are: A – *Elaeis guineensis*; B – *Cryosophila nana*; C – *Iriartea exorrhiza*; D – *Pandanus candelabrum*. Reproduced, with permission, from [127].

correlative effects such as apical dominance, and the anatomy and vasculature of the node and internode (cf. [121, 122]). These factors have been taken into account in an attempt to elucidate the regulation of root and shoot primordium development at the branch junctions of *Selaginella* stems [123]. Wochok and Sussex [123] find less auxin to be polarly transported through the ventral half of a branch junction than through the dorsal half; it is from the ventral side that the root emerges.

As long as the position and form of roots are not infinitely variable, it should be possible to classify root systems. The classification of the root systems of herbs and trees has a long history [61, 120, 124, 125, 126] of which Weber's [120] is one of the more important. His four main categories of adventitious root systems are: hypocotylar (as in *Balsamina*), nodal (as in the Ranunculaceae and Labiatae), bud (as in the Crassulaceae), and internodal (as in *Hedera, Iris* and *Salix*). This classification emphasises root location and thus enables it to encompass a wide range of plant forms.

More recently, the primary and adventitious root systems of tropical trees have been tentatively classified by Jeník [61]. The systems fall into categories (see Figs. 12 and 13) referred to as 'architectural units' [61, 127]. Probably not enough is yet known of the range of forms for this classification to be regarded as complete and its interesting implications to be fully assessed. Jeník's [61, 127] classification is perhaps likely to have some appeal to contemporary biologists since one implication of the architectural unit is that it is the expression of a particular pattern of genetic activity.

The root system units illustrated in Figs. 12 and 13 form a set of patterns in which it is possible to envisage how one type is derived from another. This interrelation of architectural units is also evident when the root systems of the Velloziaceae and Xanthorreaceae are compared with those of the palms [71]. Within the palms the adventitious roots of the Elaeis type (Fig. 12A) radiate from the stem base, while in the Iriartea and Pandanus types (Figs. 12C and 12D) they emerge from higher up on the stem (as stilt roots). The Cryosophila type (Fig. 12B) forms suppressed roots (spiny roots) on both the trunk and on the stilt roots. These types may indicate that during the evolution of the palms there has been an upward movement of the locus of root formation from the stem base towards the

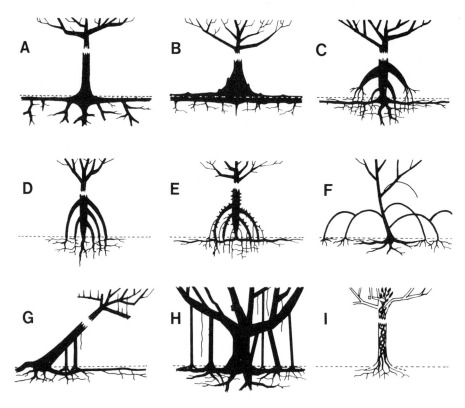

Fig. 13. Major types of root systems in dicotyledonous tropical trees. The species serving as the model for each type are: A – *Chlorophora excelsa*; B – *Piptadeniastrum africanum*; C – *Tarrietia utilis*; D – *Uapaca guineensis*; E – *Bridelia micrantha*; F – *Rhizophora mangle*; G – *Protomegabaria stapfiana*; H – *Ficus benjamina*; I – *Ficus leprieuri*. Reproduced, with permission, from [127].

crown together with an elaboration of the form and functions of the roots themselves. The same sort of migration of rooting potential in dicotyledonous tropical trees is exemplified in the sequence A, B, C, D, F, H of Fig. 13. The buttresses on the tabular root system in Fig. 13B may also be evidence of an upward extension of the rooting zone and could also represent a form in which a highly elongated (but non-emergent) root primordium develops on the trunk (see Section 2.4.).

Within the architectural units there is scope for re-arrangement of the patterning of lateral roots and for their modification. For example, in the palms, De Granville [50] proposes a scheme showing the presumed phyllogeny of ventilating tissues and organs (pneumatozones and pneumorhizae) which have become elaborated upon the simultaneously evolving adventitious root system. The same general point emerges in Jeník's [127] scheme for dicotyledonous tropical trees where ventilating roots (pneumorhizae) may have arisen secondarily on the basic unit of the adventitious root system.

Coupled with the apparent migration of rooting zones are changes in the development of the primary root system. For example, the reduction or absence of a tap root probably has some significance (perhaps even a causal interrelationship) in the development of the adventitious system (see [7, 11, 48, 127]). This may apply not only to palms and tropical trees, but also to the very different plant forms found in the Podostemaceae and Orchidaceae mentioned earlier.

Such classificatory schemes at least organize the wealth of material at hand. They may additionally provide clues to the principles operating in the evolution of root systems.

7. Evolution of roots and adventitious root systems

Adventitious roots allow plants to exploit their habitat efficiently and rapidly as well as giving them the potential to explore new and perhaps hitherto uncolonized habits (e.g. epiphytes). Often, adventitious roots re-enter areas at one time occupied by the primary root system of the plant: Kahn [128] elegantly calls this 'la reconquète de l'espace proximal'. Adventitious roots are thus effectors of species diversification and hence evolution. At the same time, this root system is a product of evolution.

Mutation must be one reason for roots developing at new locations on the plant body and so creating new classes of the architectural units referred to earlier. It has been suggested [129], on the basis of mutation studies with tomato, that the primary and adventitious root systems are under the control of separate parts of the genome. Moreover, modifications can arise within a developing root system to result in dimorphism and in specialized structures such as pneumorhizae; these are presumably also the result of mutation and/or the expression of new patterns of activity of existing genes.

What can the fossil record tell us about the evolution of roots in higher plants, and how does it influence our thinking on the nature of present-day roots systems? Palaeontology shows that the first land plants did not have roots, although they did have shoots. For example, the earliest land plants (Psilophyta), dating from the Lower Devonian (approx. 400×10^6 years ago) possessed horizontal rhizomes which linked upward-growing stems [130]. The rhizomes of *Rhynia* (Rhyniales) bore unicellular rhizoids which functionally might be regarded as precursors of root-like structures (but which are perhaps no more root-like than the root hairs of higher plants or the rhizoids of present-day moss and fern gametophytes). These plants lived in marshes, so the rhizoids presumably had as much a support function as an absorptive function. In the more highly developed plant *Asteroxylon mackeii*, a member of the Protolepidodendrales and originally found in the same deposits as *Rhynia*, the rhizomes seem to have lacked rhizoids; some rhizomes were also capable of growing downwards [130]. The evolution of a structure more like a root, particularly one attached directly to the base of an upright stem, had to wait another 50×10^6 years or so.

It was discoveries made in the mid-19th century during the building of the Manchester to Bolton railway in England which gave evidence important to the understanding of root evolution. The railway excavations unearthed five excellent specimens of fossil tree trunks which possessed at their base what are now known as 'stigmarian axes' [131]. (Originally, these axes were credited to a genus called *Stigmaria*, but with the discovery of fossil tree trunks to which such axes were joined they were realized as belonging to already-known genera (e.g. *Sigillaria*); *Stigmaria* is now a form genus.) Photographs and descriptions of these stigmarian axes from the railway site were published by Williamson [132], and more recently discovered specimens, particularly in the USA, have been illustrated and described in detail by Frankenberg and Eggert [133]. Four thick horizontal axes (some up to 80 cm in diameter) join with the base of the tree trunk; they dichotomize repeatedly and may run for great distances (10 m or more). Circular scars are found along the axes indicating that they bore fugaceous appendages; in the younger, distal portions of the axes the appendages themselves may be present. It is now generally held that the appendages are roots and that the stigmarian axis is a rhizophore [134]. In some cases the appendages resemble leaves, and it may be that the root appendage actually evolved from a leaf-like appendage from which also evolved the true leaf on aerial portions of the plant. Recalling the 'leaf-root' of *Drosera* (see section 4.4.), perhaps Goebel [9] was right when he referred to it as 'eine Rückbildung' (an atavistic structure)!

The stigmarian axes are underground portions of trees of the order Lepidodendrales which attained their greatest development in the forests of the Upper Carboniferous, approx. 300×10^6 years ago. Another interesting member of this order is *Pleuromeia*, found in Triassic deposits (approx. 225×10^6 years ago) by Mägdefrau [135]. This plant possessed four bulbous lobes at the base of the stem which may represent very much reduced rhizophore axes. These lobes bear root appendages.

We thus deduce that the first roots of the early plants were adventitious and were not necessarily in direct continuity with the stem axis. The roots probably served to supplement the absorptive function of the rhizophore. In those species with stigmarian axes, their massive size may have allowed tall stems to develop (as in *Lepidodendron* and *Sigillaria*) because they could counterbalance an increased weight of the aerial part; but where the rhizophore was reduced, as in *Pleuromeia* and in the recently discovered allied genus *Protostigmaria* [136], the roots would fulfill the same absorptive function, but they would now also assume an important support function since, in the absence of a horizontal rhizophore, they would now be the main anchor for the stem. Perhaps this fact accounts for the small height (1–2 m) of these plants, compared to the 40 m or so of some of the tree-like, rhizophore-possessing Lepidodendrales.

Pleuromeia and *Protostigmaria* seem to have affinities with present day members of the Isoetales [136]. Contemporary species of *Isoetes* have a lobed, cambium-like meristem from which roots emerge. These rhizogenic plaques are probably homologous to the lobes at the base of the stem of *Pleuromeia*. Moreover, the appendages of stigmarian axes are anatomically similar to roots of present-day *Isoetes* [137]. However, present interpretation of palaeobotanical evidence suggests that these rhizogenic plaques were not derived, by a reduction of size, from stigmarian axes. Rather, the converse seems to be the case, since it seems as though the elongated stigmarian axis was an evolutionary 'experiment' and developed on a line of the Lepidodendrales that eventually became extinct.

Another interesting genus in the Isoetales is *Stylites*, recently discovered living around the margin of Lake Caprichosa on the altiplano of Peru [138]. Here the roots arise adventitiously from a separate cambial zone lying to one side of the stele of the short stem. A fossil form, rather similar to *Stylites*, is *Nathorstiana* [139], though here the rhizogenic zone is at the base of the stem. What seems to have occurred in the genesis of *Stylites* is that the rhizogenic plaque has moved from a site around the base of the plant (as it is in *Pleuromeia* or *Nathorstiana*) to one side. Maybe this is another example of the migration of rhizogenic zones during evolution that was suggested earlier (Section 6) to have occurred in the Angiosperms. The roots of *Stylites* resemble stilt roots. It is important to note that stilt roots similar to those of present-day mangroves, and possessing lenticels on their laterals, have been described in *Amyelon iowense*, a fossil member of the Cordaitales of the Upper Carboniferous era (335×10^6 years ago) [140], and on other fossil early Gymnosperms.

Although extant modern forms of *Isoetes* have a very reduced shoot, each shoot does have a discrete shoot apex. By contrast, neither *Isoetes* nor any of the fossil Lepidodendrales have a primary root apex. The indications are that the shoot pole was the initial growing point of the plant and that rhizophore axes (stigmaria) were derived from it by dichotomous branching early in development [141]. The rhizogenic plaque of other early plants may also have arisen as a secondary structure on the basal end of the shoot.

This all too brief survey of fossil plants has focussed on those species with well preserved root systems. The examples happen to relate to species which are the forerunners of present-day Lycopsida (Lycopodiales, Selaginellales and Isoetales). In the fossil record there are glimpses of roots in the early Gymnosperms, but these have been little studied. What is remarkable is that the roots of the earliest plants were adventitious and were produced by special rhizogenic zones at the base of the stem; a primary root system did not exist. A primary root is a feature of present-day Angiosperms and Gymnosperms. Its origin must have represented a major developmental change in the way that a root organ was initiated and must have involved a change in the pattern of embryogenesis. Although the origin of the Angiosperms is something of a mystery, they nevertheless retain the ancient adventitious root system alongside the more recently acquired primary root system. What has altered, however, is the mode of root branching. Adventitious roots of primitive plants (both fossil and extant) branch dichotomously, but in the Angiosperms they now branch by means of lateral primordia as do the primary roots (see Chapters 1 and 4).

8. Conclusion

Evidence from fossils and from comparative embryology shows that the adventitious root system is a primitive character, while the primary root system is of more recent origin. In the early land plants, adventitious roots arose only on the underground parts of the plant, but in modern plants they also develop on above-ground parts. The ageotropic nature of many types of present-day adventitious roots may be another primitive character, while positive geotropism, which is a character of the primary root system, may be a more modern trait.

The ability to form adventitious roots may be related to the increased complexity of the shoots with their more highly organized system of branches and nodes; it may also be a consequence of the increased physiological complexity of the whole plant which has arisen from the evolution of novel metabolic pathways. The elevation of the sites of adventitious root formation on the stem may also be a modern trait which enables plants to colonize new habitats. Probably changes in the Earth's climate have acted as a selective force in the origin of new forms of plants and new types of adventitious root systems.

The tropical rain forest provides many examples illustrating the diversity of adventitious roots – aerial roots, stilt roots, roots associated with ant colonies, etc. This tropical ecosystem (which unfortunately is too little known or understood) is probably a site where evolution of many plant characters, including new forms of roots, has occurred. Indeed, as Corner [142] puts it: '*As a principle, every subject of botany should be studied from its unspecialized tropical aspect if it is to have its proper scientific background.*' The roots of tropical plants clearly give us great insight into the diversity of forms that are capable of being generated, and hence the range and versatility of the underlying developmental processes. All this needs to be understood more deeply since diversity of root forms raises fundamental questions that not only concern the evolution and function of this diversity, but also the means whereby it is regulated at the genetic, cellular and physiological levels.

9. References

1. Fahn, A. (1982) Plant Anatomy. 3rd Edition. Pergamon Press, Oxford, New York.
2. Nadkarni, N.M. (1981) Canopy roots: convergent evolution in rainforest nutrient cycles. Science, 214, 1023–1024.
3. Darwin, C. (1865) On the movement and habits of climbing plants. The Journal of the Linnean Society, 9, 1–118.
4. Treub, M. (1883) Observations sur les plantes grimpantes du Jardin Botanique de Buitenzorg.

Annales du Jardin Botanique de Buitenzorg, 3, 160–183.

5. Koursanov, A.L. (1956) Sur le rôle physiologique des racines aériennes chez *Ficus* sp. Bulletin de l'Institut Français d'Afrique Noir, Série A, 18, 25–34. (This is a translation into French of a paper first published in Russian in 1955 in Fiziologiya Rastenii, 2, 271–276.)

6. Lanner, R.M. (1966) Adventitious roots of *Eucalyptus robusta* in Hawaii. Pacific Science, 20, 379–381.

7. Troll, W. (1942) Vergleichende Morphologie der höheren Pflanzen. Volume 1: Vegetations-organe, Part 3, pp. 2148–2301. Verlag Gebrüder Borntraeger, Berlin-Zehlendorf.

8. Goebel, K. (1905) Organography of Plants, especially the *Archegoniatae* and *Spermatophyta*. Part 2. Special Organography. (English translation by I.B. Balfour.) Clarendon Press, Oxford.

9. Goebel, K. (1933) Organographie der Pflanzen insbesondere der Archegoniaten and Samen-pflanzen. Part 3. Samenpflanzen. G. Fischer, Jena.

9a. Weber, H. (1953) Die Bewurzelungsverhältnisse der Pflanzen. Verlag Herder, Freiburg.

10. Went, F.A.F.C. (1895) Ueber Haft- und Naehrwurzeln bei Kletterpflanzen und Epiphyten. Annales du Jardin Botanique de Buitenzorg, 12, 1–72.

11. Richards, P.W. (1952) The Tropical Rain Forest. An Ecological Study. Cambridge University Press, Cambridge.

12. Champagnat, M. & Blatteron, S. (1966) Ontogénie des organes axillaires du Cresson (*Nasturtium officinale* R.Br.). Revue Générale de Botanique, 73, 85–102.

13. Kapil, R.N. & Rustagi, P.N. (1966) Anatomy of aerial and terrestrial roots of *Ficus benghalensis* L. Phytomorphology, 16, 382–386.

14. Hinchee, M.A.W. (1981) Morphogenesis of aerial and subterranean roots of *Monstera deliciosa*. Botanical Gazette, 142, 347–359.

15. Schimper, A.F.W. (1888) Die epiphytische Vegetation Amerikas. Botanische Mitteilungen aus den Tropen, 2, 1–162.

16. Tennent, J.E. (1859) Ceylon: An Account of the Island – Physical, Historical and Topographical with Notices of its Natural History, Antiquities and Productions. 2nd Edition. Longman, London.

17. Davis, T.A. (1970) Epiphytes that strangulate palms. Principes, 14, 10–25.

18. Trelease, W. (1905) Illustrations of a 'strangling' fig tree. 16th Annual Report of the Missouri Botanical Garden, pp. 161–165.

19. Rao, A.N. (1966) Developmental anatomy of natural root grafts in *Ficus globosa*. Australian Journal of Botany, 14, 269–276.

20. Kerner von Marilaun, A. (1894) The Natural History of Plants. Volume 1. (English translation by F.W. Oliver.) Blackie and Son, London.

21. La Rue, C.D. (1952) Root grafting in tropical trees. Science, 115, 296.

22. Kirk, T. (1872) On the habit of the rata (*Metrosideros robusta*). Transactions and Proceedings of the New Zealand Institute, 4, 267–270.

23. Beddie, A.D. (1953) Root behaviour in *Metrosideros*. Wellington Botanical Society Bulletin, 26, 2–6.

24. Dawson, J.W. (1967) A growth habit comparison of *Metrosideros* and *Ficus*. Tuatara, 15, 16–24.

25. Corner, E.J.H. (1940) Wayside Trees of Malaya. Government Printing Office, Singapore.

26. Strong, D.R. & Ray, T.S. (1975) Host tree location behavior of a tropical vine (*Monstera gigantea*) by skototropism. Science, 190, 804–806.

27. King, G. (1887) The species of *Ficus* of the Indo-Malayan and Chinese countries. Part 1. Palaeomorphe and Urostigma. Annals of the Royal Botanic Garden, Calcutta, 1, p. 18.

28. Galil, J. (1984) *Ficus religiosa* L. – the tree splitter. Botanical Journal of the Linnean Society, 88, 185–203.

29. Hooker, J.D. (1854) Himalyan Journals; or, Notes of a Naturalist in Bengal, The Sikkim and Nepal Himalayas, The Khasia Mountains, etc. J. Murray, London.

30. Schimper, A.F.W. (1903) Plant Geography upon a Physiological Basis. (English translation by

106

W.R. Fischer.) Clarendon Press, Oxford.

31. Zimmermann, M.H., Wardrop, A.B. & Tomlinson, P.B. (1968) Tension wood in aerial roots of *Ficus benjamina* L. Wood Science and Technology, 2, 95–104.

32. Dressler, R.L. (1981) The Orchids. Natural History and Classification. Harvard University Press, Cambridge, Mass., London.

33. Ule, E. (1902) Ameisengarten im Amazonasgebiet. Botanische Jarbücher für Systematik, Pflanzengeschichte und Pflanzengeographie, 30 (Beiblatt No. 2), 45–52.

34. Kleinfeldt, S.E. (1978) Ant-gardens: the interactions of *Codonanthe crassifolia* (Gesneriaceae) and *Crematogaster longispina* (Formicidae). Ecology, 59, 449–56.

35. Treub, M. (1883) Sur les urnes du *Dischidia rafflesiana* Wall. Annales du Jardin Botanique de Buitenzorg, 3, 13–36.

36. Groom, P. (1893) On *Dischidia rafflesiana* (Wall.). Annals of Botany, 7, 223–42.

37. Juniper, B.E. & Jeffree, C.E. (1983) Plant Surfaces. E. Arnold, London.

38. Janzen, D.H. (1974) Epiphytic myrmecophytes in Sarawak: mutualism through feeding of plants by ants. Biotropica, 6, 237–259.

39. Gómez, L.D. (1974) Biology of the potato-fern *Solanopteris brunei*. Brenesia, 4, 37–61.

40. Ruinen, J. (1953) Epiphytosis. A second view on epiphytism. Annales Bogoriensis, 1, 101–158.

41. Johansson, D. (1977) Epiphytic orchids as parasites of their host trees. American Orchid Society Bulletin, 46, 703–707.

42. Jonsson, L. (1981) A monograph of the genus *Microcoelia* (Orchidaceae). Symbolae Botanicae Upsalienses, 23, No. 4, 1–155.

43. Benzing, D.H. (1982) Mycorrhizal infections of epiphytic orchids in Southern Florida. American Orchid Society Bulletin, 51, 618–622.

44. Menninger, E.A. (1967) Fantastic Trees. Viking Press, New York.

45. Gill, A.M. & Tomlinson, P.B. (1975) Aerial roots: an array of forms and functions. In The Development and Function of Roots (eds J.G. Torrey & D.T. Clarkson), pp. 237–260. Academic Press, London, New York, San Francisco.

46. Chipp, T.F. (1913) The reproduction of *Musanga Smithii*. Bulletin of Miscellaneous Information, Royal Botanic Gardens, Kew, 1913, 96.

47. Wallace, A.R. (1853) Palm Trees of the Amazon and their Uses. van Voorst, London.

48. Corner, E.J.H. (1978) Freshwater swamp-forest of South Johore and Singapore. Gardens' Bulletin, Supplement No. 1, 266 pp.

49. McArthur, I.C.S. & Steeves, T.A. (1969) On the occurrence of root thorns on a Central American palm. Canadian Journal of Botany, 47, 1377–1382.

50. De Granville, J.J. (1974) Aperçu sur la structure des pneumatophores de deux espèces des sols hydromorphes en Guyane, *Mauritia flexuosa* L. et *Euterpe oleracea* Mart. (Palmae). Généralisations au système respiratoire racinaire d'autres palmiers. Cahier de l'Office de la Recherche Scientifique et Technique d'Outre-Mer, Série Biologique, No. 23, 3–24.

51. Kahn, F. (1977) Analyse structurale des systèmes racinaires des plantes ligneuses de la forêt tropicale dense humide. Candollea, 32, 321–358.

52. Bieniek, M.I. & Millington, W.F. (1968) Thorn formation in *Ulex europaeus* in relation to environmental and endogenous factors. Botanical Gazette, 129, 145–150.

53. Chandler, B. (1909) Aerial roots of *Acanthorhiza aculeata*. Transactions of the Botanical Society of Edinburgh, 24, 20–24.

54. Jeník, J. & Harris, B.J. (1969) Root-spines and spine-roots in dicotyledonous trees of tropical Africa. Österreichische Botanische Zeitschrift, 117, 128–138.

55. Parija, P. & Misra, P. (1933) The 'root-thorn' of *Bridelia pubescens*, Kurz. Journal of the Indian Botanical Society, 12, 227–233.

56. Scott, D.H. (1897) On two new instances of spinous roots. Annals of Botany, 11, 327–332.

57. Treub, M. (1883) Sur le *Myrmecodia echinata* Gaudich. Annales du Jardin Botanique de Buitenzorg, 3, 129–159.

58. Chipp, T.F. (1922) Buttresses as an assistance to identification. Bulletin of Miscellaneous Information, Royal Botanic Gardens, Kew, 1922, 265–268.

59. Francis, W.D. (1931) The buttresses of rain-forest trees. Bulletin of Miscellaneous Information, Royal Botanic Gardens, Kew, 1931, 24–26.

60. Francis, W.D. (1924) The development of buttresses in Queensland trees. Proceedings of the Royal Society of Queensland, 36, 21–37.

61. Jeník, J. (1973) Root system of tropical trees 8. Stilt-roots and allied adaptations. Preslia, 45, 250–264.

62. Senn, G. (1923) Ueber die Ursachen der Brettwurzelbildung bei der Pyramiden-Pappel. Verhandlungen der Naturforschenden Gesellschaft in Basel, 35, 405–435.

63. Navez, A. (1930) On the distribution of tabular roots in *Ceiba* (Bombacaceae). Proceedings of the National Academy of Sciences of the USA, 16, 339–344.

64. Henwood, K. (1973) A structural model of forces in buttressed tropical rain forest trees. (With an appendix by H.G. Baker.) Biotropica, 5, 83–93.

65. Stokey, A.G. (1907) The roots of *Lycopodium pithyoides*. Botanical Gazette, 44, 57–63.

66. Pant, D.D. (1943) On the morphology and anatomy of the root system in *Asphodelus tenuifolius* Cavan. Journal of the Indian Botanical Society, 22, 1–26.

67. Pant, D.D. & Mehra, B. (1961) Occurrence of intracortical roots in *Bambusa*. Current Science, 30, 308.

68. Brongniart, A. (1838) Recherches sur les Lepidodendron et sur les affinités des ces arbres fossiles, précédées d'un examen des principaux caractères des Lycopodiales. Comptes rendus hebdomadaires de l'Académie des Sciences, Paris, 6, 872–879.

69. Strasburger, E. (1873) Einige Bermerkungen über Lycopodiaceen. Botanische Zeitung, 31, 81–93, 97–110, 113–119.

70. Roberts, E.A. & Herty, S.D. (1934) *Lycopodium complanatum* var. *flabelliforme* Fernald: its anatomy and a method of vegetative propagation. American Journal of Botany, 21, 688–697.

71. Weber, H. (1954) Wurzelstudien im tropischen Pflanzen I. Die Berwurzelung von *Vellozia* und *Navia* sowie der Baumfarne und einige Palmen. Abhandlungen der Akademie der Wissenschaften und Literatur Mainz – Mathematische naturwissenschaftliche Klasse, 4, 211–249.

72. Lamont, B. (1981) Morphometrics of the aerial roots of *Kingia australis*. Australian Journal of Botany, 29, 81–96.

73. Staff, I.A. & Waterhouse, J.T. (1981) The biology of arborescent monocotyledons, with special reference to Australian species. In The Biology of Australian Plants (eds J.S. Pate & A.J. McComb), pp. 216–257. University of Western Australia Press, Nedlands.

74. Weber, H. (1953) Las raíces internas de *Navia* y *Vellozia*. Mutisia, 13, 1–4.

75. Lamont, B. (1981) Availability of water and inorganic nutrients in the persistent leaf bases of the grass tree *Kingia australis*, and uptake and translocation of labelled phosphate by the embedded aerial roots. Physiologia Plantarum, 52, 181–186.

76. Coleman, W.H. (1844) Observations on a new species of *Oenanthe*. Annals and Magazine of Natural History, 13, 188–191.

77. D'Almeida, J.F.R. (1942) A contribution to the study of the biology and physiological anatomy of Indian marsh and aquatic plants. Part II. The Journal of the Bombay Natural History Society, 43, 92–96.

78. Pallis, M. (1916) The structure and history of Plav: the floating fen of the delta of the Danube. Journal of the Linnean Society (Botany), 43, 233–290.

79. Weaver, J.E. & Himmel, W.J. (1930) Relation of increased water content and decreased aeration to root development in hydrophytes. Plant Physiology, 5, 69–92.

80. Martins, C. (1866) Sur les racines aérifères ou vessies natatoires des espèces aquatiques du genre *Jussiaea* L. Académie des Sciences et Lettres de Montpellier. Mémoires de la Section des Sciences, 6, 353–370.

81. Ellmore, G.S. (1981) Root dimorphism in *Ludwigia peploides* (Onagraceae): development of

two root types from similar primordia. Botanical Gazette, 142, 525–533.

82. Hildebrand, F. (1885) Über *Heteranthera zosterifolia*. Botanische Jahrbucher für Systematik, Pflanzengeschichte und Pflanzengeographie, 6, 137–145.

83. Campbell, E.O. (1964) The restiad peat bogs at Motumaoho and Moanatuatua. Transactions of the Royal Society of New Zealand (Botany), 2, 219–227.

84. Dean, E.B. (1933) Effect of soil type and aeration upon root systems of certain aquatic plants. Plant Physiology, 8, 203–222.

85. Ellmore, G.S. (1981) Root dimorphism in *Ludwigia peploides* (Onagraceae): structure and gas content of mature roots. American Journal of Botany, 68, 557–568.

86. Lyubich, F.P. & Arbuzova, L.Ya. (1964) The biological significance of aquatic adventitious roots in *Phragmites communis* Trin. Botanicheskii Zhurnal, 49, 1299–1301 (in Russian).

87. Arber, A. (1920) Water Plants, a Study of Aquatic Angiosperms. Cambridge University Press, Cambridge.

88. Bjork, S. (1967) Ecologic investigations of *Phragmites communis*. Folia Limnologica Scandinavica, 14, 1–248.

89. Kadej, F. & Kadej, A. (1983) Die Organisation der Wurzelapikalmeristeme bei Nymphaceae und Bermerkungen über ihre Entwicklung und taxonomische Anwendung. In Wurzelökologie und ihre Nutzanwendung (eds W. Böhm, L. Kutschera & E. Lichtenegger), pp. 33–41. Bundesversuchanstalt für alpenlandische Forschung, Gumpenstein.

90. Samb, P.I. & Kahlem, G. (1983) Déterminisme de l'organogénèse racinaire de *Jussiaea repens* L. Zeitschrift für Pflanzenphysiologie, 109, 279–284.

91. Warming, E. (1883) Studien über die Familie der Podostemaceae. Botanische Jahrbucher für Systematik, Pflanzengeschichte und Pflanzengeographie, 6, 217–223.

92. Warming, E. (1881) Die Familie der Podostemaceae. Botanische Jahrbücher für Systematik, Pflanzengeschichte und Pflanzengeographie, 2, 361–364.

93. Willis, J.C. (1902) Studies in the morphology and ecology of the Podostemaceae of Ceylon and India. Annals of the Royal Botanical Garden, Peradeniya, 1, 267–465.

94. Benzing, D.H. & Ott, D.W. (1981) Vegetative reduction in epiphytic Bromeliaceae and Orchidaceae: its origin and significance. Biotropica, 13, 131–140.

95. Benzing, D.H., Friedman, W.E., Peterson, G. & Renfrow, A. (1983) Shootlessness, velamentous roots, and the pre-eminence of Orchidaceae in the epiphytic biotope. American Journal of Botany, 70, 121–133.

96. Ho, K., Yeoh, H.-H. & Hew, C.-S. (1983) The presence of photosynthetic machinery in aerial roots of leafy orchids. Plant and Cell Physiology, 24, 1317–1321.

97. Kuijt, J. (1969) The Biology of Parasitic Flowering Plants. University of California Press, Berkeley, Los Angeles.

98. Kuijt, J. (1964) Critical observations on the parasitism of New World mistletoes. Canadian Journal of Botany, 42, 1243–1278.

99. Peirce, G.J. (1893) On the structure of the haustoria of some Phanerogamic parasites. Annals of Botany, 7, 291–327.

100. Kindermann, A. (1928) Haustorialstudien an *Cuscuta*-Arten. Planta, 5, 769–783.

101. Thomson, J. (1925) Studies in irregular nutrition. No. 1. The parasitism of *Cuscuta reflexa* (Roxb.). Transactions of the Royal Society of Edinburgh. 54, 343–356.

102. Grieve, B. (1941) Studies in the physiology of host-parasite relations. II. Adventitious root formation. Proceedings of the Royal Society of Victoria (New Series), 53, 323–341.

102a. Rasa, E.A. & Esau, K. (1961) Anatomic effects of curly top and aster yellows viruses on tomato. Hilgardia, 30, 469–515.

103. Wellman, F.L. (1941) Epinasty of tomato, one of the earliest symptoms of Fusarium wilt. Phytopathology, 31, 281–283.

104. Dimond, A.E. & Waggoner, P.E. (1953) The cause of epinastic symmetry in Fusarium wilt of tomato. Phytopathology, 43, 663–669.

105. Putz, C. & Vuittenez, A. (1974) Observations de particules virales chez des betteraves présentant, en Alsace, des symptoms de 'rhizomanie'. Annales de Phytopathologie, 6, 129–138.

105a. Duhoux, E. (1984) Ontogénèse des nodules caulinaires du Sesbania rostrata (Légumineuses). Canadian Journal of Botany, 62, 982–994.

106. Bronner, R. (1969) Les tumeurs rhizogènes de Forsythia. Bulletin de la Société Botanique de France, Mémoires, 1969, 99–107.

107. Shattock, R.C. (1973) The galls on Forsythia intermedia Zab. Annals of Botany, 37, 987–992.

108. Gorecki, R.S. (1976) Gooseberry and guelder-rose tumours promote rooting. Marcellia, 39, 7–9.

109. Riker, A.J., Bamfield, W.M., Wright, W.H., Keitt, G.W. & Sagan, H.E. (1930) Studies on infectious hairy root of nursery apple trees. Journal of Agricultural Research (Washington), 41, 507–540.

110. Riker, A.J., Bamfield, W.M., Wright, W.H. & Keitt, G.W. (1928) The relation of certain bacteria to the development of roots. Science, 68, 357–359.

111. Moore, L., Warren, G. & Strobel, G. (1979) Involvement of a plasmid in the hairy root disease of plants caused by Agrobacterium rhizogenes. Plasmid, 2, 617–626.

112. White, F.F. & Nester, E.W. (1980) Hairy root: plasmid encodes virulence traits in Agrobacterium rhizogenes. Journal of Bacteriology, 141, 1134–1141.

113. Garfinkel, D.J. & Nester, E.W. (1980) Agrobacterium tumefaciens mutants affected in crown gall tumorigenesis and octopine catabolism. Journal of Bacteriology, 144, 732–743.

114. Akiyoshi, D.E., Morris, R.O., Hinz, R., Mischke, B.S., Kosuge, T., Garfinkel, D.J., Gordon, M.P. & Nester, E.W. (1983) Cytokinin/auxin balance in crown gall tumors is regulated by specific loci in the T-DNA. Proceedings of the National Academy of Sciences of the USA, 80, 407–411.

115. Pomponi, M., Spanò, L., Sabbadini, M.G. & Costantino, P. (1983) Restriction endonuclease mapping of the root-inducing plasmid of Agrobacterium rhizogenes 1855. Plasmid, 10, 119–121.

116. Huffman, G.A., White, F.F., Gordon, M.P. & Nester, E.W. (1984) Hairy-root-inducing plasmid: physical map and homology to tumor-inducing plasmids. Journal of Bacteriology, 157, 269–276.

117. Beyerinck, M.W. (1885) Die Galle von Cecidomyia Poae an Poa nemoralis. Entstehung normaler Wurzeln in Folge der Wirkung eines Gallenthieres. Botanische Zeitung, 43, 305–315, 321–332.

118. Beyerinck, M.W. (1888) Ueber das Cecidium von Nematus Capreae auf Salix amygdalina. Botanische Zeitung, 46, 1–11, 16–27.

119. Gill, A.M. (1969) The ecology of an elfin forest in Puerto Rico, 6. Aërial roots. Journal of the Arnold Arboretum, 50, 197–209.

120. Weber, H. (1936) Vergleichend-morphologische Studien über die Sproßburtige Bewurzelung. Nova Acta Leopoldina, Neue Folge, 4, 229–298.

121. Larson, P.R. & Isebrands, J.G. (1978) Functional significance of the nodal constricted zone in Populus deltoides. Canadian Journal of Botany, 56, 801–804.

122. Salleo, S., Rosso, R. & Lo Gullo, M.A. (1982) Hydraulic architecture of Vitis vinifera L. and Populus deltoides Bartr. 1-year-old twigs. II – The nodal regions as 'constriction zones' of the xylem system. Giornale Botanico Italico, 116, 29–40.

123. Wochok, Z.S. & Sussex, I.M. (1973) Morphogenesis in Selaginella. Auxin transport in the stem. Plant Physiology, 51, 646–650.

124. Freidenfelt, T. (1902) Studien über die Wurzeln krautiger Pflanzen. I. Ueber die Formbildung der Wurzel vom biologischen Geschichtspunkte. Flora, oder Allgemeine Botanische Zeitung, 91, 115–208.

125. Cannon, W.A. (1949) A tentative classification of root systems. Ecology, 30, 542–548.

126. Krasilnikov, P.K. (1968) On the classification of the root system of trees and shrubs. In Methods of Productivity Studies in Root Systems and Rhizosphere Organisms (eds M.S. Ghilarov, V.A.

110

Kovda, L.N. Novichkova-Ivanova, L.E. Rodin & V.M. Sveshnikova), pp. 196–114, Nauka, Leningrad.

127. Jeník, J. (1978) Roots and root systems in tropical trees: morphologic and ecologic aspects. In Tropical Trees as Living Systems (eds P.B. Tomlinson & M.H. Zimmermann), pp. 323–349. Cambridge University Press, Cambridge.

128. Kahn, F. (1978) Architecture et dynamique spatiale racinaires. In Physiologie des Racines et Symbioses (Proceedings of a Symposium 11–15 September 1978, Nancy, France) (eds A. Riedacker & J. Gagnaire-Michard), pp. 242–267.

129. Zobel, R.W. (1975) The genetics of root development. In The Development and Function of Roots (eds J.G. Torrey & D.T. Clarkson), pp. 261–275. Academic Press, London, New York, San Francisco.

130. Kidston, R. & Lang, W.H. (1921) An old red sandstone plant showing structure, from the Rhynie chert bed, Aberdeenshire, Part IV. Restorations of the vascular cryptogams, and discussion of their bearing on the general morphology of the Pteridophyta and the origin of the organisation of land plants. Transactions of the Royal Society of Edinburgh, 52, 831–854.

131. Hawkshaw, J. (1842) Description of the fossil trees found in the excavations for the Manchester and Bolton railway. Transactions of the Geological Society of London, 2nd series, 6, 173–176.

132. Williamson, W.C. (1887) A Monograph on the Morphology and Histology of Stigmaria ficoides. Palaeontographical Society, London.

133. Frankenberg, J.M. & Eggert, D.A. (1969) Petrified Stigmaria from North America: Part 1. Stigmaria ficoides, the underground portions of Lepidodendraceae. Palaeontographica, 128B, 1–47.

134. Stewart, W.N. (1947) A comparative study of stigmarian appendages and Isoetes roots. American Journal of Botany, 34, 315–324.

135. Mägdefrau, K. (1931) Zur Morphologie und Phylogenetische Bedeutung der fossilen Pflanzengattung Pleuromeia. Beihefte zum Botanischen Centralblatt, Zweite Abteilung, 48, 119–140.

136. Jennings, J.R., Karrfalt, E.E. & Rothwell, G.W. (1983) Structure and affinities of Protostigmaria eggertiana. American Journal of Botany, 70, 963–974.

137. Karrfalt, E.E. (1980) A further comparison of Isoetes roots and stigmarian appendages. Canadian Journal of Botany, 58, 2318–2322.

138. Rauh, W. & Falk, H. (1959) Stylites E. Amstutz, eine neue Isoetacee aus den Hochanden Perus. Parts 1 and 2. Sitzungberichte der Heidelberger Akademie der Wissenschaften – Mathematische-naturwissenschaftliche Klasse, 1959, 3–160.

139. Mägdefrau, K. (1932) Über Nathorstiana, eine Isoetacee aus dem Noekom von Quedlinburg a. Harz. Beihefte zum Botanischen Centralblatt, Zweite Abteilung, 49, 706–718.

140. Cridland, A. (1964) Amyelon in American coal-balls. Palaeontology, 7, 186–209.

141. Phillips, T.L. (1979) Reproduction of heterosporous arborescent lycopods in the Mississippian-Pennsylvanian of Euramerica. Review of Palaeobotany and Palynology, 27, 239–289.

142. Corner, E.J.H. (1946) Suggestions for botanical progress. The New Phytologist, 45, 185–192.

4. Anatomical changes during adventitious root formation

PETER H LOVELL and JULIE WHITE

Department of Botany, University of Auckland, Private Bag, Auckland, New Zealand

1. Introduction

1.1. Ease of rooting, induced and 'latent' adventitious roots

Organization is one of the most characteristic traits of living organisms. It is especially conspicuous in the orderly growth that every organism undergoes and which produces the specific forms so characteristic of a particular species [1]. The evidence for biological organization is manifested in studies in which the normal processes of growth and development are modified experimentally, e.g. by removing parts of the growing body. Thus, a 'cutting' removed from a plant, under appropriate conditions, may produce a new root system and finally an entire individual with a balanced ratio of root to shoot. These regeneration processes are present in a great range of plants. Roots that originate in locations other than from the embryo or as branches of the primary root are termed adventitious [2, 3]. They may arise spontaneously on intact plants, especially at nodes of prostrate stems and on rhizomes, or stolons or they may develop only as a response to damage when part of the plant has been deliberately or accidentally severed from the existing root system. This involves various anatomical changes associated with wound responses in addition to those involved in root formation itself. Adventitious roots are produced vigorously and rapidly in some species under appropriate conditions, but much less readily in others, and plants can be grouped according to their ease of rooting (Fig. 1).

Adventitious root production varies tremendously between species, between cultivars, with age and with the nature of the organ or part. The most basic division is between plants that will root and those that will not. Of those that will root the most important subgrouping is between those that require severance and those which do not. Plants that root vigorously when intact tend to root prolifically from nodes and this is most common in herbaceous plants such as *Veronica filiformis* and *V. persica* [4, 5] that have prostrate stems or, for example, *Agropyron repens* [6] and *Ranunculus repens* [7] with rhizomes or stolons. The climbers and species forming prop roots are also often vigorous root formers. Some intact plants will form roots if the plant is layered e.g. *Solanum andigena* [8]. A further sub-group contains those plants which possess 'pre-formed primordia' [9]. As these are initiated when the plant is intact, the early stages at least should not be complicated by wound responses. The rate of development of 'preformed primordia' is usually slow in the intact plant. They are almost always associated with woody plants and often remain as primordia for many years [9, 10]. However, if the plant part is isolated from the root system further development of preformed primordia is greatly accelerated. Thus, in this instance the early stages do not require wounding but the later stages may require it. The preformed situation thus forms an intermediate stage between those groups which require severance for primordium initiation and those which do not. There are some difficulties with the use of the term pre-formed primordia and these will be discussed later.

We suggest that the major difference is between species that require severance before primordial initiation will occur and those that do not. It is probable that the anatomical events will differ markedly between the two groups. In the no severance situation it is probable that the anatomical and physiological changes during rooting are mainly related to that process. However, this cannot possibly be true for the group in which severance is required since in addition we can expect to find a range of wound responses associated with stem severance – the exposure of stem tissue and its protection and repair, disruption of the vascular

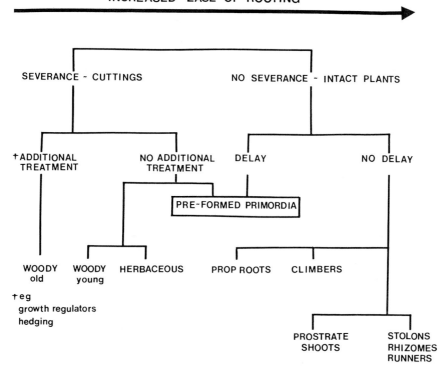

INCREASED EASE OF ROOTING

Fig. 1. The relative ease of adventitious root production in different groups of plants.

tissues with major effects both on the export of materials from the stem and the lack of connection with the root system, thereby removing the possibility of movement of metabolites and growth regulators to and from the roots. For these reasons, then, we might expect a greater level of noise to exist in adventitious root production by cuttings than by intact plants. On the other hand, examination of the processes of adventitious root production occurring under the varied conditions brought about by excision may well provide additional information about the control of the process.

Some of the plants that need to be severed from their root system before rooting will occur also require the application of growth regulators or other chemicals. It is possible that although these chemicals have effects on the anatomical events that follow not all of these may be associated with the rooting process. Unfortunately, for some species information is available only from cuttings treated with growth regulators. However, root formation occurs readily on cuttings of many herbaceous species without the need for special treatment. This is particularly true for stem cuttings although in many plants, leaves [11] and cotyledons [12, 13] root vigorously. Some species will root quickly and easily when cuttings are taken from seedlings but root only with difficulty when cuttings are taken from older plants, for example *Griselinia lucida* (14) and *Agathis australis* [15]. This is particularly true for woody plants. In some cases the anatomical events which take place prior to root production differ in cuttings taken from young and old plants e.g. *Griselinia littoralis* [16].

The majority of anatomical studies have been made on cuttings of woody species, due to their commercial importance. They are also generally more difficult to root and reasons for this have often been sought in their anatomy.

Interpretation of anatomical events described in the literature is sometimes hazardous because the terminology is not always consistent and because the illustrations do not always support the statements made by the authors. This has led to some long-standing misconceptions. Our aim is to improve our understanding by presenting an anatomical study of adventitious root formation from the first cell divisions to the development of a functional root. The anatomical events that occur prior to those leading directly to primordium formation in some species are discussed because they are essential events which determine the eventual location of the primordium.

1.2. Anatomical stages–terminology

Many authors have commented that adventitious root production can be divided into stages [16, 17, 18, 19, 20, 21], but there is a lack of agreement as to the number and nature of the stages and also in the terminology used. However, everyone supports the concept that there must be at least two stages, root initiation and root growth. Consideration of the range of possible events in different species suggests that there are a number of other stages which can be identified anatomically, but sometimes not all of these are present. For example, an important difference exists between those species which need to create a site for root primordium production and those in which the site is already present. Species in the former category will pass sequentially through all of the stages listed below whereas those of the latter type undergo only those stages that occur at the already existing potential root initiation site.

Creation of potential root initiation site.
1. Induction and activation.
2. Processes that may include
   (a) differentiation of cells producing arcs of tracheids giving rise to the creation of a site at the end of the arc [22];
   (b) continued cutting off of cells from the cambium resulting in *displacement of the site* [16].
Processes occurring at the potential root initiation site.
1. Induction and activation which involves cell division. In the case of primordia occurring *in situ* (i.e. neither 'created' nor 'displaced') division of the phloem

parenchyma, cambium or recent derivatives of the cambium usually takes place.

2. Further divisions of the cells either at the original location or at the newly identified location. These cells will be termed the *primordial initials*. However, the primordium may be not be *determined at this stage*.

3. Increase in cell number and beginning of organization. When about 1500 cells are present [22] the structure becomes organized, beginning at the apex. These are not the *root initials*. At this stage the structure is determined and is a *root primordium*. Torrey [23] considers that about 1500 cells is about the smallest cell number in the root apex of pea for it to act autonomously.

4. Further increases in complexity, and development into a root.

Thus, the cells that are activated initially may or may not be the ones that become the future primordium. The early events are characterized by the appearance of cells, with a large, centrally located nucleus and a small vacuole [24], which are capable of division [25] and may give rise to the initials of the organized root tip [26] if the primordium develops *in situ*. Both Cameron & Thomson [18] and Stangler [26] use the term root initials for these cells and we agree with them, but only for those cases where the cells do, in fact, ultimately become the initials which give rise to the tissues in the developing root. At this early stage, there is no apparent organization and the cell mass is not necessarily 'determined'. A root might not be produced. At a somewhat later stage the cells from these 'initials' often together with surrounding cells have become incorporated into an organized mass of meristematic tissue in which some differentiation has occurred. At this point it is 'determined' and we prefer to use the term 'primordium' to refer to this stage a view in line with that of Stangler [26] and Girouard [20] rather than the somewhat earlier stages as used by other authors e.g. Carlson [27] and Haissig [9]. Subsequently vascular connections are formed between the primordium and the existing vascular tissue of the plant and a functional root emerges.

There are several important points which emerge from the above discussion e.g. lack of uniformity in the use of terms such as primordia. The use of this term to identify an early stage, before the cell mass is committed may be unwise since the first cells to be 'induced' may or may not be those which ultimately become the primordial initials. When the primordium does not occur at the site of the initial activation, the cells giving rise to the primordium may be the ones first cut off from the cambium (14) or they may be another group of cells entirely [22]. A further complication is found in cuttings, especially of woody material, where the creation of new sites is more likely to occur. The narrow view would be to consider that only those processes which occur at the site of primordium development are relevant to adventitious root production and that all of the other events are wound responses. We take a wider view because of the extreme importance of the earlier events either in the creation of a potential root initiation site or in the targeting of anatomical events to already existing sites. These events are often variable from species to species both anatomically and in duration.

In a number of species, especially in intact plants, induction processes occur, but the development of the structure halts at some intermediate stage. These have been called pre-formed primordia [9] or pre-existing root initials [28] or latent adventitious root primordia [29] but in a number of cases the stage at which development was interrupted precedes any sign of organization. We feel that the terms presumptive primordial initials [30] or potential primordial initials are more appropriate because of the uncertainty over their future development.

2. Stated origin of adventitous root primordia

Wherever possible the locations given by the authors of the original papers are presented. In some cases we disagree with their findings and will discuss these special cases later.

2.1. Potential primordial initials

These occur almost exclusively in woody species and much of the work has been carried out on members of the Salicaceae (*Populus, Salix*). Studies by Carlson [36, 37], Swingle [32], van der Lek [33], Fink [29] and Haissig [31, 41] have been particularly influential in determining our thinking (Table 1). In general, the 'primordia' arise at nodes (*Salix*) associated with bud gaps (*Cotoneaster* species, *Ribes*) and leaf gaps (*Lonicera, Malus* spp.). Others are associated with ray tissue and some authors have noted that the 'primordia' only occur when the rays are abnormally broad e.g. *Thuja occidentalis, Juniperinus communis*, var. *depressa, J. horizontalis* and *J. virginiana* [42, 43] and a range of other species [29, 33].

There is a shortage of information about the frequency of occurrence of primordia at the different sites. Van der Lek [33] distinguished between nodal roots (those associated with leaf and branch traces) and internodal ones (occurring in ray tissues) but did not provide much quantitative information. One of the few studies of comparative frequency of primordia at different locations in the stem was carried out by Swingle [32] using apple but even this is, at best, semi-quantitative. He noted that, after examination of thousands of sections, the frequency of primordium initiation was directly related to the width of the band of radial parenchyma connecting the cortex with the central pith. These locations (shown diagrammatically in Fig. 2) are, in decreasing order of frequency of primordium initiation, branch gaps, leaf gaps, the pair of primary rays one on each side of each leaf trace after it has entered the stele, the secondary rays. In all cases the initial meristematic activity occurred in the parenchyma external, but close to the cambium [32]. These primordia often develop into roots over a two to three year period but the roots remain as rudimentary 'burrknots' which tend not to develop even if cuttings are made [44]. Carpenter [38] suggests that in *Citrus* the primordia form shortly after differentiation of the primary stem because their origin is close to the pith. However their origin is clearly outside the cambium region and the lignification of the ray in association with the secondary xylem could perhaps suggest an earlier origin than actually occurred. It is unlikely that the events occurring internal to the cambium are associated with the primordium.

As far as point of origin is concerned there seems to be no difference between potential primordial initials that have a protracted period of very slow or no growth and those that do not. Neither do there seem to be any anatomical differences between the two groups. In some species, layering is sufficient to induce root production in intact plants even if potential primordial initials are not present before treatment. The Cumberland black raspberry roots vigorously

Table 1. Sites of origin of potential primordial initials in plant stems. Potential primordial initials have been referred to as pre-formed primordia by Haissig, [31], as burrknots, Swingle, [32] or as rootgerms by van der Lek, [33].

Reported site of origin	Species
Wide rays	*Populus nigra* var. *pyramidalis* [33], *P. simoni* [33], *P. thevestina* [33], *P. trichocarpa* [33]
Medullary rays, three double series associated with buds	*Ribes nigrum* [33]
Nodal and connected with wide radial bands of parenchyma	*Salix amygdalina* [33]
Internodal medullary rays	*Salix alba vitellina britzensis* [33], *S. babylonica* [33], *S. discolor* [33], *S. laurifolia* [33], *S. nigricans* [33], *S. regalis* [33], *S. rotundifolia* [33]
Cambial ring in branch and leaf gap; 1° and 2° medullary rays.	*Malus* [32]
Bud gap	*Cotoneaster microphylla* [34]
Median and lateral leaf trace gaps at node	*Lonicera japonica* [28]
Parenchymatous cells in divided bud gap	*Cotoneaster dammeri* [35]
Parenchymatous secondary cells in outer part of a vascular ray[1], above a leaf or branch gap	*Salix fragilis* [36[1], 37] [1] identified in error as *S. cordata*
Medullary ray	*Citrus medica* [38]
Ray cells in 4th node below terminal leaf	*Salix fragilis* [31]
Lignified parts of new stem, precise location not determined	*Populus x robusta* [39]
Phloem ray parenchyma	*Hydrangea macrophylla* [40]
Cambial region of an abnormally broad ray	*Acer pseudoplatanus* [29], Chamaecyparis lawsoniana [29], *Fagus silvatica* [29], *Fraxinus excelsior* [29], *Juniperus communis* [29], *Populus trichocarpa* [29], *Salix alba* [29], *Taxus baccata* [29], *Thuja glutinosa* [29], *T. plicata* [29], *Ulmus carpinifolia* [29]

118

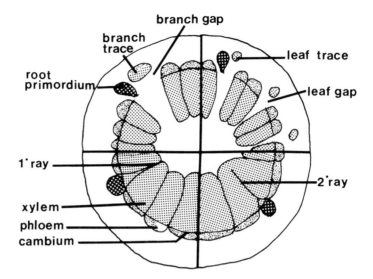

Fig. 2. Locations of potential primordial initials in intact stems: data for apple, (after Swingle [32]). The four quadrants represent the four locations where root primordia arise i.e. associated with branch traces, leaf traces, primary rays and secondary rays.

when the rapidly elongating shoot tip is layered. The roots are usually associated with leaf traces [45, 46].

2.2. *Induced primordia–woody plants*

Induced primordia form as a result of some form of manipulative treatment and the most common treatment is the taking of a cutting. Sites of origin within cuttings in a wide range of species are listed in Table 2. Primordia may arise either at nodes or internodes and the relative frequency at these locations is probably influenced by the length of internode present below the lowest node on the cutting. Occasionally root primordia occur very close to the basal end of a cutting, e.g. in *Picea abies* cuttings up to 20 primordium initials developed in the basal 1mm close to the newly formed vascular tissues [68]. However, roots tend to form a few millimetres above the base of a cutting and the distance of the node from the cut base is an important factor determining the frequency of rooting at the node. Unfortunately, since information about where the cut was made is rarely given in the literature it is often difficult to evaluate the relative importance of the two locations for most species. This point is made by Brutsch *et al.* [64] who note that in their study more roots originate from callus produced in internodal tissues or callus at the cut base than from tissues associated with leaf traces. They then make the point that this result may have been determined by the position of the cut and also by the greater emphasis that they placed on sectioning internodes

Table 2. Sites of origin of wound induced adventitious roots in woody plant stems. When other organs were used they are cited after the species name.

Reported site of origin	Species
CAMBIAL AND RAY	
Cambial and phloem portions of ray tissues	*Acanthopanax spinosa* [33], *Chamaecyparis obtusa* [50], *C. pisifera* [50], *Cryptomeria japonica* [50], *Cunninghamia lanceolata* [50], *Cupressus lawsoniana* [50], *Metasequoia glyptostroboides* [50], *Taiwania cryptomerioides* [50], *Thuja occidentalis* [50], *T. standishii* [50], *Thujopsis dolabrata* [50], *T. dolobrata* [33]
Medullary rays	*Vitis vinifera* [33]
Cambium	*Acanthus montanus* [47], *Lonicera japonica* [28]
Fascicular cambium	*Clematis similacifolia* [48]
Associated with phloem and cambium	*Coffea robusta* [49]
Phloem region usually immediately adjacent to xylem and in the cambium	*Vaccinium corymbosum* [24]
Immature multiseriate ray tissue of secondary phloem close to cambium	Rosa dilecta [26]
Phloem ray parenchyma	*Ficus pumila*, leaf bud [53], *Hedera helix*, juvenile [20]
Phloem ray parenchyma at internodes and in callus	*Hedera helix*, mature [51]
Secondary phloem in association with a ray	Apple Malling [52], *Camellia sinensis*, Brompton plum [52]
Phloem area close to the cambium	*Pistacia vera*, softwood branches [54]
Cambium and inner phloem ray also in leaf gap	*Griselinia littoralis* [16], *Griselinia lucida* [16]
BUD AND LEAF GAPS	
Outside the cambium in small groups of parenchymatous tissue within or between bundles	Rose cv Dorothy Perkins, American Pillar [55]
Parenchyma cells in bud gap	*Cotoneaster dammeri* [35]
Opposite a congerie of rays, adjacent to the leaf trace	*Pinus strobus* [56]
Bud traces (gaps)	*Cephalataxus drupacea* [50], *Larix kaempferi* [50], *Sciadopitys verticillata* [50]
Near cambial zone in bud and leaf gap parenchymatous tissue	*Malus*, M9, M26, M2 and Malling Merton 106, layering [57]
PERICYCLE	
Pericycle	*Acanthus montanus* stem, leaf, soil root, prop roots [47]
Parenchymatous cells of the bundle sheath just outside the phloem	*Phaseolus vulgaris*, leaf petiole [58]
Outside protoxylem points, generally from the cambium or from pericycle callus opposite protoxylem poles	*Abies procera*, root [59]

Table 2. Continued.

Reported site of origin	Species
Parenchymatous pericycle cells	*Dianthus caryophyllus* [26]
Interfascicular region at flanks of bundles in pericycle parenchyma	*Chrysanthemum morifolium* [26]
Parenchyma adjoining the bundles	*Boerhaavia diffusa*, petiole [60]
CALLUS – INTERNAL	
Irregularly arranged parenchymatous tissues	*Abies firma* [50], *Abies mayriana* [50], *Juniperus rigida* [50], *Picea excelsa* [50], *Sequoia sempervirens* [50], *Thujopsis dolabrata* [50]
Callus	*Larix x eurolepis* [61]
Associated with induced vascular traces	*Agathis australis*, hypocotyl [22]
CALLUS – EXTERNAL	
Callus tissues (external)	*Abies firma* [50], *Cedrus libani* [50], *Cryptomeria japonica* [50], *Ginkgo biloba* [50], *Larix kaempferi* [50] *Pinus densiflora* [50], *P. pinaster* [50], *P. silvestris* [50], *P. strobus* [50], *P. thunbergii* [50], *Podocarpus* macrophyllus [50], *Sequoia gigantea* [50], S. sempervirens [50], *Sciadopitys verticillata* [50], *Taxodium distichum* [50], *Taxus cuspidata* [50], *Thuja occidentalis* [50], *Thujopsis dolabrata* [50], *Torreya nucifera* [50], *Tsuga sieboldii* [50]
Callus tissues	*Pinus elliotii* [62], *Pinus radiata*, air layers and cuttings [18]
Bark and basal callus	*Citrus medica* [38]
Within callus at base of cutting	*Pseudotsuga menziesii* [63]
Three types of callus, basal, cambial and phloem/cortex	*Carya illinoensis* [64]
OTHER	
Hyperhydric outgrowth of the lenticels	*Tamarix aphylla* [65]
Margin of differentiating resin duct or parenchyma within the inner cortex	*Pinus radiata* [66]
Cortex and/or pith	*Guava* [67]

rather than nodes. Table 2 is extensive because most of the economically important plants fall in the 'woody' category and most of the work has been done on stem cuttings of this group. Despite the large number of species there are relatively few locations at which primordia arise. The favoured sites are rays (close to the phloem and cambium) or in bud or leaf gaps. These are not mutually exclusive for any given species e.g. in *Griselinia lucida* they form in the inner phloem ray in hypocotyl tissue from seedlings, but in leaf gaps in tissue from mature plants [16]. Other important locations for primordium origin are in the pericycle or in callus which may be produced at the base of the cutting (see especially Satoo [50] for information on gymnosperms) or in internal callus e.g. *Griselinia littoralis* [16] and *Carya illinoensis* [64]. These locations cover virtually all of the ones cited in the literature. There are some unusual locations mentioned e.g. primordium initiation occurring a few cell layers below the surface of lenticels in branch cuttings of *Tamarix aphylla* [65]. Wally *et al.* [67] also suggest that adventitious roots may develop under a lenticel or in the outer cortex in guava. Other workers have shown that lenticels are often closely related to 'storage' rays [69, 70]. Since rays are a common location for adventitious root initiation (Table 2 above) roots may often emerge via a lenticel if one happens to be present in an appropriate place [36, 37]. The margin of differentiating resin ducts has been mentioned as a possible location in *Pinus radiata* [66]. Resin ducts have been shown to develop in the rooting zone of the cutting at the same time as developing primordia in *Agathis australis* but their structure is somewhat different [71]. A report of adventitious root initiation occurring in the pith in guava is not supported by any illustration in their paper [67].

2.3. Induced primordia–herbaceous plants

Table 3 catalogues the sites where induced root primordia have been found. Much of the work has been carried out on young plants using hypocotyl tissue [21, 73, 79, 81, 84], cotyledons [73, 80] and leaf tissue or petioles [58, 75, 82]. Thus, the list has a different bias from those in Tables 1 and 2 which are almost exclusively stem tissue.

In almost all cases, irrespective of the species or material the origin of the primordium is close to vascular tissue (Table 3). Some of these locations are shown diagrammatically (Fig. 3). There are few physical obstacles to root initiation and development in the material illustrated. Where incomplete rings of fibre bands occur e.g. *Phaseolus vulgaris* [84] it is noticeable that roots originate in sectors where fibre bands are not present. In *Phaseolus* hypocotyls the roots develop in four distinct longitudinal rows parallel to and between the four pairs of vascular bundles. Application of auxin does not alter the location. Roots are initiated from parenchyma cells outside the cambium on the same radius as the pericycle fibres [84]. It is also common for four rows of roots to be produced in hypocotyl cuttings of *Phaseolus aureus* [79]; they are associated with the cotyledonary and primary leaf vascular traces. These roots usually develop near to the cut base but occasionally arise up to 20 to 30 mm above it. Hypocotyls of tomato root within 10 mm of the base [81]. The first events reported are anticlinal divisions in the cell layers close to the phloem. In general, initial events are usually close to vascular bundles or in layers close to the interfasicular region; occasionally sites further out into the cortex have been reported. In segments of rhizomes of *Helianthus tuberosus* cultured *in vitro*, differentiation of islets of phloem and tracheids occurs after 4 d, followed by the development of a cambium which, in turn, gives rise to root primordia [86]. Apart from the absence of primordia immediately internal to pericycle fibres the origins of primordia appear to be affected very little by hypocotyl or stem structure in herbaceous plants (Fig. 3).

Both Wilson [72] and McVeigh [76] indicate that adventitious roots may have epidermal origins. Wilson suggests that they arise in nodal regions in the axils of branch buds in stem cuttings of *Rorippa* while McVeigh [76] observed that primordia are produced from cells of the lower epidermis of cut laminae and petioles close to, but not from, wound callus tissue in *Crassula*. In cut laminae the activity is greatest in the midrib region. Sometimes outgrowths from petioles develop both root and shoot primordia.

Table 3. Sites of origin of wound induced adventitious roots in herbaceous plants.

Reported site of origin	Species
Epidermal and cortical cells in the axil of a bud	*Rorippa austriaca*, stem [72]
Between fibrovascular bundles in cells of the pericycle	*Coleus blumei*, branch tip [27]
Parenchymatous tissue of pericycle, phloem and pith	*Linum usitatissimum*, hypocotyl, [73]
Parenchymatous tissue surrounding veins	*Linum usitatissimum*, cotyledon [73]
Interfascicular cambium occasionally fascicular cambium at edge of bundle	*Begonia maculata*, stem [74], *B. semperflorens*, stem [74]
Parenchyma between leaf traces	*Saintpaulia ionantha*, petiole [75]
Near veins – exact tissue not determined.	*Saintpaulia ionantha*, lamina [75]
Epidermis	*Crassula multicava*, leaf [76]
Interfascicular cambium	*Tropaeolum majus*, stem [77]
Parenchymatous cells just outside phloem and medullary rays	*Phaseolus vulgaris*, isolated leaf petiole [58]
In parenchyma alongside vascular bundle in line with border between xylem and phloem	*Phaseolus vulgaris*, petiole [78]
Four sites associated with vascular tissue	*Phaseolus aureus*, hypocotyl [79]
Cells located between xylem and phloem at sides of vascular strands	*Sinapis alba*, cotyledon petiole [80] *Raphanus sativus*, cotyledon petiole [80]
Between two vascular bundles from specialized phloem parenchyma	*Vigna*, hypocotyl [21]
In pericycle and endodermal cells related to primary xylem and phloem	*Lycopersicon esculentum*, hypocotyl [81]
Main veins in external phloem parenchyma. Minor veins in sheath parenchyma	*Lycopersicon esculentum*, leaf discs [82]
Interfascicular region between endodermis and interfascicular cambium	*Azukia angularis*, stem [83]
In four rows parallel to and between four pairs of vascular bundles	*Phaseolus vulgaris* cv Brittle Wax, hypocotyl [84]
Interfascicular cambium	*Pelargonium x hortorum* cv Yours Truly, stem [85]

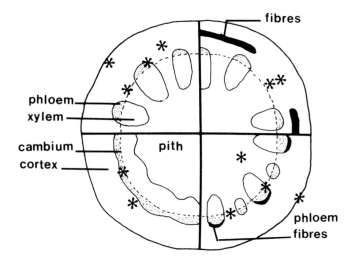

Fig. 3. Reported locations of origin of adventitious roots in herbaceous hypocotyls and stems (*). The four quadrants represent (counter-clockwise) discrete bundles without fibre caps, vascular tissue with narrow rays, bundles with phloem fibre caps, and pericyclic fibres in incomplete rings outside the bundles.

3. Initiation and development of adventitious roots

3.1. 'Preformed primordia'

Root initials develop in intact stems of brittle willow at node 4 which is in the transition region which separates the zones of rapid growth and maturation [31]. They are located in leaf gaps (Fig. 4A) where ray cells develop large nuclei, prominent nucleoli and dense cytoplasm. These groups of cells are increased by the addition of cambial derivatives (Fig. 4B). The cells increase in number (Fig. 4C) as a result of cell divisions occurring in all planes but with little enlargement of the daughter cells. After severance [36] further divisions give rise to a structure which has the appearance of a very early stage primordium which is beginning to push out into the cortex (Fig. 4D). It is close to the interfascicular cambium but shows little organization at this stage. In the absence of severance the structure may remain at a primordium initial stage over the winter or even for a number of years. In *Salix fragilis* the primordium forms at the site of the initial cell divisions.

3.2. Formation of primordia in situ

This occurs in hypocotyl cuttings of both *Griselinia littoralis* and *G. lucida* [16]. It is not certain whether the initial cells originate directly from the cambium or from

124

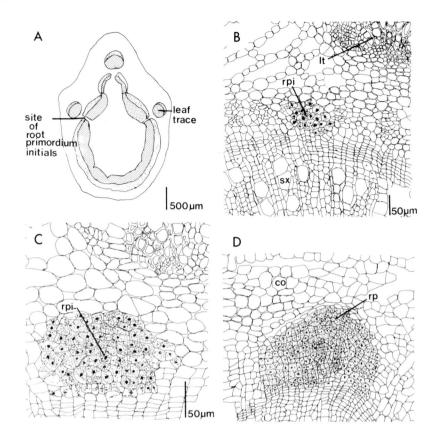

Fig. 4. The initiation and development of root primordia in intact stems of *Salix fragilis*. (A) A transverse section through a node showing the location at which root primordium initials develop in a leaf gap. Root primordium initial, early stage (B) and later stage (C). Further cell division and the beginning of organization occurring after severance (cutting taken from one year-old wood). A-C (after Haissig [31]) and (D) (after Carlson [36]) which is approximately the same scale as (B). A scale was not given by the author.
Abbreviations: co, cortex; lt, leaf trace; rp, root primordium; rpi, root primordium initial; sx, secondary xylem.

Fig. 5. Root initiation in cuttings of *Griselinia*. Hypocotyl cuttings taken from 3 month old seedlings (A-C) of *G. lucida* show that root primordia developed in the region where the first cell divisions occurred. (A) Cell divisions of recent derivatives from the cambium give rise to a root primordium initial in association with a phloem ray; (B) the outermost cells continue to divide while those closest to the xylem do not divide again and remain undifferentiated at this stage. These cells will differentiate later to form vascular connections; (C) root primordium close to emergence. The apex is highly organized and covered by a root cap. Vascular connections are differentiating between the root and stem vascular systems. Cuttings from mature *Griselinia littoralis* plants (D-G) – sections taken approximately 5 mm from the cut base. (D) Cross section showing that over 50 per cent of the

125

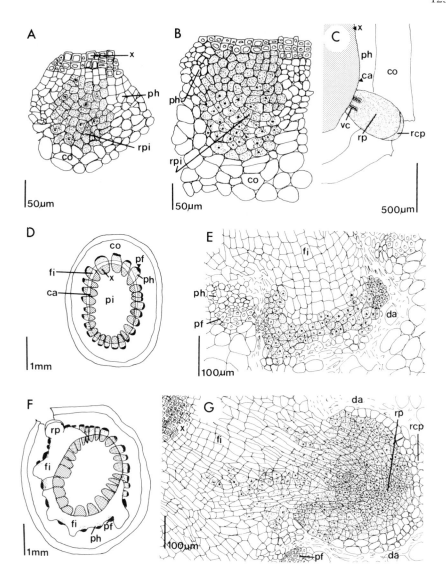

cambium has produced files of cells to the outside; (E) the files of cambial derivatives push the phloem and the fibre caps outwards and sidewards causing cell damage. The outermost cells of the files are dividing or preparing for division; (F) although a substantial amount of cambial activation has taken place only one root primordium is present in this cross section; (G) a root primordium which is becoming organized at the apex. A root cap is present and cell division is occurring in strands back towards the xylem (after White & Lovell [16]).

Abbreviations: ca, cambium; co, cortex; da, damaged cells; fi, files of cambial derivatives; pf, phloem fibres; ph, phloem; pi, pith; rcp, root cap; rp, root primordium; rpi, root primordium initials; vc, vascular connections; x, xylem.

adjacent immature derivatives but they form a group of identifiable cells situated close to the cambium (Fig. 5A). The cells closest to the cambium enlarge but do not divide further at this stage and the outer cells are those that continue to divide and are the root primordium initials (Fig. 5B). By the time that the root primordium has traversed the cortex it is almost fully organized and vascular connections have been established (Fig. 5C). In this case the root primordium initials are the same cells as those involved in the first divisions and the primordium is initiated at the location of those divisions quite close to the cambium.

3.3. Formation of primordia at a point distant from the first cell divisions

3.3.1. Continued division of the cambium to produce an internal callus. In cuttings taken from mature plants of *Griselinia littoralis* [16] the initial events that occur include activation of a large part of the cambium in a region about 5 mm from the cut base (Fig. 5D). The areas of maximum proliferation are associated with leaf traces and in these regions the cambium cuts off files up to 22 cells in number on the phloem side (Fig. 5E, F). There are no corresponding divisions on the xylem side. The files of cells push outwards into the cortex and this results in displacement and damage to cells. Under certain circumstances e.g. when the files of cells force outwards between phloem fibre caps a further wave of cell division begins in the cells that were the first to be cut off from the cambium. These cells, by this time, are at about the same radius as the fibre caps (Fig. 5E). The most active area of cell division is in the region between the fibre caps although some cells closer to the cambium side of the active region also divide. The primordial initials form in the area of actively dividing cells and a root primordium develops (Fig. 5G). The primordia can be readily distinguished by changes in the plane of cell division, the developing spherical structure and the beginning of organization at the apex. The stimulus for the second wave of cell division appears to be associated with constriction and cell damage as the files of cells come into contact with the fibre caps. Other regions of very active cambial activity give rise only to files of cells and not to root primordia (Fig. 5F). By the time of root emergence, the strands developing from the primordium to the vascular tissue have differentiated into vascular connections. Thus, in mature tissue of *Griselinia littoralis* the root primordial initials are the first cells to be cut off from the cambium. Because the cambium continues to cut off cells and because the further division of those cells is delayed, the root primordium is initiated at the same radius as the fibre caps and not close to the cambium. Similar events are illustrated in other species although the authors may not have made specific reference to them or noted that they might have some implications for the rooting process. Thus, proliferation of divisions of the cambium giving rise to files of cells prior to the formation of root primordium initials can be seen in Dorothy Perkins rose [55], *Pinus strobus* [56], mature *Hedera helix* [51], hybrid larch [61] and for a range of other woody species in a paper by Satoo [50] in which she refers to irregularly arranged parenchymatous tissue. Davies *et al.* [53] do refer to the production of files of cells radiating out from the cambium as an event occurring prior to rooting in both juvenile and adult cuttings of *Ficus pumila*. In the absence of time course studies these early events could be missed and the situation misinterpreted so that the initial events could be thought to occur some distance from the cambium.

3.3.2. The formation of root primordia in association with induced vascular tissue.

In order for a root to be functional it is essential that vascular connections are established between the developing root and the stem vascular system. Although the final link-up is usually made only as the root primordium is about to emerge from the stem tissues there is an increasing body of information which indicates that the origin of the primordium is associated or even determined by the presence of induced vascular tissue. This is a characteristic of 'woody' material. The sites of origin of the root primordia may be either in the stem tissue or alternatively in callus produced at the base, e.g. in air layered *Pinus radiata* trees, root primordia do not form in the original stem tissues, they develop in callus produced by proliferation of cells from the cambium, phloem and xylem parenchyma (Fig. 6A). Cameron and Thomson [18] noted that the formation of root primordia is preceded by lateral extensions of induced vascular tissue. The 'callus xylem' on occasions is shown as vascular strands extending from the stem tissues into the callus. In an extensive study of induced vascular development in cuttings of *Agathis australis* it was found that the first event to occur in the sub-base of cuttings is the differentiation of tracheids and phloem in the interfascicular region [22, 71]. These cells are similar in size and shape to the adjacent parenchyma and form short files of vascular tissue which advance by cell division and differentiation. The files arc outwards and downwards (Fig. 6B) and root primordia ultimately form at the end of the arc much closer to the base of the cutting and well out in the cortex (Fig. 6C, E). At the time of emergence there is continuity of vascular tissue from the stem vascular system to that of the primordium (Fig. 6D). However, all of the vascular tissue except the final connection develops basipetally and arises prior to primordium initiation. The link is made as a result of vascular development from the primordium to the end of the vascular strands. In this example primordium initiation is substantially displaced both in space and time from the first anatomical events. The primordium location is 'determined' by the vascular strands. Brutsch *et al.* [64] found that cuttings of *Carya* (pecan) produce large amounts of callus tissue. They note that it is difficult to identify the initiation stage of root primordia partly because of its short duration and also because of the extensive callus development. Root primordia appear to originate in the vicinity of tracheids which have formed in callus produced within the stem. The tracheids are usually discrete bands or bundles. However, when callus is produced at the base although primordia develop in association with induced vascular tissue (Fig. 6F) they suggest that the vascular tissue may not link up with

128

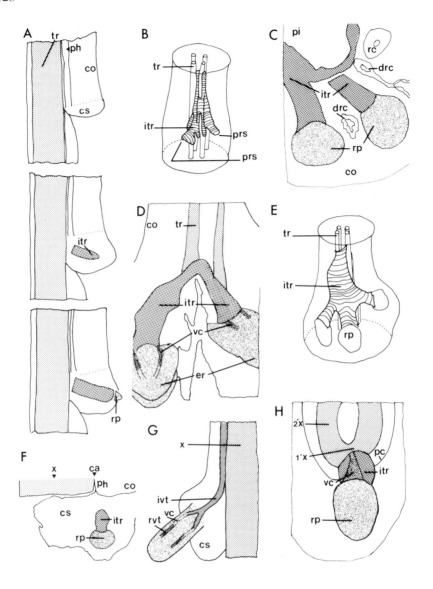

the vascular bundles of the stem. This would imply that the vascular tissue had been induced in the callus tissue itself. Data for root production on pruned roots [59] also show the association of the newly developed primordium with induced vascular tissue (Fig. 6H).

Another variant on the theme is illustrated by root formation in callus which formed at the base of cuttings of hybrid larch [61]. The initiation of root primordia within the callus is preceded by the initiation of tracheids about 3 mm above the cut base. The tracheids then arc down through the stem and callus tissues. Organized areas of cells (tracheid nests) develop at the end of the induced

Fig. 6. A diagrammatic representation of the development of root primordia in association with induced vascular tissue. (A) *Pinus radiata* – root primordia produced in callus after stem girdling (air layering) mature trees. Callus is produced on both the vertical and exposed cortical faces (upper), subsequently the callus mass increases, induced vascular tissue develops (middle) and later root primordia are initiated (lower) at the ends of the vascular strands; (after Cameron & Thomson [18]). (B-E) *Agathis australis* – root production in hypocotyl cuttings. The three dimensional reconstructions (B) and (E) were made from serial sections (after White & Lovell [22]); (B) strands of induced vascular tissue are initiated close to the vascular system about 3 mm from the cut base and develop downwards and outwards through the cortex; (C) root primordia are initiated at the ends of these induced vascular strands which are connected to the hypocotyl vascular tissue; (D) vascular connections between the primordium and the induced vascular tissue are formed at about the time of root emergence; (E) reconstruction showing the location of the root primordia in relation to the earlier events. (F) *Carya illinoensis* – callus has developed from the cambium, phloem and cortex. Differentiation within the callus gives rise to induced tracheids and a root primordium develops in association with them (after Brutsch, Allan & Wolstenholme [64]). (G) *Larix × eurolepis* – longitudinal section through the base of a cutting showing a callus mass with an emerged root. The root is connected via induced vascular strands to the main vascular system (after John [61]). (H) *Abies procera* – an early stage in the regeneration of a lateral root on a pruned diarch root. Induced tracheids are produced on both sides of the primordium. Vascular connections are present between the root primordium and these tracheids. The lateral root primordium arises in the cambial region opposite the protoxylem pole (after Wilcox [59]).

Abbreviations: ca, cambium; co, cortex; cs, callus; drc, degrading resin canal; er, emerged root; itr, induced tracheids and phloem; ivt, induced vascular tissue; pc, pericycle; ph, phloem; pi, pith; prs, potential root initiation site; rc, resin canal; rp, root primordium; rvt, root vascular tissue; tr, tracheid; vc, vascular connections; x, xylem; $1°X$, primary xylem; $2°X$, secondary xylem.

vascular tissue in the callus and it is from these areas that the root primordia form. Vascularization of the callus commonly occurs prior to root formation [9] and it may be an essential pre-requisite. However, the induction of vascular tissue alone does not mean that root primordia will necessarily be initiated. Sometimes massive amounts of vascular development may take place in the basal regions of stems of cuttings e.g. *Agathis australis* [71], or in basal callus [87] but root primordia are not produced.

4. Development of root primordia and later stages of root formation

In *Agathis australis* the first indication that root primordium initials may develop into emerged roots is when the end of the induced vascular strand is in the mid cortex and the cells immediately ahead of it divide but do not differentiate (Fig. 7A, B). These cells remain unchanged anatomically and eventually form the final vascular connection. Prior to this the neighbouring cells in the cortex are induced to divide repeatedly (Fig. 7C, D). Cell division is not confined to one plane and a roughly spherical group of cells is produced. By the time that the root primordium has about 1500 cells the beginning of organized cell arrangement can be seen at the apex. Periclinal divisions give rise to three rows of cells and these become the vascular and columella initials (Fig. 7E). Initials for the cortex and root cap are

130

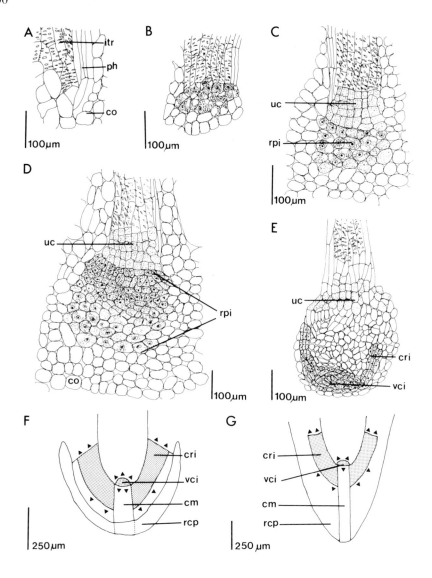

Fig. 7. The development of a root primordium in association with vascular strands in the sub-basal zone of a hypocotyl cutting of *Agathis australis*. (A) The basipetal end of an induced strand of vascular tissue which has terminated in the cortex (see Fig. 6B). (B) Cells abutting the ends of the strand are preparing to divide or are dividing. (C) Cells that have divided do not divide again at this stage, later they will differentiate to form vascular connections. The root primordium initials develop from cortical cells neighbouring the undifferentiated cells. (D) Further cell divisions increase the cell mass and the root primordium initial is clearly delimited. (E) The apex of the developing primordium is organized and the root initials can be seen. (F) A late stage primordium. The apex is highly organized and the root initials are present. (G) Root tip of an emerged root (after White & Lovell [22]).

Abbreviations: cm, columella; co, cortex; cri, cortex and root cap initials; itr, induced tracheids and phloem; ph, phloem; rcp, root cap; rpi, root primordium initials; uc, undifferentiated cells; vci, vascular and columella initials.

also present. The zone of undifferentiated cells is still present. The apex is organized (Fig. 7F) with a root cap consisting of three layers of cells and the columella forming in a central position. By the time the root emerges (Fig. 7G) the root apex resembles a seedling root tip. The files of cells produced by the cortical and root cap initials can now be easily identified. Differentiation of the uncommitted cells takes place at about the time of root emergence. These cells form the vascular connection between the primordium and the hypocotyl vascular tissue.

This sequence of events in the later stages, after the formation of the root primordium initials is fairly similar for most species for which information is available, although the processes may differ in detail. As noted earlier there are differences between species in the earlier stages *i.e.* those determining the location of the root primordium and also whether cortical cells are incorporated into the primordium (as in *Agathis*) or not (as in *Griselinia*).

5. The importance of physical barriers

There has been a continuing debate in the literature over whether or not certain anatomical structures present physical barriers to adventitious root production. Restriction of rooting could occur as a result of obstruction by sclerenchyma bands, secretory canals, resin canals, or large volumes of induced vascular tissue occupying space in the sub-basal region which is thus unavailable for primordium initiation. Alternatively, the presence of obstacles could inhibit primordium initiation through some mechanism other than reduction of available space or the initiation of the primordium could occur but its outgrowth might be prevented. Some of the 'barriers' that occur are illustrated in Fig. 8. Studies of *Phaseolus vulgaris* and three varieties of *Hibiscus rosa-sinensis* showed that etiolated plants have less structural tissue and relatively more parenchyma in the stems, and that rooting is more vigorous in cuttings taken from etiolated material [88]. Studies of woody species that are difficult to root suggest that poor adventitious rooting ability may be associated with the presence of a ring of sclerenchymatous tissue [52, 89, 90, 91]. It was also found that rooting in cuttings of *Ilex aquifolium* is aided by wounding which disrupts the sclerenchyma band [91]. In *Dianthus caryophyllus* [26] and *Coffea robusta* [49] root primordia are initiated, grow down inside the fibre band and emerge through the cut base. There is a discontinuous fibre band in *Phaseolus vulgaris* hypocotyls and root production occurs in four files. The root initiation areas alternate with the arcs of sclerenchyma fibres [84]. Secondary resin canals modify the direction of growth of induced vascular tissue in *Agathis australis* [71] and, to that extent, affect root primordium initiation. However, Kachecheba [92], after studying several species of *Hibiscus* which had varying amounts of fibres, came to the conclusion that there is no general relationship between the presence of tissues that might be considered as barriers and poor adventitious root production.

132

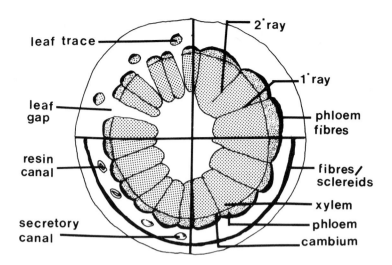

Fig. 8. Diagrammatic representation of physical barriers that may be present in cuttings from woody stems. Barriers include phloem fibre caps, bands of fibres/sclereids, secretory canals and resin canals.

In general, we support the view of Sachs *et al.* [93] who believe that ease or difficulty of rooting is associated primarily with root initiation rather than subsequent development. Davies *et al.* [53] argue that the poorer rooting of mature material of *Ficus* has a physiological basis and not an anatomical one despite the increase in sclereids, perivascular fibres and laticifers as material passes from juvenile to mature. Cuttings from mature *Griselinia lucida* have complete rings of sclerenchyma fibres. When a cutting roots it does so in association with the leaf traces where the sclerenchyma band is discontinuous [16]. There is no evidence that primordia are initiated inside fibre bands and are then restricted from breaking through during subsequent development. In general, primordia are not formed in these locations.

One of the few pieces of work indicating that anatomical structures could prevent the outgrowth of primordia through the tissues is for the emergence of roots through callus in *Populus balsamifera* [94]. When cuttings are grown at a high pH, the callus mass that develops at the base is composed of very small, compact cells. Even when roots are initiated they do not break through the callus. Wu & Overcash [95] also argued that the hard, compact mass of callus which forms in red raspberry hybrid cuttings could be a cause of the lack of root emergence.

6. Timing, variation and overall pattern of events

Species can be divided into four categories based on characteristics of the primordia: *i.e.* 'pre-formed' primordia, primordia that develop *in situ*, primordia that develop in the sub-base but not at the site of the initial cell divisions and those that develop in basal callus. It is not possible to make a quantitative assessment of the relative frequency of the four categories because the information is based on a very biased sample. Herbaceous species are easy to section and may be over-weighted, some authors e.g. [29, 33, 50] have surveyed a considerable number of similar species of woody plants giving rise to an over-representation of those groups. However, irrespective of category, the initial cell divisions almost always occur in the cambium/interfascicular area, or the ray or in bud or leaf gaps or some combination of these. There may be alternative locations possible and the outcome may depend, in cuttings, on the distance below the node at which the cut was made.

Root primordia are usually associated with nodes in intact plants but arise in a more diverse array of locations in cuttings. The initiation of root primordia at nodes may be a consequence of the distinctive anatomical and physiological features of nodes. Nodes are often sites where xylem differentiation is initiated and also may be regions of locally high auxin concentration [96, 97]. Auxins may be involved in a number of events associated with adventitious root production. For example, auxins were shown to be the factor limiting the differentiation of tracheary elements in *Coleus* pith [98] and they may also be implicated in the production of the induced files of tracheids that are pre-requisites for rooting in cuttings from a number of woody species. In *Griselinia* the greatest regions of cambial activation are always associated with traces from young leaves which are probably exporters of auxins [16]. The application of auxins can result in increased numbers of roots and extension of the normal rooting zone [78, 99]. Reduction in the basipetal transport of auxin from the expanding leaf zone has also been associated with fewer cells per root primordium [41].

The time taken for root production varies greatly from species to species. It is difficult to compare the 'preformed primordium' situation with induced primordia because with the former type it is difficult to state precisely when the inductive stimulus is received, even if the primordia develop at a specific node, e.g. node 4 in *Salix fragilis* [31], and the primordia may also remain 'dormant' for a protracted period.

The lag phase, *i.e.* the period between taking a cutting and the first anatomical event, is usually the most variable in terms of time. It varies not only from species to species but also it is often affected by the time of year and the type of cutting taken. The lag phase for *Agathis australis* varies from 1–24 weeks depending on age and time of year. Herbaceous plants usually have a lag phase of a few hours e.g. *Azukia*, [83], or 1–2 d e.g. *Phaseolus aureus* [79, 100], *Phaseolus vulgaris* [101], *Taraxacum officinale* [102]. Cuttings from woody plants normally take longer e.g. 3 d for *Coffea robusta* [49] and for *Chrysanthemum morifolium* [26] and 5–7 d for *Dianthus caryophyllus* and *Rosa dilecta* [26], or very much longer e.g. 10–36 weeks for cuttings from mature plants of *Griselinia lucida* [16]. The information on the timing of the early stages is much less comprehensive for material that takes longer to root.

134

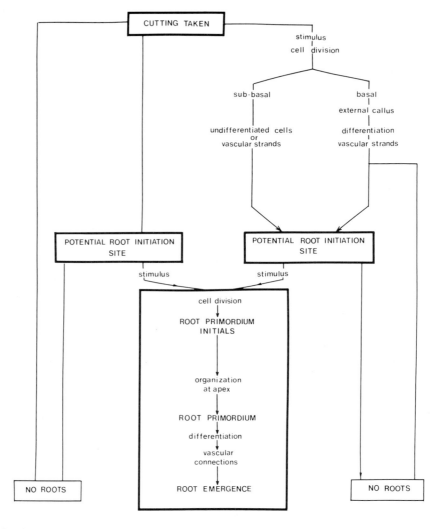

Fig. 9. Flow diagram summarizing the main sequences of events in a cutting from initial activation to root production. When a potential root initiation site is already present the initial cell divisions lead to root production *in situ.* When a site is not present alternative routes leading to the creation of a site are shown.

The alternative pathways that may be present are shown in a flow diagram (Fig. 9). There are several possibilities. A cutting may or may not root. It may fail to root because no anatomical events occur even if a potential root initiation site exists or even if anatomical changes do take place they may result only in vascular development and not in a root primordium. The reasons for the very variable time period for production of a root primordium are two-fold, firstly the variable duration of the lag phase as already mentioned and secondly if a potential root initiation site does not exist in the cutting the subroutines shown in Fig. 9 may be necessary and these may take days or weeks. The situations shown include sub-basal e.g. *Griselinia littoralis* [16] and *Agathis australis* [22] and basal e.g. *Carya illinoensis* [64]. Completion of the early events does not guarantee that root production will occur. Until root primordium initials are produced it is unwise to suggest that a root will eventuate because the cellular events have not reached a stage of 'determination'. After the root primordium initial has formed it is rare for the process not to continue to root production [15].

When roots form *in situ,* i.e. the potential root initiation site is at the location of the initial cell divisions, the early events usually take only a short time, about 1–2 d or less. Once a root initiation site is present or has been established and the stimulus resulting in cell division has been received the remaining stages are much more standardized. These are shown in the large rectangle of Fig. 9. Once the process becomes 'determined' the time period from the root primordium initials phase to root emergence is much less variable from species to species than the earlier stages, e.g. *Chrysanthemum morifolium,* 7–12 d [26]; *Pinus radiata* hypocotyl, 9 d [66]; *Hedera helix* mature plants, 7–14 d [51]; *Larix, Picea, Pinus* and *Torreya,* 7–17 d [50]; mature *Griselinia littoralis* and *G. lucida,* 15 d [16]. Information of this type is not common because in many investigations only the time that has elapsed between the cutting being taken and root emergence is recorded.

There are many unresolved problems. For example, it is not known whether the stimulus that is associated with the initial events in the creation of a root primordium site is the same as that which initiates the cell divisions at that site. At a slightly later stage there is a lack of firm information about the role of neighbouring cells in the development of a primordium. About half of the reports state that neighbouring cells, e.g. cortical cells, are incorporated into the developing primordium and half indicate that they are not. This may reflect species difference. The final vascular connection from the primordium to the original vascular tissues never forms until the primordium has at least developed to the root initial stage and generally the link up does not occur until the apex of the developing root is completely formed and the root is emerging. The vascular connections develop back from the primordium differentiating in sequence until the final connection is made. By the time that the root emerges it should usually be functional.

The anatomical processes that occur at the root initiation site are fundamentally similar for all species examined and also the later events closely resemble those that take place in lateral root formation. In adventitious root production the anatomical changes differ only in minor respects such as the length of the vascular connections and whether or not additional cortical cells are incorporated into the primordium. The precise cell arrangements at the root apex will, of course, differ to some extent.

However, there is substantial variation between species, and even between cuttings from young and old plants of the same species, in the events leading to the creation or locating of a potential root initiation site. We believe that it is these

136

processes that need our detailed attention in future because they may hold the key to our understanding of the physiological basis for adventitious root production with profound implications for applied research in forestry, horticulture and agriculture. In order to achieve this it is essential that more wide-ranging, dynamic studies of the anatomical processes involved be undertaken.

7. References

1. Sinnott, E.W. (1961) Cell and Psyche, The Biology of Purpose. Harper & Brothers, New York.
2. Esau, K. (1953) Plant Anatomy. John Wiley and Sons, New York.
3. Hayward, H.E. (1938) The Structure of Economic Plants. MacMillan, New York.
4. Harris, G.R. & Lovell, P.H. (1980) Growth and reproductive strategy in *Veronica* spp. Annals of Botany, 45, 447–458.
5. Harris, G.R. & Lovell, P.H. (1980) Adventitious root formation in *Veronica* spp. Annals of Botany, 45, 459–468.
6. Palmer, J.H. & Sagar, G.R. (1963) Biological flora of the British Isles No. 93 *Agropyron repens* (L.) Beauv. Journal of Ecology, 51, 783–794.
7. Ginzo, H.D. & Lovell, P.H. (1973) Aspects of the comparative physiology of *Ranunculus bulbosus* L. and *Ranunculus repens* L. II. Carbon dioxide assimilation and distribution of photosynthates. Annals of Botany, 37, 765–776.
8. Booth, A. (1959) Some factors concerned in the growth of stolons in potato. The Journal of the Linnean Society of London, Botany, 56, 166–169.
9. Haissig, B.E. (1974) Origins of adventitious roots. New Zealand Journal of Forestry Science, 4, 299–310.
10. Shapiro, S. (1958) The role of light in the growth of root primordia in the stems of lombardy poplar. In Physiology of Forest Trees (ed K.V. Thimann), pp. 445–465. Ronald Press, New York.
11. Hagemann, A. (1932) Untersuchungen auf Blattstecklingen. Gartenbauwissenschaft, 6, 69–195.
12. La Rue, C.D. (1933) Regeneration in mutilated seedlings. Proceedings of the National Academy of Science, 19, 53–63.
13. Lovell, P.H., Barratt, E. & Moore, K.G. (1970) The rooted excised cotyledon as an experimental system. The New Phytologist, 69 1185–1187.
14. White, J. & Lovell, P.H. (1984) Factors influencing adventitous root production in cuttings of *Griselinia littoralis* and *Griselinia lucida*. Annals of Botany, 53, 443–446.
15. White, J. & Lovell, P.H. (1984) Variation in the lag phase and uniformity of the length of the rooting phase prior to adventitive root production in cuttings of *Agathis australis*. Physiologia Plantarum, 61, 459–463.
16. White, J. & Lovell, P.H. (1984) The anatomy of root initiation in cuttings of *Griselinia littoralis* and *Griselinia lucida*. Annals of Botany, 54, 7–20.
17. Argles, G.K. (1959) Root formation and root development in stem cuttings: a re-examination of certain fundamental aspects. Annals of Applied Biology, 47, 626–628.
18. Cameron, R.J. & Thomson, G.V. (1969) The vegetative propagation of *Pinus radiata*: root initiation in cuttings. Botanical Gazette, 130, 242–251.
19. Dore, J. (1965) Physiology of regeneration in cormophytes. In Encyclopaedia of Plant Physiology 15, (2), (ed W. Ruhland), pp. 1–91. Springer-Verlag, New York.
20. Girouard, R.M. (1967) Initiation and development of adventitious roots in stem cuttings of *Hedera helix*. Anatomical studies of the juvenile growth phase. Canadian Journal of Botany, 45, 1877–1882.
21. Sircar, P.K. & Chatterjee, S.K. (1973) Physiological and biochemical control of meristemization

and adventitious root formation in *Vigna* hypocotyl cuttings. The Plant Propagator, 19, 17–26.

22. White, J. & Lovell, P.H. (1984) Anatomical changes which occur in cuttings of *Agathis australis* (D. Don) Lindl 2. The initiation of root primordia and early root development. Annals of Botany, 54, 633–646.

23. Torrey, J.G. (1963) Cellular patterns in developing roots. Symposia of the Society for Experimental Biology 17, 285–314.

24. Mahlstede, J.P. & Watson, D.P. (1952) An anatomical study of adventitious root development in stems of *Vaccinium corymbosum*. Botanical Gazette, 113, 279–285.

25. Fahn, A. (1974) Plant Anatomy. Second Edition. Pergamon Press, Oxford.

26. Stangler, B.B. (1956) Origin and development of adventitious roots in stem cuttings of chrysanthemum, carnation and rose. Cornell University of Agricultural Experimental Station Memoirs, 342, 1–24.

27. Carlson, M.C. (1929) Origins of adventitious roots in coleus cuttings. Botanical Gazette, 87, 119–126.

28. Sandison, S. (1934) The rooting of cuttings of *Lonicera japonica*. A preliminary account. The New Phytologist, 33, 211–217.

29. Fink, S. (1982) Adventitious root primordia – the cause of abnormally broad xylem rays in hard- and softwoods. International Association of Wood Anatomists, 3, 31–38.

30. Halperin, W. (1966) Alternative morphogenetic events in cell suspensions. American Journal of Botany 53, 443–453.

31. Haissig, B.E. (1970) Preformed adventitious root initiation in brittle willow grown in a controlled environment. Canadian Journal of Botany, 48, 2309–2312.

32. Swingle, C.F. (1927) Burr knot formation in relation to the vascular system of the apple stem. Journal of Agricultural Research, 34, 533–544.

33. van der Lek, H.A.A. (1925) Root development in woody cuttings. Mededelingen Landbouwhogeschool Wageningen, 28, 211–230.

34. Clark, W.A. (1933) Vegetative propagation of *Cotoneaster*. Transactions and Proceedings of the Botanical Society of Edinburgh, 31, 255–261.

35. Wolfe, F.H. (1934) The origin of adventitious roots in *Cotoneaster dammeri*. Botanical Gazette, 95, 686–694.

36. Carlson, M.C. (1938) The formation of nodal adventitious roots in *Salix cordata*. American Journal of Botany, 25, 721–725.

37. Carlson, M.C. (1950) Nodal adventitious roots in willow stems of different ages. American Journal of Botany, 37, 555–561.

38. Carpenter, J.B. (1961) Occurrence and inheritance of preformed root primordia in stems of citron (*Citrus medica* L). Proceedings of the American Society for Horticultural Science, 77, 211–218.

39. Smith, N.G. & Wareing, P.F. (1972) The distribution of latent root primordia in stems of *Populus × robusta*, and factors affecting the emergence of preformed roots from cuttings. Forestry, 45, 197–209.

40. Molnar, J.M. & LaCroix, L.J. (1972) Studies of the rooting of cuttings of *Hydrangea macrophylla*: enzyme changes. Canadian Journal of Botany, 50, 315–322.

41. Haissig, B.E. (1972) Meristematic activity during adventitious root primordium development. Plant Physiology, 49, 886–892.

42. Bannan, M.W. (1941) Vascular rays and adventitious root formation in *Thuja occidentalis* L. American Journal of Botany, 28, 457–463.

43. Bannan, M.W. (1942) Notes on the origin of adventitious roots in the native Ontario conifers. American Journal of Botany, 29, 593–598.

44. Priestley, J.H. & Swingle, C.F. (1929) Vegetative propagation from the standpoint of plant anatomy. United States Department of Agriculture Technical Bulletin, 151, 1–98.

45. Sudds, R.H. (1934) A study of the morphological changes and the origin of roots in tip-layered

138

Cumberland raspberry plants. Proceedings of the American Society for Horticultural Science, 32, 401–406.

46. Sudds, R.H. (1935) The origin of roots in several types of red and black raspberry stem cuttings. Proceedings of the American Society for Horticultural Science, 33, 380–385.

47. Taylor G. (1926) The origin of adventitious growths in *Acanthus montanus*. Transactions and Proceedings of the Botanical Society of Edinburgh, 29, 291–296.

48. Smith, E.P. (1928) A comparative study of the stem structure of the genus *Clematis* with special reference to anatomical changes induced by vegetative propagation. Transactions of the Royal Society of Edinburgh, 55, 643–664.

49. Reaño, P.C. (1940) Histological study and observations on the effect of some synthetic growth substances on stem tip cuttings of coffee. Philippine Agriculturist, 29, 87–99.

50. Satoo, S. (1956) Anatomical studies on the rooting of cuttings in coniferous species. Bulletin of the Tokyo University Forests, 51, 109–158.

51. Girouard, R.M. (1967) Initiation and development of adventitious roots in stem cuttings of *Hedera helix*. Anatomical studies of the mature growth phase. Canadian Journal of Botany, 45, 1883–1886.

52. Beakbane, A.B. (1969) Relationships between structure and adventitious rooting. Combined Proceedings of the International Plant Propagators' Society, 19, 192–201.

53. Davies, F.T. Jr., Lazarte, J.E. & Joiner, J.N. (1982) Initiation and development of roots in juvenile and mature leaf bud cuttings of *Ficus pumila* L. American Journal of Botany, 69, 804–811.

54. Al Barazi, Z. & Schwabe, W.W. (1982) Rooting softwood cuttings of adult *Pistacia vera*. Journal of Horticultural Science, 57, 247–252.

55. Carlson, M.C. (1933) Comparative anatomical studies of Dorothy Perkins and American Pillar roses. I. Anatomy of canes. II. Origin and development of adventitious roots in cuttings. Contributions from the Boyce Thompson Institute for Plant Research, 5, 313–330.

56. Delisle, A.L. (1940) Histological and anatomical changes induced by indole-acetic acid in rooting cuttings of *Pinus strobus*. American Journal of Botany, 27, Supplement p3S.

57. Doud, S.L. & Carlson, R.F. (1977) Effects of etiolation, stem anatomy and starch resources on root initiation of layered *Malus* clones. Journal of the American Society for Horticultural Science, 102, 487–491.

58. Gregory, F.G. & Samantarai, B. (1950) Factors concerned in rooting responses of isolated leaves. Journal of Experimental Botany, 1, 159–193.

59. Wilcox, H. (1955) Regeneration of injured root systems in noble fir. Botanical Gazette, 116, 221–234.

60. Mitra, G.C. & Bose, N. (1957) Rooting and histological responses of detached leaves to β-indole-butyric acid with special reference to *Boerhaavia diffusa* Linn. Phytomorphology, 7, 370–381.

61. John, A. (1978) An anatomical study of root initiation in stem cuttings of hybrid larch. The New Phytologist, 81, 111–116.

62. Mergen, F. & Simpson, B.A. (1964) Asexual propagation of *Pinus* by rooting needle fascicles. *Silvae Genetica,* 13, 133–139.

63. Bhella, H.S. & Roberts, A.N. (1975) Seasonal changes in origin and rate of development of root initials in Douglas-fir stem cuttings. American Society for Horticultural Science, 100, 643–646.

64. Brutsch, M.O., Allan, P. & Wolstenholme, B.N. (1977) The anatomy of adventitious root formation in adult-phase pecan (*Carya illinoensis* (Wang) K. Koch) stem cuttings. Horticultural Research, 17, 23–31.

65. Ginzburg, C. (1967) Organization of the adventitious root apex in *Tamarix aphylla*. American Journal of Botany, 54, 4–8.

66. Smith, D.R. & Thorpe, T.A. (1975) Root initiation in cuttings of *Pinus radiata* 1. Developmental sequence. Journal of Experimental Botany, 26, 184–192.

67. Wally, Y.A., El-Hamady, M.M., Boulos, S.T. & Abu-Amara, N.M. (1981) Rooting experiments on guava using hardwood stem cuttings. Egyptian Journal of Horticulture, 8, 77–86.
68. Dalgas, K.F. (1973) Anatomical studies on cuttings of Norway spruce (Picea abies (L.) Karst.) undergoing the rooting process. Forest Tree Improvement, 5, 1–20.
69. Wetmore, R.H. (1926) Organization and significance of lenticels in dicotyledons. I. Lenticels in relation to aggregate and compound storage rays in woody stems, lenticels and roots. Botanical Gazette, 82, 71–90.
70. Wetmore, R.H. (1926) Organization and significance of lenticels in dicotyledons. II Lenticels in relation to diffuse storage rays of woody stems. Botanical Gazette, 82, 113–135.
71. White, J. & Lovell, P.H. (1984) Anatomical changes which occur in cuttings of Agathis australis (D. Don) Lindl l. Wounding responses. Annals of Botany, 54, 621–632.
72. Wilson, C.L. (1927) Adventitious roots and shoots in an introduced weed. Bulletin of the Torrey Botanical Club, 54, 35–38.
73. Crooks, D.M. (1933) Histological and regenerative studies on the flax seedling. Botanical Gazette, 95, 209–239.
74. Smith, A.I. (1936) Adventitious roots in stem cuttings of Begonia maculata and B. semperflorens. American Journal of Botany, 23, 511–515.
75. Naylor, E.E. & Johnson, B. (1937) A histological study of vegetative reproduction in Saintpaulia ionantha. American Journal of Botany, 24, 673–678.
76. McVeigh, I. (1938) Regeneration in Crassula multicava. American Journal of Botany, 25, 7–11.
77. Smith, A.I. (1942) Adventitious roots in stem cuttings of Tropaeolum majus L. American Journal of Botany, 29, 192–194.
78. Gramberg, J.J. (1971) The first stages of the formation of adventitious roots in petioles of Phaseolus vulgaris. Proceedings Koninklijke Nederlandse Academie van Wetenschappen, 74, 42–45.
79. Chandra, G.R., Gregory, L.E. & Worley, J.F. (1971) Studies on the initiation of adventitious roots on mung bean hypocotyl. Plant and Cell Physiology, 12, 317–324.
80. Moore, K.G. & Lovell, P.H. (1972) Rhizogenesis in detached cotyledons. Physiologie Végétale, 10, 223–235.
81. Aung, L.H., Bryan, H.H. & Byrne, J.M. (1975) Changes in rooting substances of tomato explants. Journal of the American Society for Horticultural Science, 100, 19–22.
82. Coleman, W.K. & Greyson, R.I. (1977) Analysis of root formation in leaf discs of Lycopersicon esculentum Mill. cultured in vitro. Annals of Botany, 41, 307–320.
83. Mitsuhashi-Kato, M., Shibaoka, H. & Shimokoriyama, M. (1978) Anatomical and physiological aspects of developmental processes of adventitious root formation in Azukia cuttings. Plant and Cell Physiology, 19, 393–400.
84. Friedman, R., Altman, A. & Zamski, E. (1979) Adventitious root formation in bean hypocotyl cuttings in relation to IAA translocation and hypocotyl anatomy. Journal of Experimental Botany, 30, 769–777.
85. Cline, M.N. & Neely, D. (1983) The histology and histochemistry of the wound-healing process in Geranium cuttings. Journal of the American Society for Horticultural Science, 108, 496–502.
86. Gautheret, R.J. (1969) Investigations on the root formation in the tissues of Helianthus tuberosus cultured in vitro. American Journal of Botany, 56, 702–717.
87. Reines, M. & McAlpine, R.G. (1959) The morphology of normal, callused and rooted dwarf shoot of slash-pine. Botanical Gazette, 121, 118–124.
88. Herman, D.E. & Hess, C.E. (1963). The effect of etiolation upon the rooting of cuttings. The International Plant Propagators' Society Combined Proceedings, 13, 42–62.
89. Beabane, A.B. (1961) Structure of the plant stem in relation to adventitious rooting. Nature, London, 192, 954–955.
90. Goodin, J.J. (1965) Anatomical changes associated with juvenile to mature growth phase transition in Hedera. Nature, London, 208, 504–505.

91. Edwards, R.A. & Thomas, M.B. (1980) Observations on physical barrier to root formation in cuttings. The Plant Propagator, 26, 6–8.

92. Kachecheba, J.L. (1975) Anatomical aspects of the formation and growth of roots in stem cuttings of some species of *Hibiscus*. I. Stem anatomy and its relation to the formation and growth of roots. Horticultural Research, 14, 57–67.

93. Sachs, R.M., Loreti, F. & De Bie, J. (1964) Plant rooting studies indicate sclerenchyma tissue is not a restricting factor. California Agriculture, 18, 4–5.

94. Cormack, R.G.H. (1965) The effects of calcium ions and pH on the development of callus tissue on stem cuttings of balsam poplar. Canadian Journal of Botany, 43, 75–83.

95. Wu, L.L. & Overcash, J.P. (1971) Anatomical structure of red raspberry hybrid cuttings rooted under mist. Journal of the American Society for Horticultural Science, 96, 437–440.

96. Bruck, D.K. & Paolillo, D.J. Jr. (1984) Anatomy of nodes vs. internodes in *Coleus*: the nodal cambium. American Journal of Botany, 71, 142–150.

97. Bruck, D.K. & Paolillo, D.J. Jr. (1984) Anatomy of nodes vs. internodes in *Coleus*: the longitudinal course of xylem differentiation. American Journal of Botany, 71, 151–157.

98. Comer, A.E. (1978) Pattern of cell division and wound vessel member differentiation in *Coleus* pith explants. Plant Physiology, 62, 354–359.

99. Thimann, K.V. (1972) The natural plant hormones. In Plant Physiology Vol VIB Physiology of Development: The Hormones (ed F.C. Steward), pp. 3–332. Academic Press, London, New York, San Francisco.

100. Blazich, F.A. & Heuser, C.W. (1979) A histological study of adventitious root initiation in mung bean cuttings. Journal of the American Society for Horticultural Science, 104, 63–67.

101. Oppenoorth, J.M. (1978) The influence of colchicine on initiation and early development of adventitious roots. Physiologia Plantarum, 42, 375–378.

102. Khan, M.I. (1973) Anatomy of regenerating root segments of *Taraxacum officinale* Web. Pakistan Journal of Botany, 5, 71–77.

5. Metabolic processes in adventitious rooting of cuttings

BRUCE E. HAISSIG

USDA-Forest Service, North Central Forest Experiment Station, Forestry Sciences Laboratory, P.O. Box 898, Rhinelander, WI, 54501, USA

1. Introduction

A cutting is in a thermodynamically unfavourable state from the time it is prepared [1]. Part of the physical and physiological support, the root system, is gone. Only a new root system can restore 'whole plant' thermodynamics and sustain life. Altered metabolism in a cutting regenerates the root system.

The present review addresses metabolic studies of adventitious root primordium initation and development. In addition, enough physiology has been included to develop the primary topics. Generalities rather than specifics are sometimes presented because past research has not supplied a precise understanding of most metabolic subjects [2]. The text discusses primarily research published during the last 10 years. I have reviewed the earlier literature on a previous occasion [3][1]. In the present chapter I sometimes summarize work from the foregoing paper and cited it as a source of references to save space.

Much more is known about the initiation, development, and metabolism of non-adventitious roots than about adventitious roots. However, as Zobel [4] has indicated there is a general lack of information about root system development. Gross anatomical and developmental similarities in adventitious root development between species [5, 6, 7, 8] suggest that the supporting metabolism is similar. If so, metabolic information should be at least partially transferrable between species, which I have assumed in the present review unless contrary evidence was available.

However, the present review is exclusively limited to studies of adventitious rooting. Lateral rooting has been considered by McCully [9] and in Chapter 2. The present review is also restricted to adventitious rooting in excised shoots, stems, and leaves ('cuttings'), except where other information seemed particularly relevant to studies with cuttings or was specially needed for a cohesive presentation. Organogenesis and related metabolic studies with *in vitro* cultures have recently been reviewed [2, 10]. Terminology concerning anatomical differentiation during the formation of adventitious roots (see also Chapter 4) is consistent with [5]. 'Primordium initiation' refers to cellular differentiation (often called 'de-differentiation') that directly yields the root primordium initial cells; 'Primordium development' refers to division of the initial cells and of cells adjacent to the initial cells that may become incorporated into the primordium after the initial cells appear. 'Rooting' refers to the generation of adventitious roots by cuttings, without reference to any specific phase of anatomical or biochemical differentiation.

2. Experimental limitations

Before addressing the literature, it is appropriate to list and consider some of the difficulties in studying the metabolism of adventitious rooting. An understanding of past difficulties and limitations may allow structuring of future metabolic research to improve ways in which evidence is obtained or interpreted. In addition, the information should help the reader interpret the conclusions of various authors.

(1) It seems that adventitious rooting has a unique metabolism [3, 11] as will be discussed below. However, the metabolism of adventitious rooting is not unique enough to be quickly, easily, and unambiguously distinguished from the metabolism of 'normal' whole plant development. Therefore, it has proven challenging

---

[1] This chapter is based partly on computerized literature searches of the Dialog (Palo Alto, CA) data bases Agricola (US Department of Agriculture), Biosis Previews (BioSciences Information Service), and CAB Abstracts (Commonwealth Agricultural Bureaux) covering 1972 to mid 1984.

to experimentally isolate the metabolic peculiarities of adventitious rooting.

(2) Adventitious rooting may occur ordinarily at more than one ontogenetic point, even in the same species. For example, in most monocotyledonous species, the entire root system is adventitious [7], whereas in many dicotyledonous species only part of the root system may be adventitious [7]. In other species preformed adventitious root primordia normally develop in stems [5] but in certain species, profound perturbation of normal development is required before adventitious rooting occurs, if at all, in the intact plant [12, 13] or its severed parts [14]. Furthermore, all or most types of adventitious rooting can occur in the same plant [7]. Thus, there may be no generally occurring, single status of whole plant or whole organ development that can be unequivocally linked to the metabolic peculiarities of adventitious rooting.

(3) It is difficult or impossible to understand the metabolism of rooting without a comprehensive knowledge of the associated anatomical differentiation [15, 16]. Completed adventitious rooting results in functional roots with the usual anatomical characteristics [7]. However, specific anatomical changes during rooting may differ markedly between species [14, 17] and often between tissues of the same species (compare bean callus [18] with bean cuttings [19]). Moreover, the time course and anatomy of rooting may depend on the age (maturity) of the stock plant [14, 17, 20, 21], physiological condition of the stock plant, propagation environment, chemical treatment, etc. [22, 23]. Studies concerned with the initiation of primordia primarily describe biochemical differentiation during the period from severance of the cutting to anatomical differentiation of the first primordial (initial) cells [5]. This period of biochemical differentiation may be very brief [19] or require weeks or months [14, 20, 21]. In many studies it is not clear which developmental stage of 'rooting' was sampled and, therefore, what aspect of anatomical differentiation was chemically described.

(4) Gaspar and co-workers [24] have described adventitious rooting as consisting of an 'inductive' phase and an 'initiative' phase (see later discussion). Adventitious root initials form only from cells that have been induced to become 'competent' [25]. Studies of the metabolism of primordium *initiation* attempt to define biochemically the establishment of this competent state, which is presently obscure. Therefore, such studies must consider when the inductive phase occurs in relation to the beginning of experimental assessments. As an extreme example, both the inductive and early initiative phases may have occurred before cuttings are prepared in those species that contain preformed adventitious root primordia. In comparison, the inductive phase cannot occur in cuttings that initiate primordia in callus until the callus is forming. This may require days or weeks. Thus, study of the metabolism of adventitious root primordium initiation must be placed within a known schematic of developmental anatomy (from (3) above) and developmental physiology, which vary with species of cutting, stock plant age, etc. (see Chapter 4). Construction of the anatomical and physiological schematic may be greatly aided by defining biochemical markers for successive

144

developmental periods [T. Gaspar, Laboratory of Fundamental and Applied Hormonology, Botanical Institute, University of Liège, Belgium, personal communication], but that has rarely been accomplished. Perhaps enzyme activities could serve as such markers (see Sections 8 and 11).

(5) Most means for measuring biochemical differentiation provide only 'averages' for the rooting zone. Most cells of the rooting zone of a cutting do not initiate primordia. It is presently unclear how biochemical differentiation in the rooting zone as a whole specifically relates to primordium initiation. Thus, understanding metabolism requires description of biochemical differentiation in the whole rooting zone and in the initial cells of root primordia [26], and probably in the entire cutting. Most information concerning metabolism during rooting describes the rooting zone but not events in the precise location of primordium initiation. At present, histochemical tests offer the only hope of describing biochemical differentiation within root primordium initials and their progenitor cells [3].

(6) Many studies of metabolism during rooting have used indirect approaches. A good example has been the use of inhibitors of protein and nucleic acid metabolism during rooting [3]. In such work it is often difficult to isolate the primary action of an inhibitor from any secondary actions that may also profoundly influence the metabolism of rooting [3].

(7) The biochemical data reported in the rooting literature is often statistically unreliable. It may be unclear whether experiments were replicated in time and whether appropriate statistical tests were applied. Inadequate replication within experiments and graphic presentation of data that lacks statistical information (e.g. without variances) may sometimes be sufficient to display a trend in chemical concentration or enzyme activity, but often fails to resolve quantitative differences between treatments.

3. Genetic effects

There is plausible evidence for genetic control of adventitious rooting. Consistent differences have been reported between and within genera (e.g. *Populus*) in adventitious rooting ability [27, 28, 29], while preformed adventitious root primordia occur only and unfailingly in certain species [5]. It is also known that genetic mutations affect rooting ability [4], and that plant regeneration in microculture depends on the presence of specific genes [30]. An excellent specific example of genetic dependence is found in jack pine (*Pinus banksiana*). Adventitious rooting varies markedly between cuttings from half-sib families within good- and poor-rooting groups. There also exist pronounced familial differences in rooting response to applied auxin [D.E. Riemenschneider, USDA-Forest Service, North Central Forest Experiment Station, Forestry Sciences Laboratory, Rhinelander, WI 54501, USA, personal communication]. However, it remains unknown how many genes are involved, what the gene products are or what metabolic pathways and pathway controls are involved. The theory of cellular totipotency suggests that all higher plant cells have the genetic capability to initiate roots although this has proven difficult to demonstrate since the capacity for rooting often remains unexpressed [10]. This indicates that one or more factors needed to support adventitious rooting are lacking or ineffective. The nature of these deficiencies is also unknown. Moreover, there is no basis for assuming that a given genetic block in one species would be the same in another species. Presumably, the deficiencies include a failure to establish or maintain a metabolism that specifically supports adventitious rooting. This may be why many investigators have chosen metabolism as a prime subject for exploration.

4. Water relations

Adventitious rooting often occurs under conditions of water stress in cuttings and in the aerial organs of intact plants [31], but cause-effect relations between water stress and rooting are unclear. In particular, the role of water in the metabolism of rooting has not received enough discussion, possibly because the importance of water in turgor maintenance and as a reactant-product and cellular solvent is so obvious. Nevertheless, separation of the cutting from its water-supplying root system would seemingly influence volume, turgor pressure, and water content of cells, and upset physiological integrity [32]. Metabolic changes in cuttings have indeed been demonstrated as a result of water stress. Water deficits may have additional, as yet unstudied metabolic influences, as will be discussed below.

Water stress imposed on the stock plant may influence the subsequent rooting of cuttings. For example, Rajagopal and Andersen [33] tested rooting of pea (*Pisum sativum*) cuttings under low irradiance and low water stress (see also Chapter 7). The cuttings were obtained from stock plants grown under either low or high irradiance, and then water stressed by exposure of their roots to polyethylene glycol solutions. The authors found that cuttings from non-stressed, high-irradiance stock plants rooted poorly, compared to non-stressed, low-irradiance stock plants (but see [34] for conflicting evidence for *Chrysanthemum*). Brief water stress of stock plants increased rooting of cuttings from the high-irradiance treatment but not that of cuttings from the low-irradiance treatment [35]. Prolonged water stress of stock plants reduced rooting of cuttings regardless of stock plant irradiance [35].

In the foregoing experiments, the difference in rooting between pea cuttings from non-stressed, high-irradiance stock plants and non-stressed, low-irradiance stock plants may also have been an effect of water stress. Similar tests with pea have indicated that leaf temperatures may differ by several degrees between irradiance treatments, which would result in different levels of water stress [B. Veierskov, Royal Veterinary and Agricultural University, Department of Plant

Physiology and Anatomy, Thorvaldsensvej 40, DK1871 Copenhagen V, Denmark, personal communication]. Increased irradiance might increase or decrease leaf temperature, depending upon the amount of evaporative cooling that occurs. Stock plant irradiance and tissue temperature also influence auxin transport and, therefore, rooting of pea cuttings [36].

Water stress in Norway spruce (*Picea abies*) cuttings may reduce rooting [37]. Leakey [38] similarly found that low leaf water potentials of lateral shoots on stock plants of *Triplochiton scleroxylon* were related to less extensive rooting. On the contrary, Orton [39] found that water stressing of *Chysanthemum* cuttings increased root formation. Therefore, water stress influences rooting, but the type of response depends on general environmental influences to which stock plants and cuttings are exposed.

The rate of water loss from a cutting is determined by the vapour pressure gradient between leaves and the surrounding air, and by the conductance of the leaf to water vapour [40]. Stomatal aperture and to a lesser extent, cuticular conductance control leaf conductance [40]. Level of irradiance seems to be a key factor in determining the amount of water stress in a cutting [33, 35, 37, 41]. However, Loach [42] indicated that, in addition to current day's radiation, number of days from insertion of the cutting and previous day's leaf water potential or radiation must be considered.

In addition, the CO_2 level to which cuttings are exposed may influence water relations through effects on stomatal aperture [43]. Uptake of inorganic ions may also influence water relations in cuttings through osmoregulatory influences. Eliasson [41] conducted tests indicating that pea cuttings may require a source of inorganic ions with which to adjust their osmotic balance and thereby maintain turgor. He found that pea cuttings were best maintained in nutrient solution, and that tap water or deionized water containing calcium was superior to deionized water alone [41]. However, in the foregoing tests, ion uptake could have indirectly influenced osmoregulation via influences on carbohydrate concentrations.

Orton [39] demonstrated with *Chrysanthemum* cuttings that stomata remained fully open for only 10 min after cuttings were prepared and that stomata were fully closed by 30 min. Similarly, Eliasson and Brunes [44] reported that stomata closed rapidly after preparation of hybrid aspen (*Populus tremula* × *tremuloides*) cuttings. Stomatal resistance was also shown to increase in water-stressed *Populus robusta* cuttings [45]. Stomatal closure, which reduces leaf gas exchange and water loss, also reduced rates of net photosynthesis in cuttings of various species [40, 44, 46]. In concurrence with the above findings, the best rooting of *Cornus* and *Rhododendron* cuttings occurred under conditions that produced the highest leaf conductance [40]. However, reduced rates of net photosynthesis in cuttings often do not result in decreased endogenous carbohydrate levels (see Section 8) because lower rates of carbohydrate production are more than compensated for by the absence of root system carbohydrate consumption [47].

Available evidence indicates that water stress in cuttings influences rooting, partly through changes in carbohydrate metabolism [33, 35]. Carbohydrate metabolism in the rooting zone may be influenced because of the different distribution of growth in stressed vs. non-stressed cuttings [39]. Water stress of whole plants that is severe enough to produce wilting may also influence accumulation of other metabolites, such a proline [32], but it is unknown whether such changes regularly occur in cuttings. One recent study found marked accumulation of proline and other amino acids in the rooting zone of mulberry (*Morus alba*) cuttings [48].

Water stress in cuttings may also influence hormone metabolism. Sivakumaran and Hall [45] reported that abscisic acid (ABA) and ethylene levels, but not indole-3-acetic acid (IAA) levels, increased in leaves and stems of water-stressed *Populus robusta* cuttings. Rajagopal and Andersen [35] also implicated ABA as influencing rooting in water-stressed pea cuttings. ABA and ethylene have been shown to influence rooting, but it is unclear how they modulate metabolism during rooting.

Water stressing of cuttings may have less direct effects on the metabolism of rooting than those stated above for carbohydrates and hormones. For example, water stressing of cuttings may reduce translocation [J.R. Potter, USDA, SEA-AR Ornamental Plants Research Laboratory, 3420 SW Orchard St., Corvallis, OR 97331, USA, personal communication]. In addition, cellular solvent capacity must be preserved in order to maintain metabolic integrity [49]. As Atkinson [49] has indicated, the maintenance of low cellular metabolite concentrations may be the most important problem relating to metabolic control, because excessive metabolite concentrations can result in non-enzymic side reactions. Cellular structure itself offers certain strategies for preserving the solvent capacity of cellular water [49]. However, the rooting of cuttings poses a particularly challenging situation for the proper maintenance of cellular solvent capacity and, therefore, metabolic control. Firstly, as stated above, the preparation of cuttings causes water stress and probably reduces cellular ability to maintain optimum solvent capacity. Secondly, the tissues of cuttings, especially the rooting zone, may accumulate rather large concentrations of soluble sugars, soluble nitrogenous compounds, and phenolics (see Sections 8, 9, and 11). Simple sugars, amino acids, organic acids, and inorganic ions are known osmotica in plant cells [32, 50, 51, 52]. Organic solutes may also influence protein solubility and hydration by their effects on water structure [53].

Under conditions of low water availability and increasing concentrations of osmotically active solutes, it may be impossible for a cutting to maintain normal cellular osmotic balance [32], resulting in profound metabolic changes [49]. Much additional information is needed to ascertain whether part of the control of rooting resides in metabolic changes that result from variations in cellular solvent capacity, water structure, and solute concentration. However, it seems safe to hypothesize that such changes occur and profoundly influence the metabolism of rooting when cuttings are separated from their normal supply of water and kept in desiccating environments of low relative humidity, intense irradiance, or warm temperatures.

5. Oxygen effects

Air-moisture relations of the rooting medium must reside within critical limits for the successful propagation of cuttings [54]. In keeping with the foregoing statement, the general promotive effects of oxygenation on rooting of cuttings have been repeatedly demonstrated ([55, 56, 57] and Table 1). In general, the normal physiological activities of higher plants, such as root growth, require O_2 [58 and references therein]. Anaerobic metabolism, which may increase under O_2 stress, is often inadequate to support normal growth [58]. Oxygen may reach the rooting zone in several ways, for example through the rooting medium into the lower stem, by photosynthetic generation of O_2 in the stem, or by basipetal movement of O_2 in intercellular space from the upper stem and leaves [55, 56, 57, 58]. However, not all the specific metabolic needs for O_2 during rooting are clear. Oxygen serves a major role during rooting as a respiratory electron acceptor [59]. However, O_2 may also have other functions. For example, Pingel [60] found a positive correlation in *Tradescantia albiflora* between O_2 uptake by nodes, IAA oxidase (IAAox) activity, and rooting.

Oxygen influences the biochemistry of mitosis, which is a primary event in root primordium initiation and development [3, 5]. Amoore [61, 62] studied the dependence of respiration and mitosis on O_2 concentration in excised pea roots. Respiration in pea root tips was limited by slow diffusion of O_2 through tissues. Anaerobisis reduced energy availability by 99% [62]. Amoore [62] also found that all stages of mitosis required O_2. However, visible stages of mitosis were less O_2 dependent, compared to the onset of mitosis (i.e. interphase-prophase). Thus, low O_2 (0.001% v/v) concentrations allowed the completion of active mitoses at reduced rates, but further mitoses did not occur [62]. Amoore [62] also proposed that, prior to visible stages of mitosis, there is a stage with a lower O_2 requirement followed by a stage with a higher O_2 requirement, both compared to the visible stages. Thus, the DNA synthesizing S-phase was probably most affected by O_2

Table 1. Effect of aeration on growth and rooting of pea cuttings. Initial shoot length was about 7 cm. Mean ± S.E. From Table 3 of [63].

Aeration	Final shoot length (cm)	Root length (cm)		No. roots per cutting	
		Day 8	Day 13	Day 8	Day 13
−	27 ± 0.4	2.2 ± 0.2	7.3 ± 0.3	11.0 ± 0.7	19 ± 1.0
+	30 ± 0.4	2.4 ± 0.2	15 ± 0.2	10.6 ± 0.7	25 ± 1.2

deficiency. The results also suggest mitosis is less sensitive to O_2 deficiency than root elongation or respiration. Amoore's finding indicate that the initiation of root primordia in cuttings is partly controlled by O_2 concentration within the rooting zone, which is usually a solid or liquid medium. Therefore, O_2 concentration to which the tissues are externally exposed will partly depend on the O_2 concentration of the substrate liquid and the O_2 diffusion coefficient of the medium (i.e. the facility with which O_2 moves through the medium to the tissue) [56]. The O_2 diffusion rate in a liquid (e.g. water) is much slower than in air. Thus, the availability of O_2 to the base of the cutting where roots are forming is primarily determined by the air/water ratio in the medium [56]. Successful rooting may therefore depend on the rate at which O_2 moves to the roots through gas-filled, interconnected pores of the rooting medium (see Chapter 7 and [57]). In liquid media, the concentration of dissolved O_2 will be the principal determinant of O_2 availability.

Therefore, the metabolism of rooting may depend on the physical environment in which the cuttings are rooted, e.g. liquid vs. solid substrate, misting vs. plastic tent, etc. However, it is uncertain which stages of rooting are most influenced by tissue O_2 concentration. Tests with pea yielded similar root primordium initiation in cuttings rooted in unaerated vs. aerated solution ([63] and Table 1). However, root growth and number of roots per cutting late in the test were much greater if the rooting solution was aerated. These results indicate that the stages of rooting before or during development of the first root primordia were relatively less sensitive to O_2 concentration than were root development and secondary initiation of root primordia. However, such results could also indicate that aeration is needed for primordium initiation only after internal O_2 concentrations have dropped below some critical level.

Other evidence also suggests that the development of established primordia is clearly O_2 dependent, for example, the rooting of poplar or willow (*Salix*) cuttings that contain preformed root primordia in aerated and unaerated water. Aeration produces substantially faster root development ([55] and Haissig, unpublished). Kordan [64] has shown for adventitious roots of rice (*Oryza sativa*) seedlings that primordium development but not initiation requires O_2. Primordium development arrested by anaerobisis was found by Kordan [64] to restart under aerobic conditions. Comparison of Amoore's [61, 62] and Kordan's [64] results indicates that the period of de-differentiation preceding primordium initiation occurs during the premitotic states that initially can proceed in low O_2 concentrations and then require higher O_2 concentrations for completion. However, the influences of O_2 concentration on primordium initiation in cuttings require much additional study.

6. Mineral nutrition

The metabolism of rooting may be influenced by mineral nutrition of the stock plant or of its cuttings during propagation, or both. Over 40 years ago, Pearse [65] observed relations between specific stock plant mineral deficiencies and rooting of *Vitis* cuttings. Rooting of cuttings (56%) from fully nourished stock plants was surpassed if stock plants were nitrogen deficient (91%) but decreased if stock plants were deficient in potassium (42%), phosphorous (37%), magnesium (35%), or calcium (0%). Thus, of the deficiencies that Pearse [65] tested, nitrogen deficiency had the greatest promotive effect and calcium deficiency had the greatest negative effect on rooting. Haun and Cornell [66] also observed that excessive nitrogen deterred rooting of cuttings from *Pelargonium* stock plants.

In general, sufficient or excessive nitrogen nutrition for shoot growth of stock plants appears to diminish subsequent rooting of cuttings, compared to mild nitrogen deficiency ([67] and see Section 9). However, depending on species and environmental conditions [41, 66, 68], rooting may be increased if stock plants are properly nourished, i.e. supplied with a combination of low to moderate nitrogen together with amounts of other macro- or micro-nutrients ordinarily required for satisfactory plant vigour. In particular, fertilization of stock plants with potassium and phosphorous may be beneficial if optimum nitrogen levels for stock plant shoot growth are retained [66, 68].

The proper mineral nutrition of stock plants has not consistently been linked to modulation of specific endogenous organic compounds, although carbohydrates and tryptophan have been implicated [68]. Most probably, mineral nutrition influences a variety of endogenous biochemical responses that are sensitive to or cause variations in stock plant growth (see Chapter 7). As discussed elsewhere in this chapter, excessive vegetative growth of the stock plant and cutting may diminish or preclude rooting. Therefore, mineral nutrition of the stock plant that maximizes vegetative development may deter rooting, and further increases in mineral nutrition may be even more harmful [B. Veierskov, Royal Veterinary and Agricultural University, Department of Plant Physiology and Anatomy, Thorsvaldsensvej 40, DK-1871 Copenhagen V, Denmark, personal communication].

After separation from the stock plant, cuttings have a fixed mineral nutrient pool, except for any minerals absorbed from the rooting medium or irrigation solution. The mineral nutrient pool in the cutting may even decline as a result of leaching during propagation [69], an effect exacerbated by warmer temperatures [70]. Mineral deficiencies could presumably result because of low ion uptake or leaching. Such deficiencies may perturb metabolism. However, mineral deficiencies in the rooting zone may be overcome by redistribution of minerals within the cutting. For example, Eliasson [41] suggested that applied nutrients may not generally be needed during rooting because endogenous nutrients are basipetally transported from the shoot. He also suggested that applied calcium and boron may influence rooting because the endogenous pool of these minerals is poorly transported in the phloem [41]. In tests with *Chrysanthemum morifolium* cuttings, measurements of nitrogen, phosphorous, and potassium indicated a redistribution in particular of potassium, from mature to developing tissues, including the rooting zone [71]. However, Blazich and Wright [72] found that nitrogen, phosphorous, potassium, calcium, and magnesium were not redistributed in holly (*Ilex crenata*) cuttings. In a subsequent study with holly cuttings, Blazich *et al.* [69] tested the influences of applied indole-3-butyric acid (IBA) on mineral distribution during propagation. They found that nitrogen, phosphorous, potassium, calcium, and magnesium were not redistributed to the rooting zone of untreated or IBA-treated cuttings during primordium initiation. Minerals were redistributed to the upper stem from leaves and rooting zone, and this took place only after primordium initiation and budbreak. IBA treatment enhanced movement of nitrogen, phosphorous, and potassium to the rooting zone [69].

There is some evidence that supplying specific nutrients to cuttings during rooting influences primordium initiation and development (nitrogen is discussed in detail in Section 9). Eliasson [41] found that rooting of cuttings from inadequately nourished pea stock plants was improved by exposing cuttings to nutrient or boric acid solution. Calcium appeared to be the only nutrient required for root elongation in the foregoing tests [65]. Relative to Eliasson's observations, calcium has been shown to activate peroxidase [73], which may be an essential enzyme for rooting. In another test, Reuveni and Raviv [74] found that mineral content of leaves was not positively related with rooting of 10 avocado clones. However, manganese concentration was negatively correlated with rooting in their tests. Manganese has often been associated with enhancing the activity of oxidative enzymes active during rooting (see Sections 11 and 12).

Boron is one of the most frequently mentioned minerals in rooting studies (see also Chapters 6 and 7). In one study, mung bean stock plants were grown in light or darkness, and cuttings were propagated under the light regime used for their respective stock plants [75]. Cuttings from light-grown stock plants did not root, whether or not IBA-treated, unless they were propagated in boric acid solutions [75]. This experiment established the obligatory role of boron in rooting of light-grown mung bean cuttings. Calcium chloride treatment interacted with boric acid treatment to promote rooting, if boric acid concentration was low. Middleton *et al.* [75] concluded that IBA-induced rooting did not proceed beyond the pre-primordial stage in the absence of boron (also indicating that IBA influenced the very early states of primordium initiation). These authors also proposed that boron may be inactivated by binding with phenolics in light-grown cuttings, which have higher phenolic levels than cuttings grown in the dark (see Section 11). Middleton *et al.* [76] conducted further tests with the boron-sensitive mung bean cuttings from light-grown plants. They found that: (1) IBA treatment promoted increases in free sugar levels in leaves and promoted movement of ^{14}C to hypocotyls from leaves treated with ^{14}C-sucrose; and (2) boric acid treatment promoted translocation of sugars and [14]C (from ^{14}C-sucrose) to hypocotyls. However, the authors noted that the ^{14}C translocated to hypocotyls may not have been in sucrose but in another root-promoting factor. In addition, they proposed that boron may be bound by phenolics with *cis*-hydroxyl configurations, which appear in higher concentrations in light- than in dark-grown cuttings (see Section 11 and Chapter 6). However, as stated below, compounds other than phenolics may bind boron. In an extension of the foregoing study, Jarvis and Booth [77] tested the influences of boron on the rooting of light-grown mung bean cuttings. They also found that rooting depends on exogenous boric acid. The authors concluded that boron initiates or maintains transport of carbohydrates, perhaps *myo*-inositol or precursors, from leaves to the rooting zone. The role of applied boron in modifying rooting may have a much more complex metabolic basis than its effect on carbohydrate transport. Lewis [6, 78] recently summarized information and theorized concerning the influences of boron on growth and development in vascular plants. These papers should be read by anyone interested in studying the potential roles of boron on the metabolism of rooting. According to Lewis [6], boron may influence the control of: (1) growth and differentiation at the whole plant level, (2) membrane permeability and sugar transport and (3) enzymes involved in the metabolism of carbohydrates, phenolics, lignin, auxins, and nucleic acids. In particular, wounding of plant tissue results in boron binding and alterations in hormone and phenolic metabolism similar to effects produced by boron deficiency [78]. Boron is bound by compounds that contain *cis*-hydroxyl configurations, of which carbohydrates (but not sucrose) and *o*-diphenols are important representatives. Deficiency of boron results in poor lignification and xylem differentiation, which causes phenolic lignin precursors to accumulate [6]. Lewis [6] proposed and explained metabolic interactions between lignification, auxin, boron, and peroxidase activity. This simple scheme suggests the complexity of interpreting boron-related modifications of rooting according to any single, isolated metabolic process.

Overall, the literature suggests that root primordium initiation is not markedly influenced by excesses or deficiencies of any particular mineral nutrient, with the possible exception of boron in light-grown mung beans, and of nitrogen in many species (see Section 9). However, this conclusion is based on limited evidence.

Some time ago, Good and Tukey [71] suggested that good rooting depends on adequate mineral nutrition before and during rooting. The most direct approach to discovering what constitutes adequate mineral nutrition during primordium initiation involves rooting of cuttings from stock plants that have intentionally been made deficient in certain minerals. This approach has been little used in recent years [65, 66, 67]. Such experiments may better define the specific mineral nutrient requirements during rooting than studies involving the application of mineral nutrients since excess may be one of the greatest deterrents to rooting in many species.

7. Temperature relations

The temperature at which stock plants and cuttings are maintained influences the metabolism of rooting. Based on tests with peas, Veierskov et al. [79] suggested that the growth temperature of the stock plant, rather than photosynthesis, was most important in determining initial carbohydrate content of cuttings (see Section 8). Various studies have considered the influence of temperature on rooting. For example, rooting of *Sinapis alba* cotyledons was found by Moore et al. [80] to be best at 25–30°C, inhibited below 25°C (10, 13, or 20°C), and eliminated above 30°C (35 and 40°C). These authors suggested that temperature may have influenced translocation of supportive and inhibitory factors, and also mitosis in the rooting zone. Carbohydrate may be one of the supportive factors because the accumulation-metabolism of carbohydrate within pea cuttings has been found to depend on temperature [81]. Veierskov and Andersen [81] found that low molecular weight carbohydrates and starch accumulated in cuttings maintained at 15°C, but no starch accumulated at 25°C. As will be discussed in Section 8, carbohydrate content and catabolism in cuttings have important implications for rooting.

In experiments with *Camellia japonica* and *Chrysanthemum morifolium*, Ooishi et al. [82] found that temperature effects on cuttings may be mediated through carbohydrate metabolism. Rooting occurred in direct relation to temperature: 16%, 36%, and 87% rooted at 17°C, 23°C, and 30°C, respectively. *Chrysanthemum* cuttings rooted earlier at 23°C or 30°C, but root development was not maintained. Respiration rate in the rooting zone of both species was positively associated with rooting. Peak respiration rate during rooting occurred earlier and was higher in both species with increasing temperature [82]. The decline in rooting with time at high temperature observed with *Chrysanthemum* cuttings [82] probably related to differential temperature effects on the initiation of primordia and their subsequent development. Dykeman [83] tested rooting of *Chrysanthemum* and *Forsythia* cuttings at 25°C and 30°C. More rapid rooting and more roots per cutting were obtained at 30°C, but root elongation, root diameter, and root hair development were superior at 25°C. Thus, higher temperatures favoured primordium initiation whereas lower temperatures favoured root development. The beneficial influence of higher temperature on initiation may be due to the related increase in respiration [82] and catabolism of simple sugars that would have been stored in starch at lower temperatures [81]. However, data for *Chysanthemum* [82, 83] and *Camellia* [83] indicated pronounced varietal differences in the response of primordium initiation and development to temperature.

8. Carbohydrate metabolism

Like all developmental processes, adventitious rooting is endergonic. This is indicated by increases in total cutting mass that can accompany rooting before root system development ([14, 17, 84] and Fig. 1). Respiration rate may also increase during rooting [82], especially in auxin-treated cuttings [85]. Stimulation of rooting by uncouplers of oxidative phosphorylation has been attributed to the resultant increase in respiration rate [86, 87]. However, inhibitors or uncouplers of oxidative phosphorylation may inhibit rooting under some conditions [3].

Apparently, the demands for energy and carbon skeletons to support rooting vary between species and depend somewhat on the type of cutting. For example, woody cuttings may require weeks or months to root during which time they can produce large amounts of callus and undergo substantial increase in mass [17, 21, 84]. In contrast, many herbaceous cuttings initiate primordia quickly, with little accretion of mass. Nevertheless, all types of cuttings may be in an equally tenuous energy state when first severed from the stock plant root system. Carbohydrates have most often been considered to be the principal source of energy and carbon skeletons during rooting because in most cuttings carbohydrates are present in greater concentrations than alternative energy sources such as lipids [3]. Nevertheless, lipid metabolism in certain species is also involved in primordium development, and possibly in initiation [88], which suggests that further studies of lipid metabolism during rooting are warranted in lipid-storing species. Substantial evidence suggests that cuttings root best under conditions that yield optimum internal total non-structural carbohydrate (TNC) concentrations before and during rooting [38, 74, 89, 90, 91]. In part, carbohydrate concentrations in cuttings may be influenced by auxin treatment, which can enhance mobilization of carbohydrate in leaves and upper stem, and increase transport to the rooting zone [76, 81, 92, 93, 94, 95, 96, 97]. However, the optimum carbohydrate conditions in stock plants and cuttings have not been defined. Thus, after over 50 years of intensive investigations with cuttings, controversy still exists concerning the metabolic control of rooting that may be effected by total or specific carbohydrate concentrations. Conflicting discussions are found in the literature even for the most studied species, such as pea [41, 46, 79, 81, 90, 98, 99, 100].

Five reasons may explain the ongoing controversies concerning the roles of carbohydrates during rooting of cuttings. Firstly, many years of investigation have been devoted to carbohydrate-rooting relations involving either total soluble carbohydrate or TNC [3]. Recent evidence suggests that total soluble carbohydrate and starch levels in cuttings may be positively related to rooting but not through any cause-effect relation (e.g. [84] and references therein). For example, total carbohydrate (Fig. 2A) accumulated in upper and basal stems of jack pine seedling cuttings during propagation. However, accumulations were equal in upper and basal stems, but only basal stems formed roots. Total carbohydrate

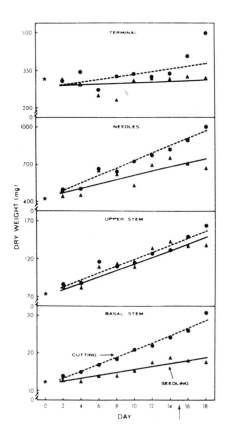

Fig. 1. Dry weight of terminals, needles, upper stem, and basal 1 cm of stem during the first 18 d propagation of cuttings from 90 d-old *Pinus banksiana* seedlings. Terminals were composed of the terminal bud with some subtending needles (see Fig. 3 in [17]). Measurements from intact seedlings grown in the same environment are also shown. Stars mark the initial weight of each tissue source. The arrow on the time-axis indicates the onset of root primordium initiation. Regression coefficients are significant (P<0.05), except for seedling terminals. Regression coefficients from the same tissue sources are different only for needles (P<0.01) and basal stems (P<0.001). Comparisons between tissue sources of cuttings indicate differences (P<0.05) between all regression coefficients except terminal vs upper stem. From [84].

levels in cuttings may be more related to root growth than to root initiation [100]. Concentrations of individual carbohydrates may have a much more direct relation to rooting [84, 97]. For example, reducing sugar (Fig. 2B) and sucrose (Fig. 3A) but not starch (Fig. 3B) levels differed between upper (non-rooting) and basal (rooting) stems of jack pine seedling cuttings during propagation. In these cuttings, the ratio of reducing sugar to starch was a particularly sensitive indicator of differential carbohydrate partitioning during rooting (Fig. 4).

156

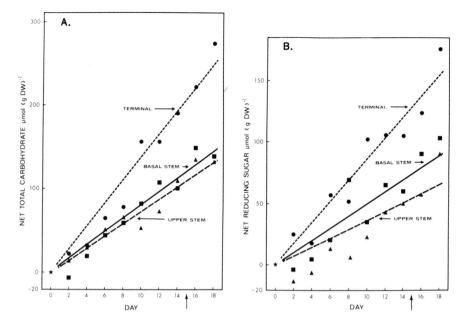

Fig. 2. (A) Net total carbohydrate and (B) net reducing sugar in terminal, upper stem, and basal 1 cm of stem during the first 18 d propagation of cuttings from 90 d-old *Pinus banksiana* seedlings. Net concentrations were obtained by subtracting data for seedlings, yielding on initial value of 0 (star). Seedlings and cuttings were grown in the same environment. The arrow on the time-axis indicates the onset of root primordium initiation. All regression coefficients in (A) and (B) are significant (P<0.001). In (A) the regression coefficient for the terminal differs (P<0.001) from coefficients for upper stem and basal stem. Regression coefficients for upper stem and basal stem do not differ significantly. In (B) all regression coefficients differ from each other (P<0.001). From [84].

A second reason why the role of carbohydrates remains controversial arises from studies of the influence of irradiance on rooting. Such studies often yield results that are difficult to interpret in a single major way. Recent evidence indicates that the influences of different intensities or durations of irradiance, although giving pronounced effects on endogenous carbohydrate levels, may also modify stock plant and cutting development. This can produce secondary effects that markedly influence rooting independently of carbohydrate status (see Chapter 7 and [79, 101, 102] and B. Veierskov, Royal Veterinary and Agricultural University, Department of Plant Physiology and Anatomy, Thorvaldsensvej 40, DK1871 Copenhagen V, Denmark, personal communication). Hansen and Eriksen [98], for example, proposed that auxin, growth inhibitor, and carbohydrate levels may be influenced by irradiance treatments. In addition, high irradiance may lessen rooting by hastening maturation [101], imposing water stress [B. Veierskov, personal communication], or lessening etiolation [J.R. Potter, USDA, SEA-AR Ornamental Plants Research Laboratory, 3420 SW Orchard St., Corvallis, OR 97331, USA, personal communication].

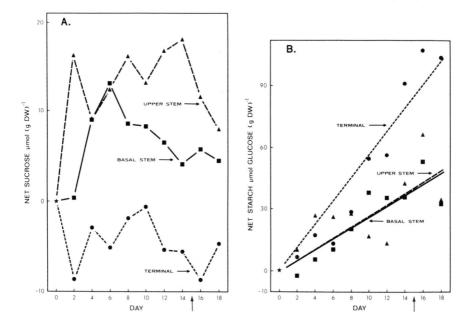

Fig. 3. (A) Net sucrose and (B) starch in terminal, upper stem, and basal 1 cm stem during the first 18 d propagation of cuttings from 90 d-old *Pinus banksiana* seedlings. Net concentrations were obtained by subtracting data for seedlings, yielding an initial value of 0 (star). Seedlings and cuttings were grown in the same environment. The arrow on the time-axis indicates the onset of root primordium initiation. In (A) mean net concentrations of sucrose differ ($P<0.05$) between tissue sources. In (B) the regression coefficient for terminals differs ($P<0.001$) from those for upper stem and basal stem. Regression coefficients for upper stem and basal stem do not differ significantly. From [84].

A third cause of the difficulty in understanding carbohydrate relationships in rooting is that authors may have drawn erroneous conclusions from experiments in which carbohydrates have been supplied to stock plants and cuttings. It is often assumed that the applied carbohydrates are taken up unchanged, are not metabolic inhibitors, are not themselves physico-chemically active, and are transported within the cutting in normal carbohydrate pathways. These assumptions may sometimes be invalid. Applied sucrose can be converted to reducing sugars on uptake [102, 103]. In addition, applied sucrose has been shown to inhibit stock plant and root development [41]. Sucrose, glucose, and mannose, for example may inhibit enzyme activities [32, 103, 104]. Certain sugars may be toxic or antagonize the effects of beneficial sugars [105]. Applied sugars also have osmotic influences [106, 107]. And, sugars are often applied at concentrations that greatly exceed physiological levels, even when possible concentrating effects in plant tissues are disregarded ([100] and B. Veierskov, Royal Veterinary and Agricultural University, Department of Plant Physiology and Anatomy, Thorvaldsensvej 40, DK1871 Copenhagen V, Denmark, personal communication).

158

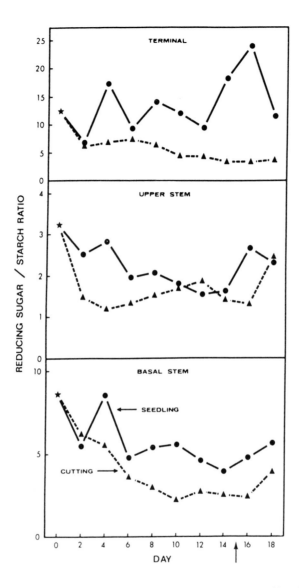

Fig. 4. Reducing sugar/starch concentration ratios in terminal, upper stem, and basal stem during the first 18 d propagation of cuttings from 90 d-old *Pinus banksiana* seedlings and in the parent seedlings. Seedlings and cuttings were propagated in the same environment. The arrow on the time-axis indicates the onset of root primordium initiation. Ratios differed between seedlings and cuttings in terminal ($P<0.001$), upper stem ($P<0.02$), and basal stem ($P<0.02$). Ratios in different tissues of the cuttings were significantly different ($P<0.05$) from each other. From [84].

Finally, cuttings uptake and translocate sugar mostly in the xylem, which is wholly abnormal distribution. Each of the foregoing influences might indirectly modify rooting, i.e. apart from the intended carbohydrate treatment effect.

A fourth difficulty arises from the widely held view that the sole importance of carbohydrates in cuttings resides in their being sources of energy and carbon skeletons. This importance cannot be denied, for example auxin treatments that enhance rooting may concomitantly increase respiration rate [82] and thus promote depletion of carbohydrate reserves [3, 17]. However, carbohydrates that accumulate during rooting may have an impact over and above those of providing carbon and energy. Carbohydrate concentration of the cytosol might influence osmoregulation, cellular solvent capacity, and other physico-chemical phenomena that influence metabolism (see Section 4). Thorpe and others have proposed important osmotic effects of starch degradation products and of free sugars in the medium on organogensis in *in vitro* cultures ([10] and references therein). Therefore, differential partitioning of carbohydrate into soluble, readily metabolized forms, soluble storage forms, and insoluble storage forms may aid in determining the roles of carbohydrates in rooting of cuttings [84]. In studies other than rooting, Huber [108] indicated that partitioning of carbon between starch and sucrose in the leaf may influence relative growth of shoots vs. roots. Ackerson [109] suggested for cotton that starch is involved in the regulation of cellular non-osmotic volume and, therefore, in osmotic adjustment (see Section 4).

Fifthly, physiological status of the stock plant influences carbohydrate metabolism in cuttings by causing variations in the amounts and types of substrates available for metabolism. Carbohydrate conditions of the stock plant are important [90, 96]. Any carbohydrate deficiencies may not be rectified through later environmental manipulation, for example by increasing irradiance to stimulate photosynthesis. In addition, rooting occurs so quickly in some herbaceous species that there may be insufficient time for major endogenous adjustments in carbohydrates.

Leafy cuttings are usually propagated under conditions that would favour fast rates of net photosynthesis, i.e. a large quantum flux density of extended duration, low water stress, and, sometimes, elevated CO_2 concentration [110]. Nevertheless, such treatments may not maintain photosynthesis at or above that for the intact plant. Rates of net photosynthesis in leafy cuttings are reported to be much lower than in corresponding stock plants and to decline during rooting [44, 46]. Photosynthesis is not an absolute requirement for rooting. This is amply demonstrated by the rooting of etiolated tissue cultures and of dark-propagated cuttings [e.g. 111]. Such observations may have discouraged much needed investigation of photosynthesis during the rooting of leafy cuttings. However, dark-propagated cuttings may not always root, as Eliasson [41] demonstrated by showing that rooting of pea cuttings increased up to the maximum irradiance used. In accord with Eliasson's findings, recent evidence suggests that the availability of current photosynthate may be important for rooting of some species of leafy cuttings. For example, Davis and Potter [46] found that rooting of leafy pea cuttings was reduced by about 50% if photosynthesis was decreased to the compensation point by manipulating light intensity, CO_2 concentration, or by applying an anti-transpirant. The treatments used to reduce the rates of net photosynthesis also decreased carbohydrate in the rooting zone before root emergence. In contrast, Breen and Muraoka [112] reported that less than 5% of photosynthetically fixed ^{14}C was translocated to the basal rooting zone of plum cuttings before rooting. The rooting zone of plum cuttings accumulated photosynthetically fixed ^{14}C only after roots were formed [113]. However, as Davis and Potter [46] noted, current photosynthate may only be needed when shoot growth establishes a sink that competes for carbohydrates to the detriment of rooting. Possibly, accumulation of photosynthetically fixed ^{14}C in the upper part of plum cuttings [112, 113] satisfied an upper sink and obviated competition with the rooting zone below. Although there is much to be learned about the roles of current photosynthate during the rooting of leafy cuttings, the observations of Breen and Muraoka [112, 113] and Davis and Potter [46] underscore the need to understand the metabolism of rooting on a whole cutting basis, rather than only in the rooting zone.

In addition to the above direct evidence, indirect evidence for several species indicates that current photosynthesis contributes substantially to carbohydrate pools throughout leafy cuttings, particularly in the basal rooting zone. For example, carbohydrate accumulation during rooting has often been reported [46, 84, 95, 97, 100, 101, 114]. Such accumulations probably occur, even under conditions of reduced photosynthesis, because cuttings lack the root system that may account for about 40% of whole plant respiration [47]. In addition, starch has been shown to accumulate in rooted bean leaf cuttings as a result of phosphorous deficiency [115], which might also occur during rooting under normal circumstances, particularly if there is active shoot growth (see Section 6). Carbohydrate accumulation in leafy cuttings often occurs first in the basal regions and thereafter in more aerial parts, although there are exceptions, e.g. plum cuttings. These differing observations may indicate that carbohydrate accumulates in basal zones to some specific level, after which carbohydrates are redirected to other regions. According to this hypothesis, further carbohydrate would not accumulate in the basal regions of cuttings, if the basal regions were already saturated with carbohydrate when cuttings were removed from the stock plant. Sucrose, glucose, fructose, sugar alcohols, and starch have been most frequently reported to accumulate in cuttings, under conditions where carbohydrates accumulated [46, 84, 95, 97, 100, 101, 114]. These observations suggest that sucrose not immediately directed to biosynthesis of other compounds or to respiration undergoes a partitioning into soluble and insoluble storage carbohydrate. Under circumstances where carbohydrates do not accumulate, concentrations of starch, reducing sugars, and sugar alcohols usually decline [112, 113, 116], indicating consumption.

Pronounced metabolism of carbohydrate in the rooting zone of cuttings is

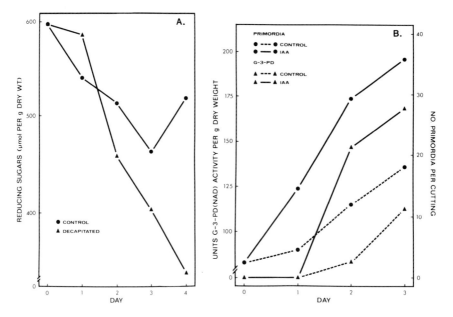

Fig. 5. Reducing sugars, number of adventitious root primordia and glyceraldehyde 3-phos-phate(NAD) dehydrogenase activity [G-3-PD(NAD)] in the basal 2 cm of the hypocotyl of *Phaseolus vulgaris*. In (A) and (B) leafy and leafless (decapitated) cuttings were propagated in low light intensity (approx. $30 \mu mol m^{-2} s^{-1}$) in vials that for the first 24 h contained $25 mmol l^{-1}$ MES-NaOH buffer (pH 6.0) with or without $50 \mu mol l^{-1}$ indole-3-acetic acid. Thereafter, vials contained double de-ionized water that was replaced daily. Differences between treatments are significant (P<0.01). From [116].

indicated by observations that soluble carbohydrates or TNC disappear from the basal regions of cuttings under circumstances that hinder photosynthetic carbon fixation ([116] and Fig. 5A) or where basipetal translocation does not occur [112, 113]. In part, TNC is metabolized to form insoluble macromolecules (most likely, cellulose, hemicellulose, lignins, proteins, nucleic acids). As an example, Breen and Muraoka [113] reported that up to 75% of photosynthetically incorporated ^{14}C was not extractable in starch or simple sugars. Carbohydrate catabolism during rooting probably follows the well-known pathways of intermediary metabolism. Thus, aerobic degradation of carbohydrates via the Embden-Meyerhoff-Parnas (EMP) pathway of glycolysis and the pentose phosphate (PP) pathway are the likely sources of energy in cuttings, i.e. complete or partial catabolism of glucose 6-phosphate, which is readily derived from simple sugars and starch [3]. Starch may be hydrolyzed by amylases (EC 3.2.1.1 and 3.2.1.2), which may or may not change quantitatively and qualitatively during rooting [3, 117, 118]. However, the phosphorylation of free glucose liberated by amylases would consume adenosine triphosphate (ATP) [3]. Starch hydrolysis by amylases seems to be an inefficient use of cellular energy because glucose 1-phosphate can be produced directly from starch by the action of phosphorylase (EC 2.4.1.1).

Glucose 1-phosphate is efficiently and readily converted to glucose 6-phosphate [3]. Subsequent partial degradation of glucose 6-phosphate would, in addition to energy, supply the necessary carbon skeletons to support rooting.

Rooting appears to depend strongly upon nucleic acid and protein biosynthesis ([3] and see Section 10). Primary requirements of nucleic acid biosynthesis (i.e. pentose and NADPH) are met through operation of the pentose phosphate pathway [116]. However, the requirement for NADPH might also be met from the conversion of phosphoenol pyruvate (PEP) to pyruvate, which then enters the citric acid cycle [26]. This conversion involves reactions catalyzed by PEP carboxylase (EC 4.1.1.31), malate dehydrogenase (EC 1.1.1.37), and malic enzyme (EC 1.1.1.40) in the following reaction sequence:

$$PEP + CO_2 \rightarrow oxaloacetate$$
$$oxaloacetate + NADH \rightarrow malate + NAD$$
$$malate + NADP \rightarrow pyruvate + NADPH + CO_2$$

Protein synthesis requires amino acids, and ATP as an energy source. The requirement for ATP can be met through oxidation of glucose by the EMP pathway ([116] and references therein), which also yields NADH. NADH can be converted to NADPH as shown above, or by NAD(P) transhydrogenase (EC 1.6.1.1). Overall, the metabolic requirements of rooting may be met by interrelated rather than separate operation of the EMP and PP pathways [116]. Interrelated operation of the EMP and PP pathways during rooting has been suggested in tests with bean (*Phaseolus vulgaris* cv. Top Crop) cuttings [116]. In the foregoing tests, the activities of two EMP pathway enzymes [phosphofructokinase, (EC 2.7.1.11) and glyceraldehyde 3-phosphate(NAD) dehydrogenase, (EC 1.2.1.12)] and two PP pathway enzymes [glucose 6-phosphate dehydrogenase, (EC 1.1.1.49) and 6-phosphogluconate dehydrogenase, (EC 1.1.1.44)] were found to increase in cuttings not treated with auxin. Only activity increases in 6-phosphogluconate dehydrogenase appeared to be unrelated to rooting. Enzyme activity increases associated with rooting precede or occur concomitantly with primordium development and enhanced metabolism of reducing sugars (compare Figs 5A and 5B). Results obtained for glyceraldehyde 3-phosphate(NAD) dehydrogenase with bean cuttings were generally confirmed with cuttings of jack pine [116]. In tests with both bean and jack pine, glyceraldehyde 3-phosphate(NAD) dehydrogenase activity was found to increase substantially as a result of auxin treatment, which promoted rooting in both species ([116] and Fig. 5B).

In summary, the foregoing research indicates that there are relations between the carbohydrate status of stock plants and the rooting response of cuttings taken from them. However, carbohydrate status of the stock plant alone has proven inadequate for interpreting rooting response. It is not clear whether relations between carbohydrate levels in stock plants and rooting are direct or indirect. This uncertainty arises because carbohydrate status of the stock plant also relates directly to overall developmental history. The developmental history of the stock plant influences its degree of cellular differentiation and overall maturation, both of which may profoundly influence rooting (see also Chapter 7). Influences of stock plant maturity and differentiation can probably be illustrated by rooting studies using an age sequence of woody stock plants or etiolated herbaceous stock plants. Similarly, it is unclear what carbohydrate changes during propagation of cuttings mean. The literature suggests that varying carbohydrate levels in cuttings during propagation are either a controlling influence over rooting or the direct manifestation of some controlling influence. At present, some evidence suggests that changed carbohydrate levels are a manifestation of auxin action. This observation may be important because auxin positively influences the earliest stages of primordium initiation [25, 119]. However, auxin-carbohydrate relations during propagation of cuttings may only have the appearance of importance because other, possibly more important relations have yet to be identified. The influence of carbohydrate metabolism in rooting will only become clear through further study of translocatable carbohydrates and their partitioning into other forms. Such work would benefit from the use of radiotracers and experimental modification of specific carbohydrates in stock plants and cuttings.

9. Nitrogen metabolism

Various low molecular weight nitrogenous compounds have strong but often variable influences on rooting [3]. Purines and pyrimidines are among such nitrogenous compounds, but their influences on rooting are discussed later (see Section 10). The effects of cytokinins and indole auxins are treated elsewhere in this volume (Chapters 7 and 8). Thus, the present discussion is primarily limited to inorganic nitrogen, amino acids, polyamines, and oligopeptides, which have been collectively termed 'nitrogen' for simplicity.

Precise endogenous and exogenous relations between nitrogen and rooting have not been established with cuttings [3]. The need for nitrogen during rooting is seemingly acute because for example, amino acid, and purine and pyrimidine synthesis are associated with protein and nucleic acid anabolism. Some evidence suggests that, at least in part, amino acids that support rooting arise by *de novo* synthesis [3]. Elevated respiration rates during rooting [82] alone may indicate an increased need for nitrogen because protein nitrogen levels (e.g. root tips) can be proportional to respiratory activity [26]. Carbon/nitrogen (C/N) ratios have long been used to estimate influences of nitrogen, and of carbohydrates and nitrogen, on rooting [3]. It is generally held that high C/N ratios favour rooting [3, 110]. However, contrary evidence has frequently appeared. In a recent example, Leakey [38] calculated C/N ratios in lateral shoots of *Triplochiton scleroxylon* stock plants with differing numbers of nodes. Shoots from tall (high C/N ratio) plants rooted poorly, compared to shoots from short (low C/N) ratio plants.

Thus, good rooting was associated with low C/N ratios, which is contrary to the generalization that high C/N ratios favour rooting. Based on other tests, it was recently concluded that C/N ratio in pea stock plants was not related to subsequent rooting of cuttings [100]. Evidence and conclusions concerning the importance of C/N ratios to the rooting of cuttings often conflict [3]. Carbon/nitrogen ratios in themselves may be unreliable predictors of rooting ability for several reasons [3]. Rooting is influenced by vegetative development of the cutting. Vegetative development may depend as much upon the absolute magnitude of carbohydrate and nitrogen concentrations as upon their ratios [107]. Endogenous nitrogen and carbohydrate levels in various tissues of cuttings may change during propagation. Redistribution, biosynthesis, etc. would then change the C/N ratios. The form of nitrogen (e.g. nitrate, amino acids, type of amino acids) and carbohydrate (e.g. simple sugars, starch) may be as important as the total amounts present. Other conditions in the stock plant and cutting (e.g. hormone levels, degree of maturity), and environmental influences during propagation may influence overall requirements for the amount and type of nitrogen and carbohydrate.

As stated above, soluble nitrogen in both inorganic and organic compounds may become redistributed to the rooting zone of cuttings during propagation. Auxin treatment may also enhance the redistribution and use of nitrogen in cuttings [3]. However, as with the redistribution and accumulation of carbohydrates, there appears to be no fixed pattern of spatial change of nitrogen within cuttings of different species, or under different test conditions. For example, it has been reported that amino acids accumulated in the bases of cuttings [3] but in a recent test with jack pine seedling cuttings, no difference was found in total amino acids of the needles, upper stems, or basal rooting zones of untreated cuttings [97]. In addition, naphthaleneacetic acid (NAA) treatment resulted in a greater accumulation of total amino acids, compared to untreated cuttings, but only in the non-rooting upper stem. In the same test carbohydrate accumulated in all tissues of untreated and NAA-treated cuttings [97]. Lack of accumulation of amino acids does not necessarily mean that they have not been synthesized, or basipetally transported, or both. Amino acid concentrations may simply remain constant under varying rates of protein synthesis. In another recent experiment, Suzuki and Kohno [48] found that developing callus and roots of mulberry cuttings markedly accumulated 11 amino acids. In the previous study, growing portions of the rooting cuttings (buds, callus, and roots) accumulated total nitrogen, whereas total and protein nitrogen levels decreased in wood and, more so, in bark. Thus, the authors concluded that the rooting and other growing zones accumulated nitrogen as amino acids from protein hydrolyzed in bark or, to a lesser extent in wood, and from soluble nitrogen in wood and bark [48]. The authors proposed that storage protein was rich in arginine, and that arginine was transported to the rooting zone after protein hydrolysis. Their data indicate pronounced metabolism of arginine to other amino acids in the rooting and other growing zones [48].

Of inorganic nitrogen sources, nitrate levels have seemed the most closely related to rooting [3]. Recently, Hyndman *et al.* [107] also found that nitrate levels to which cultured rose shoots were exposed were positively related with rooting [120]. In addition, applied sucrose seemed to overcome the negative effects of high nitrogen levels on rooting [107]. Welander [121] observed that IAA-induced rooting was improved in sugar beet (*Beta vulgaris*) by applying more ammonium and nitrate nitrogen (which raised endogenous nitrogen concentrations) and by supplying more sucrose (which depressed endogenous nitrogen concentrations). In this test, it was clearly impossible to isolate a specific nitrogen effect on IAA-induced rooting [121]. In addition to ammonium and nitrate nitrogen, endogenous and exogenous amino acids may influence rooting. In tests with *Pelargonium* petiole explants, Welander [102] discovered several interesting relations. Rooting increased with the nitrogen and sucrose supply, and with increasing stock plant irradiance. When stock plant irradiance was decreased, endogenous nitrogen, and reducing sugar and sucrose decreased. Decreased irradiance yielded the highest initial endogenous levels of lysine, arginine, and ornithine, whereas high-irradiance gave the greatest amounts of γ-aminobutyric acid, proline, alanine, glutamic acid, glutamine, aspartic acid, and asparagine. Welander also noted that inhibition of rooting was related to increased endogenous concentrations of arginine, alanine, aspartic acid, asparagine, glutamic acid, and glutamine during propagation. Thus, the foregoing tests indicated complicated interactions between irradiance, endogenous nitrogen levels and endogenous carbohydrate levels. But, neither the foregoing research nor other studies yielded general conclusions concerning the importance of specific amino acids or their endogenous levels to the promotion or inhibition of rooting (e.g. compare [48], [102], and [122]).

Polypeptides containing specific amino acids have recently been proposed as initiators of organogenesis [123]. According to this theory, oligopeptides that induce rooting are of shoot origin. The root-promoting oligopeptides would contain tryptophan or phenylalanine, to mimic the core structure of IAA or phenylacetic acid, plus an acidic amino acid '. . . *which keeps the free acidic group within a definite position to the aromatic ring'*. Klämbt [123] also suggested that IAA mimics the function of the root-inducing oligopeptide. Kinetin was proposed as the inhibitor of rooting in this scheme. Thus, oligopeptide-cytokinin balance would determine whether or not rooting would occur [123]. This interesting theory lacks direct experimental evidence, but suggests a need for specific amino acids as a basis for root-inducing hormonal compounds. The root-inducing effects of auxin-amino acid conjugates have introduced similar possibilities (see Section 12).

There has been a recent interest in the influences of the polyamines putrescine, spermidine, and spermine on rooting [124]. Putrescine is synthesized from L-ornithine and L-arginine. The synthesis of spermidine and spermine additionally requires L-methionine [125]. L-arginine and L-ornithine have the respective

166

analogs L-canavanine and L-canaline [126, 127]. Methylglyoxal-bis(guanyl-hydrazone) (MGBG) blocks spermidine and spermine biosynthesis by inhibiting adenosylmethionine decarboxylase (EC 4.1.1.50) activity [127]. Friedman *et al.* [127] tested the effects of applied polyamines on rooting of mung bean hypocotyl cuttings in the presence of IBA. They found that applied polyamines did not increase rooting [124]. However, L-canavanine and L-canaline inhibited rooting; effects that could be reversed by applying L-arginine or L-ornithine, respectively. Treatment with MGBG also inhibited rooting. Concentrations of extracted polyamines were higher in IBA-treated than in untreated hypocotyl, epicotyl, and leaves. The data implicates the biosynthesis of polyamines as a necessity for IBA-induced rooting [127]. As discussed above, it was found that stock plant irradiance, which best promoted rooting of *Pelargonium* explants, yielded the highest initial endogenous levels of the polyamine precursors arginine and or-nithine [102]. In addition, arginine was the primary amino acid to accumulate in the rooting zone of mulberry cuttings [48]. But in these studies no attempt was made to determine if arginine or ornithine were polyamine precursors. Jarvis *et al.* [128] also studied the influences of polyamines on rooting, but in mung bean cuttings. The authors concluded that polyamines and their metabolism were involved in the early stages of rooting. Their evidence included observations that MGBG treatment inhibited rooting and resulted in lower levels of spermine and spermidine, while giving the expected increase in putrescine. Treatment with IBA also enhanced biosynthesis of spermine, spermidine, and putrescine in association with more profuse rooting. Furthermore, treatment with MGBG blocked IBA-induced increases in endogenous spermine and reduced increases in endogenous spermidine. It seems spermine levels in cuttings are particularly important to rooting [128]. As Friedman *et al.* [127] concluded, there appears to be a relation between auxin-induced rooting and endogenous polyamine concen-trations, but the metabolic role of polyamines is unclear. Kaur-Sawhney *et al.* [129] concluded from tests with isolated oat protoplasts that polyamine treatment promoted DNA synthesis and the inception of mitotic activity, both of which were arrested in the G1 phase of the cell cycle before polyamine treatment. Polyamines associate with nucleic acids [124, 130] and may stabilize membranes and participate in maintaining the structure and function of ribosomes [124, 126] which could influence rooting. However, polyamines may merely store and supply nitrogen [124]. This seems reasonable because both spermine and sper-midine can be enzymically oxidized to produce ammonium ions [126]. In addition, the preparation of cuttings initiates basal cellular injury, which may be increased further by auxin treatment. Auxin dosages that best promote rooting of cuttings may be nearly lethal to cells in the rooting zone. Thus, polyamine synthesis may occur in injured or dying cells of the rooting zone, but without direct relation to the metabolism of rooting. However, many other possibilities exist and are addressed in a recent review on polyamines [125].

Literature concerning relations between nitrogen and rooting implies that often, possibly always, low molecular weight nitrogenous compounds affect rooting by influencing stock plant and cutting development (e.g. growth rate, growth correlations, maturity). Growth correlations within cuttings have often been noted as strongly influencing rooting. For example, Reuveni and Adato [89] suggested that rooting ability in date palm (*Phoenix dactylifera*) depended upon large carbohydrate reserves and little competition from vegetative growth. Nitrogen levels in cuttings seem to be very influential in modifying competitive balances between developing organs. The effects of nitrogen on development are most probably related to the general metabolism of carbohydrates, nucleic acids, and proteins. The redirection of metabolic resources, prominently carbohydrates, may or may not benefit rooting, depending on the particular physiological status of the cutting. Redirection of carbohydrate metabolism by nitrogen treatment has been suggested by experiments in which applied nitrogen depressed EMP enzyme activity in leaves [3]. Rooting may depend on maintenance of nitrogen metabolism that does not stimulate cutting shoot development so much that rooting is placed at a competitive disadvantage for carbohydrates, mineral nutrients, and hormones.

10. Nucleic acid and protein metabolism

Cells of adventitious root primordia synthesize DNA, RNA, and proteins [3]. Protein complement in the rooting zone can change quantitatively and qualitatively during rooting [3, 131, 132]. For example, activities of specific enzymes have been shown to vary by histochemical tests [133] and by more gross assays [3, 22, 116]. The need for nucleic acid synthesis during rooting has been indirectly demonstrated by the positive effects of applied purine and pyrimidine bases. Bhattacharya and Nanda [134] found that, in the presence of IAA and sucrose, treatment of mung bean cuttings with adenine, guanine, cytosine, and thymine stimulated rooting. Jordan *et al.* [135] also reported that adenine and cytosine treatments resulted in significantly more rooting of *Prunus avium* hypocotyls, while guanine and thymine were less effective. Similarly, the pentose sugars of RNA (ribose) and DNA (deoxyribose) have been found to promote rooting of mung bean cuttings when applied in the presence of sucrose and IAA [136]. These results suggest that metabolic changes which bring about rooting depend on the availability of specific active proteins [132] and associated nucleic acid synthesis. Auxin treatments have been shown to influence nucleic acid and protein metabolism during rooting [3]. For example, Tripathi and Schlosser [137] found that IAA treatment of cabbage (*Brassica*) leaf cuttings promoted rooting and enhanced RNA and protein content, and to a lesser extent, DNA content. Similarly, they reported that reduced rooting as a result of fungicide treatment was associated with diminished DNA, RNA, and trichloroacetic acid-soluble protein, and increased buffer-soluble protein. Therefore, one might expect chemical inhibition of nucleic acid or protein metabolism to inhibit rooting. However, this has not always proved to be the case and a given inhibitor can yield different responses, depending upon species and other treatments [3]. The effects of inhibitors of nucleic acid and protein synthesis on rooting have been termed 'paradoxical' [87], which truly fits the existing evidence.

Actinomycin D treatment has frequently been used to determine whether the inhibition of DNA-dependent RNA synthesis influences rooting. For example, Kantharaj et al. [138] found a 90% inhibition of RNA synthesis in bean hypocotyl segments after treatment with actinomycin D. A rather extensive study has recently been made concerning the influences of actinomycin D on rooting of *Azukia* bean epicotyl cuttings [15, 16, 139]. In these studies, actinomycin D treatment promoted rooting if it was applied within 8 h after cuttings were prepared [15]. The authors studied the root promoting effect in comparison with cell divisions leading to primordium initiation. Rooting of the *Azukia* bean cuttings occurred in two phases, a preparatory phase and a root-forming phase [15]. The preparatory phase consisted of 5 subphases [15, 16, 139]: (1) subphase A, no cell divisions (0–8 h); (2) subphase B, transverse cell divisions (8–16 h); (3) subphase C1, longitudinal cell divisions (16–24 h); (4) subphase C2, cell divisions occurring after C1; and (5) subphase D, cell enlargement. Rooting of the *Azukia* bean cuttings occurred in the basal zone in 12 potential sites containing 10–20 cells adjacent to vascular bundles [15]. The authors found that, as a result of actinomycin D treatment, subphases A, B, and C1 were delayed about 4 h, and that subphases C2 and D were delayed more than 4 h [139]. However, the number of primordia that developed past subphase C1 was greater after actinomycin D treatment [139]. Thus, actinomcyin D treatment delayed the initiation of longitudinal divisions but increased the number of primordia with longitudinally dividing cells [139]. IAA treatment promoted longitudinal cell divisions and development of primordia with longitudinally divided cells [16]. In contrast, actinomycin D treatment suppressed rooting of *Populus nigra* cuttings [117, 131]. Actinomycin D treatment also suppressed rooting of bean leaf cuttings, particularly if the treatment was applied within 6 h of preparing cuttings [140].

Purine and pyrimidine analogs that modify normal nucleic acid synthesis have also been shown to suppress rooting [3, 131]. Blazich and Heuser [141] used 6-methyl purine to inhibit RNA synthesis in mung bean cuttings. They found that 6-methyl purine treatment inhibited rooting in the presence or absence of NAA, without visible injury to cuttings. Rooting was completely inhibited if 6-methyl purine was applied up to 12 h after NAA treatment, with lessening inhibition thereafter [141].

Jarvis et al. [142] tested effects on rooting of mung bean cuttings by cordycepin (3′-deoxyadenosine), which inhibits transcription and poly adenylation. They found cordycepin treatment stimulated rooting if it was applied within 0–4 h (10μg ml^{-1}) or 8–20 h (25μg ml^{-1}) of IBA treatment. Cordycepin treatment did not influence rooting if it was applied before IBA. Cordycepin treatment inhibited rooting of non-auxin-treated cuttings if applied within 0–24 h [142]. Sawhney *et al.* [143] reported interactions between auxins and rifampicin, which inhibits RNA synthesis. In their tests, combined rifampicin and IAA treatment promoted rooting better than if either compound was applied alone. However, similar promotion was not obtained if the auxin was IBA or NAA [143]. Rifampicin binds to RNA polymerase only in bacteria [143]. However, rifampicin has recently been shown to reduce RNA and protein content in higher plant tissue [144].

Recent studies have tested the effects of protein synthesis inhibitors on rooting of cuttings. Cycloheximide has often been used although it may also influence DNA synthesis [140]. Dhaliwal *et al.* [145] found that cycloheximide treatment stimulated rooting of *Impatiens balsamina* cuttings and induced formation of new peroxidase isozymes. Rooting was also promoted in *Azukia* bean cuttings when cycloheximide was applied 0–16 h after cuttings were prepared [15]. However, in another test, cycloheximide treatment of IAA-treated bean leaf cuttings completely inhibited rooting when the inhibitor was applied with or after auxin treatment [140]. Rooting of *Populus nigra* cuttings was also suppressed by cycloheximide treatment [131, 145].

Kantharaj *et al.* [138] found that protein synthesis increased within 30-min of exposing bean hypocotyl segments to IBA, followed by the first cell division within 24 h. Cycloheximide treatment inhibited rooting and protein synthesis if applied during the 30 min IBA treatment [138]. Similarly, Blazich and Heuser [141] found that cycloheximide treatment of mung bean cuttings inhibited rooting, especially in NAA-treated cuttings. However, visible damage to mung bean cuttings occurred at cycloheximide concentrations that inhibited rooting, which was also true for actinomycin D and puromycin treatments.

The older literature contains very few autoradiographic investigations of nucleic acid and protein anabolism during rooting [3]. The situation has changed little in the last 10 years, but recently at least one important study has been conducted with mung bean cuttings [146]. In that test, preparing the mung bean cuttings stimulated cell divisions over the full length of the hypocotyl, but only in some areas were primordia subsequently initiated [146]. The authors found that ^{3}H-thymidine and ^{3}H-uridine were incorporated along the whole hypocotyl, whether or not cell divisions were related to primordium initiation. NAA treatment of the cuttings promoted rooting but not ^{3}H-thymidine or ^{3}H-uridine incorporation, which further indicated that incorporation was not specific to cells that would form primordia. Nuclei in the potential rooting zone became labeled first with ^{3}H-uridine (nucleolus, 2 h) and then with ^{3}H-thymidine (14–26 h). By 8 h, ^{3}H-uridine was also incorporated in the cytoplasm and cell walls. Cell divisions were first observed at 23–26 h, in close agreement with ^{3}H-thymidine incorporation. The ^{3}H-uridine and ^{3}H-thymidine were specifically incorporated into RNA and DNA, respectively, based on control evidence obtained with nuclease enzyme treatments of tissue sections. The use of uridine and thymidine as tracers of nucleic acid synthesis did not in itself influence rooting [146].

The authors cited in this section have reached some primary conclusions about nucleic acids and proteins during rooting. Nanda and Bhattacharya [147] concluded that rooting of *Populus nigra* stem segments was associated with two actinomycin D-sensitive, low molecular weight RNA's. Based on tests with bean leaf cuttings, Oppenoorth [140] proposed that two types of DNA-dependent RNA synthesis occurred. One was termed an early, very actinomycin D-sensitive type, and the other a less actinomycin D-sensitive type. The early RNA's appeared to have a faster turnover than the later type, suggesting that they were messenger RNA and ribosomal RNA, respectively. Tripepi *et al.* [146] also concluded that messenger RNA synthesis was followed by ribosomal RNA synthesis in mung bean hypocotyl cuttings. However, root-promoting treatments may initially delay such RNA synthesis [142]. Oppenoorth [140] concluded that protein synthesis is required for de-differentiation leading to primordium initiation, and that the proteins involved have short half-lives and high turnover rates. In tests with bean hypocotyl segments, Kantharaj *et al.* [138] concluded that IBA influenced 'post-transcription' and translation in a step that was vital to rooting. They further concluded that a soluble factor in addition to IBA was required for early protein synthesis [138].

These conclusions agree generally with previous studies indicating that at the beginning of the rooting process but after an initial lag phase, the promotive influence of auxin on primordium initiation and development is effected through quantitative and qualitative changes in protein synthesis [3]. It was also shown that inhibitors of nucleic acid synthesis may lengthen the lag phase and thereby enhance root primordium initiation and development in some species, and that primordium initiation may require messenger RNA synthesis. Auxins can certainly enhance RNA synthesis and thus root primordium initiation through some action at the chromosomal level. Transcription of messenger RNA is likely to be involved. Furthermore, some factor other than auxin is apparently needed to trigger maximum RNA synthesis during initiation of primordia [3]. The research discussed in this section has extended the knowledge of nucleic acid and protein metabolism during rooting by demonstrating the need for sequential messenger RNA and ribosomal RNA synthesis, and for protein metabolism in cells that will initiate root primordia. Further, rooting-supportive nucleic acid and protein metabolism seems, at least partly, to be influencing the first cellular divisions leading to root primordium initiation. The research has also indirectly indicated that the conflicting influences on rooting by inhibitors of nucleic acid synthesis probably relate to the specific status of nucleic acid and protein metabolism in those particular cells that initiate primordia when a cutting is prepared, and to the amount of non-specific chemical injury that a cutting sustains as a result of inhibitor treatment. Presumably, the metabolic status of cells that will initiate root primordia depends on the species and type of cutting, status of maturation, and other aspects of physiology.

11. Phenolic metabolism

There is substantial evidence that applied phenolics may influence rooting (see also Chapter 6). Various theories to explain their effects have been proposed [122]. These include modification of IAA oxidase (IAAox) activity, stimulation of auxin synthesis, liberation of endogenous auxin, and formation of covalently bonded auxin-phenolic conjugates. However, each of these theories lacks adequate supporting evidence, as indicated by the fact that certain of them are mutually exclusive [122].

Identification of specific phenolic-rooting relations has proven difficult because no single structural type of phenolic consistently influences rooting [122, 148]. Results may not be substantiated when the same species of cutting is tested in different laboratories (compare [149] with [148, 150, 151, 152]), and rooting responses to a given phenolic may differ between species [148, 149, 150, 153]. Thus, it is not easy to identify an advantageous starting point to study the effects of phenolics on the metabolism of rooting. This may be why the metabolism of simple and complex phenolics during rooting has received little attention. However, it remains likely that anabolism of simple and complex phenolics is important in rooting if only because aromatic amino acids are involved in the synthesis of proteins and of phenylpropanoids concerned in lignin biosynthesis.

Unlike carbohydrates, there have been few studies of total soluble endogenous phenolic levels before or during the progress of rooting. However, some research suggests that low endogenous levels of simple phenolics in the stock plant favour rooting of cuttings [154, 155], for example because low levels sometimes occur during active shoot growth [154]. Breaking of bud dormancy and active (but not excessive) shoot growth have often been associated with a physiology that promotes rooting.

Levels of endogenous phenolics have been shown to increase in propagules during rooting [155], which is consistent with observations that phenolics accumulate as a result of wounding [78]. In jack pine seedling cuttings a pronounced basipetal movement of soluble phenolics has been observed during propagation and before rooting (Fig. 6). The data suggest synthesis of phenolics during rooting because accumulation occurred in each part of the seedling cuttings. Further, the time of initial increase and final concentration in the various tissues indicate synthesis of phenolics in the terminals (bud plus some subtending needles), followed by basipetal transport to and greatest accumulation in basal stem tissue. In jack pine seedling cuttings, only basal stems initiate roots [20, 21]. The basipetally transported phenolics may be involved in lignification of tracheid nests that form in basal callus tissue where root primordia initiate [20, 21]. In comparison, it has been found that phenylalanine ammonia lyase (EC 4.3.1.5, PAL) activity reaches a maximum in bean callus when nodules of xylem and phloem are forming [18]. PAL catalyzes the conversion of phenylalanine to *trans*-cinnamic acid, which is used in lignin biosynthesis. Tripathi and Schlosser [137]

172

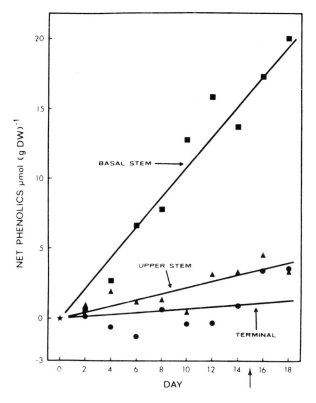

Fig. 6. Net soluble (Folin) phenolics in terminal, upper stem, and basal 1 cm of stem during the first 18 d propagation of cuttings from 90 d-old *Pinus banksiana* seedlings. Net concentrations were obtained by subtracting data for seedlings, yielding and initial value of 0 (star). Seedlings and cuttings were propagated in the same environment. The arrow on the time-axis indicates the onset of root primordium initiation. All regression coefficients are significant (P<0.05) and differ (P<0.05) from each other.

found that the systemic fungicide methyl benzimidazol-2-yl carbamate, which may inhibit PAL activity, inhibited rooting. In general, however, lignin synthesis during rooting has been a much neglected subject.

Lignin biosynthesis involves three major steps [6]: (1) shikimate pathway activity yielding phenylalanine; (2) conversion of phenylalanine to special alcohols; (3) polymerization of the alcohols. Lignin production results from the enzymic dehydrogenative polymerization of *trans-p*-coumaryl, *trans*-coniferyl, and *trans*-sinapyl alcohols in reactions that require peroxidase, hydrogen peroxide, and free radical formation [6, 73]. Conifer (guaiacyl) lignin primarily contains coniferyl alcohol; angiosperm (guaiacyl-syringyl) lignin also contains sinapyl alcohol [6]. Physiological circumstances that limit lignification result in accumulation of lignin-precursor phenolics [6]. Inadequate lignification during rooting would influence vascularization needed to support developing roots, and would have profound consequences on the metabolism of rooting and root development (see also Chapter 4 and [6, 78]).

Varied PAL activity during rooting may also be associated with biosynthesis of compounds other than lignin. *Trans*-cinnamac acid is a precursor for the biosynthesis of flavonoids, phenolic esters, benzoic acids, etc., which may influence rooting [122, 156]. Prominent among these compounds are anthocyanins, which are water-soluble flavonoids [156]. Anthocyanin biosynthesis is incompletely understood but initially requires *trans*-cinnamic acid and malonyl coenzyme A [156]. Over 20 years ago, Bachelard and Stowe [157] observed a positive relationship between rooting of *Acer rubrum* cuttings and the concentration of leaf anthocyanins (cyanidin glycosides) that developed during propagation. However, enhanced anthocyanin biosynthesis does not always characterize metabolism that improves rooting. For example, Lee [156] found that misting of *Euonymus alatus* stock plants inhibited development of red foliage, and delayed dormancy and leaf abscission. Cuttings from misted stock plants rooted better than cuttings from non-misted plants. Compared to non-misted stock plants, leaf tissues of misted stock plants had lower concentrations of anthocyanins, total sugars, soluble nitrogen, and potassium, and higher concentrations of starch, protein nitrogen, PAL, flavonols, flavans, leucoanthocyanins, and total phenolics. In particular, misting leached compounds (e.g. sugars, potassium, and flavonoids) that were essential for anthocyanin biosynthesis. Treating cuttings from non-misted stock plants with rutin increased rooting to that of cuttings from misted stock plants [158]. Rutin is a flavonol similar to those that accumulated in leaves of misted plants [156]. These results do not explain how precursors of anthocyanin biosynthesis, or anthocyanins themselves, may modify rooting of cuttings. However, they may explain why degree of maturation of the stock plant can influence rooting. Anthocyanin accumulation has been associated with reduced stock plant growth [e.g. 41] and slow growth of callus tissue cultures [K.E. Wolter, USDA-Forest Service, Forest Products Laboratory, Gifford Pinchot Drive, Madison, WI 53705, USA, personal communication]. Therefore, maturation may be modifying endogenous levels of anthocyanin precursors that are essential for the rooting of cuttings. This connection may also explain why variable rooting results are often obtained in tests with applied phenolics. The activity of a specific phenolic in rooting may depend on its particular chemical type, in relation to the qualitative and quantitative availability of phenolics, which is maturation-dependent.

The metabolism of simple phenolics during rooting has been demonstrated through studies of the oxidoreductases catechol oxidase (COx, EC 1.10.3.1, commonly known as polyphenol oxidase or phenolase) and peroxidase (POx, EC 1.11.1.7). COx's are a group of copper-containing proteins (MW = 120,000) that use O_2 to oxidize various substituted catechols [159]. COx may also catalyze reactions of monophenol oxygenases [159]. Therefore, COx will catalyze the conversion of *o*-diphenols to *o*-diquinones and, at a substantially lower rate, also catalyze the *o*-hydroxylation of monophenols [160]. COx has been termed a *'physiological enigma'* [160], but is becoming better understood. Recent evidence suggests that COx is a plastidic enzyme that exists in a latent form on thylakoid membranes. COx is not involved in biosynthesis of phenolic compounds, but functions as a phenol oxidase in senescent or damaged tissue. Wounding, such as that incurred in the preparation of cuttings, releases phenolics from vacuoles and exposes them to COx [160]. Kominsky [161] has suggested that COx activity accompanies cellular proliferation and that POx activity indirectly influences rooting through effects on cellular differentiation. In one recent study of POx and COx activity, Dhawan and Nanda [162] found that coumarin treatment of *Impatiens balsamina* cuttings resulted in increased activities of both enzymes after cuttings were made. In the foregoing tests, the increased enzyme activities were associated with earlier cell divisions that preceded primordium initiation in coumarin-treated as opposed to untreated cuttings.

Bassuk and Howard [163] vacuum extracted the phenolic phloridzin and COx from apple cuttings. The authors found that treatment with the COx-phloridzin reaction product(s) plus IBA stimulated rooting in mung bean cuttings and apple hardwood cuttings. Thus, it appeared that the enzymic oxidation of phloridzin led to formation of a rooting cofactor. The role of such rooting cofactors or auxin synergists has been discussed elsewhere [122]. In tests with carrot, Habaguchi [164] found that increases in COx activity preceded rooting in untreated callus cultures. In addition, there was a positive relation between rooting induced chemically by cyclic adenosine monophosphate (cAMP) and the resulting COx activity. Habaguchi [165] concluded that cAMP treatment of carrot callus cultures resulted in increased COx activity because of increased transcription for COx. Habaguchi [164] also compared activity responses toward various substrates of COx's obtained from non-rooting (2,4-D medium) and rooting (IAA medium) carrot callus cultures. Differences were found in COx activity toward dopamine (large difference), caffeic acid (moderate difference), and catechol (little difference), suggesting that the COx's from the two tissue sources were different. Almost no activity was demonstrated by COx from either tissue source toward cresol or dopa. Other differences between these two COx's were found in activity responses to copper ions, dithiothreitol, ethylenediamine tetra acetic acid, temperature, and pH. The chromatographic behaviour of the two COx's was also different [164]. In a subsequent study, Habaguchi [166] tested 45-fold purified COx from the rooting carrot callus cultures. The heat-labile enzyme (MW range of 110,000–120,000) oxidized *o*-diphenols but not monophenols. At pH 6-7, pyrocatechol, dopamine, caffeic acid, and chlorogenic acid were substrates. And, the initial velocity of the COx-catalyzed reaction was enhanced by copper, cobalt, and manganese ions, and lessened by dithiothreitol, diethyldithiocarbamate, *p*-chloromercuryphenyl sulfonic acid, and glycylglycine [166].

POx's are haemoproteins (MW range = 40,000–50,000, plant sources) that oxidize various electron donors (e.g. NAD, NADP, fatty acids, ferrocythochrome c, iodide, glutathione, chloride) usually with hydrogen peroxide as the acceptor [159]. However, POx's also have been reported to oxidize with O_2 as the acceptor, and to decarboxylate, halogenate, hydroxylate, and polymerize [73]. For example, POx's oxidize IAA by using O_2 (see Section 12 and [73]). At very low hydrogen peroxide concentrations POx may catalyze the formation of quinones from phenolics, thus mimicking catalysis by COx [160]. As a consequence, some studies may not have adequately distinguished between those COx and POx activities, which can be accomplished if catalase or alcohols are added to COx reaction mixtures [160]. Not all known POx's have been reported in higher plants. POx's commonly reported in the rooting literature are of the group (EC 1.11.1.7) that oxidize unnamed donors [159]. POx's are associated with the cell wall and endocellular membrane systems (e.g. Golgi, endoplasmic reticulum, vacuoles). Numerous isoperoxidases have been reported. However, all isoperoxidases may not be native molecules but arise from interaction of a few isoforms, or from one or more isoforms interacting with other molecules [73]. Perhaps immunochemical methods will aid in analysis of POx isoforms [167]. Recent reviews provide additional detailed information concerning COx [160] and POx [73].

Decreased POx activity in leaves has been associated with increased rooting in some tests [168]. POx activity in leaf tissue of *Cichorium intybus* was higher in light-grown than in dark-grown tissue, and endogenous phenolic levels varied inversely with POx activity [169, 170]. POx activity in the leaf tissue was reversible with light regime. Based on tests with pre-oxidized caffeic and chlorogenic acid, it appeared that guaiacol-based POx activity in the leaf tissue was controlled by the endogenous levels of oxidized phenolics [169]. These observations have recently been confirmed and related to rooting activity [155, 171]. Druart *et al.* [155] described rooting in terms of an '*inductive*' phase before primordium initiation, and an '*initiative*' phase during which primordia were initiated. The initiative phase began when the test organs were transferred to a medium that stimulated rooting. Axillary shoot cuttings that originated from apple meristem tip cultures were exposed only to light (L) or darkness (D) during the inductive phase, and each of these types of cultures were then exposed only to L or D during the initiative phase. Rooting, based on percentage of rooted cultures, increased in the order: LL, DL, LD, and DD. The authors concluded that darkness in the initiative phase produced the best rooting [111]. The DD treatment resulted in cultures that had, compared to light regime treatments, the highest POx activity and the lowest endogenous phenolic levels at the start of the initiative phase. DD treatment also resulted in the largest decrease in POx activity and largest increase in endogenous phenolics later in the initiative phase [169]. In subsequent rooting trials with *Cynara scolymus*, Moncousin and Gaspar [171] found a positive relationship between the amount of POx activity induced by treatments with boron, auxins, or vitamin D_2, and the extent of their promotive effect on rooting. The authors concluded that rooting was promoted by treatments which resulted in a faster increase in POx activity during the inductive phase, or a faster decrease in

POx activity during the initiative phase. Similarly, total POx activity declined markedly during rooting of cacao (*Theobroma cacao*) cuttings [172]. The authors concluded that rooting was determined by a balance between auxin synthesis and catabolism, in which POx participated. Moncousin and Gaspar [171] also suggested that rooting was controlled by an '*inverse variation*' in auxin level during the inductive and initiative phases that resulted from the decline in total POx activity and concomitant decline in IAA-oxidizing basic POx isozymes.

Appearance of POx isozymes has been positively related to rooting in several species. Chandra *et al.* [173] found that the rooting zone of mung bean cuttings contained three POx isozymes and that four additional POx isozymes appeared during the initial 48 h of propagation. Appearance of new isoperoxidases was associated with differentiation of parenchyma cells that were involved in the initiation of root primordia [174]. Rooting has also been associated with appearance of one or a few new POx isozymes in *Impatiens balsamina* [175], mung bean [176], *Populus nigra* [177, 178], *Salix tetrasperma* [179], and tobacco [180, 181]. Variations in anodic and cathodic POx isozymes have also been noted during rooting [168, 180]. However, POx isozymes may also diminish in activity or disappear during rooting [e.g. 172]. Nevertheless, most authors have suggested that POx activity or isozyme changes during propagation influence rooting through effects on either lignification or IAA catabolism, whether or not the POx changes have been quantitative or qualitative, or increasing or decreasing during propagation. The relationship between POx isozymes and IAA oxidation is discussed in Section 12.

In summary, phenolics seem to directly affect rooting, perhaps as cofactors with auxin, or have less direct effects that relate to auxin, lignin, and anthocyanin metabolism. However, these effects of phenolics during rooting are only putative. Substantial research is needed concerning the metabolism of endogenous phenolics during rooting, and of the metabolic fates of applied phenolics. It is presently difficult to interpret the results of application studies. In particular, it is necessary to define the metabolism of simple phenolics that are substrates for lignification and anthocyanin biosynthesis, as opposed to non-substrate phenolics. In addition, it is necessary to explore roles of peroxidation apart from auxin oxidation. The role of COx also needs defining because it is unclear whether COx activity directly relates to rooting or is an ancillary, generalized wounding response.

12. Auxin metabolism

Much evidence suggests that together with unidentified factors, auxin is a principal endogenous promoter of root primordium initiation [17, 25, 119, 122 and Chapter 6]. Treatment with applied auxins has frequently been shown to stimulate rooting in cuttings from stock plants that are not so mature as to have lost their natural rooting ability [14, 182]. It is not known if it is only unmodified auxins that participate in triggering primordium initiation. Perhaps auxins are changed chemically, e.g. partially oxidized or conjugated with other small or large molecular weight compounds before they act physiologically [73, 122, 145, 183]. There is substantial evidence that auxins can be oxidized enzymically and non-enzymically. For example, Tang and Bonner [184] extracted an IAA oxidase enzyme (IAAox) from pea epicotyls that required one mole of O_2 per mole of substrate, liberated one mole of CO_2 per mole of substrate, and did not destroy the indole nucleus. Neither IBA nor indole-3-propionic acid (IPA) was appreciably oxidized in 3 h [184]. Subsequently, Wagenknecht and Burris [185] extracted similar IAAox's from bean roots and etiolated pea epicotyls that also catalyzed the oxidation of IBA and IPA. The catalysis was sensitive to the partial pressure of O_2 [185]. Kenten [186] tested a highly purified horseradish POx that catalyzed the oxidation of IBA and IPA by O_2 in the presence of manganese ions. An obligatory intermediate role of hydrogen peroxide was suggested because catalase inhibited the oxidation [186]. Wheat leaf extracts, catalase, and horseradish POx have also been shown to decarboxylate and oxidize IAA, in the presence of manganese and monohydric phenols (phenol, o-cresol, 2,4-dinitrophenol), or resorcinol [187].

Available evidence suggests that POx contains a special catalytic site for IAA oxidation [73]. The most cationic POx isozymes appear to be most catalytically effective in IAA oxidation. The following partial reaction sequence has been suggested: IAA + O_2 → indole-3-aldehyde or 3-methylene oxindole + CO_2 + H_2O. Indole-3-aldehyde is the dominant product at high enzyme concentrations [73]. However, oxindole-3-acetic acid has also been found to be a major catabolic product in maize (*Zea mays*) seedlings [188]. IAA oxidation apparently requires the POx heme or manganese, and is promoted by small amounts of hydrogen peroxide [73]. Monophenols may enhance and polyphenols may diminish POx-based IAA oxidation, phenomena sometimes used to explain the promotive effects of applied phenolics on rooting [122].

Auxins can also be oxidized non-enzymically. Brennan and Frenkel [189] reported non-enzymic oxidation of IAA in the presence of as little as 4 μmol l^{-1} hydrogen peroxide, with catalysis promoted by Fe ions. Indole-3-acetic acid, IBA, and NAA may also undergo riboflavin-sensitized photo-oxidation (see references in [190]). For example, root system development by shoot explants of *Eucalyptus ficifolia* was influenced by IBA concentration, which was decreased in the rooting medium by riboflavin-mediated destruction [190].

Auxins may also become bound by forming conjugates, for example with amino acids and sugar alcohols [191]. Ethylene may enhance such binding of auxin [192]. Naturally occurring bound auxins have been shown to promote rooting [193] and to influence embryogenesis [194]. Synthetic phenyl esters, phenyl thioesters, and phenyl amides of IAA and IBA have also been shown to promote rooting of bean and jack pine cuttings more effectively than the free acids [14,

195]. In comparison, the phenyl ether of IAA (phenyl tryptophyl ether) did not promote rooting of bean cuttings [182]. Bound auxins can be hydrolyzed [191, 195] and are protected from enzymic oxidation [191]. The interested reader should refer to a recent review concerning the chemistry and physiology of bound auxins [191].

There is substantial evidence that POx activities in cuttings are related to rooting, and some studies have related the POx activities directly or indirectly to IAA oxidation (see Section 11). In mung bean cutting extracts, gel electrophoresis revealed the presence of a major multifunctional (POx, COx, and IAAox) protein complex during the early stages of rooting [196]. IAA and catechol treatment enhanced activity of the multifunctional protein complex [196]. Chibbar et al. [197] and Brunner [198] also found that POx-IAAox activity increased during rooting of bean cuttings. Rooting was associated with increased POx-IAAox activity in Tradescantia albiflora [60]. In Tradescantia, phenolic promoters and inhibitors of IAAox stimulated rooting, with promoters being most effective [60]. Bansal and Nanda [199] compared IAAox activities in species with differing rooting potential. Cuttings of Salix tetrasperma, Populus robusta, Hibiscus rosa sinensis, and Eucalyptus citriodora cuttings rooted 100%, 100%, 30%, and 0%, respectively. IAAox activity was greatest in the best rooters and lowest in the poorest rooters. A Eucalyptus protein was found to inhibit rooting of mung bean cuttings and also to inhibit IAAox and COx activity in vitro in a concentration-dependent manner [199]. In addition to total IAAox activity, the number of IAAox isozymes may increase during rooting, for example in Impatiens balsamina [145], Populus nigra [131], and mung bean [200]. However, POx-IAAox activity may not always increase during rooting. Quoirin et al. [201] reported for cherry (Prunus) that total POx activity decreased in leaves and stems during rooting; activity of anodic isozymes decreased whereas activity of cathodic (probably IAAox) isozymes increased. Based on their trials, Quoirin et al. [201] concluded that IBA treatment reduced activity of cathodic peroxidases, especially those thought to be IAAox's, thus sparing root-promoting IAA from oxidation. Total POx activity was also found to decrease during the rooting of Asparagus [202]. In that study, the activity of two anodic POx's increased; activity of four cathodic POx's decreased, with all activity changes occurring in both the upper and rooting region of the cutting. Recently, pronounced differences in rooting response between apple clones to the applied phenolic phloroglucinol have been attributed to differences in auxin metabolism (e.g. oxidation, conjugation) as opposed to auxin uptake [148, 151, 152, 203].

The foregoing research suggests that rooting depends on a high metabolic capability for auxin oxidation before the initiation of primordia, with a relatively lower ability for auxin oxidation during primordium initiation. Presumably, endogenous auxin levels must rise during primordium initiation. This is consistent with observations that applied auxins may stimulate rooting. On the whole, however, endogenous auxin levels during primordium initiation require further study. Current evidence supports previous findings that primordium initiation depends more upon auxin than do subsequent phases of primordial development [25, 119]. Nevertheless, there is no reasonable explanation for the high auxin oxidizing capability that may be required before rooting. Without an explanation of a specific physiological requirement for auxin oxidizing capability, it seems necessary to explore other reasons for quantitative and qualitative variations in POx-IAAox activity that accompany rooting (see Section 11). In addition, it is unclear whether auxin oxidation products themselves or auxin conjugates, or both, influence rooting directly.

13. Conclusions

Except for the metabolism of lipids, substantial progress has been made during the last 10 years in increasing our general understanding of metabolic events during adventitious rooting. However, we presently face the dilemma of abundant factual information together with much uncertainty concerning specific metabolic controls over adventitious rooting. Only more rigorous, precisely targeted experimentation will yield the needed insights. Additional progress in understanding the metabolism of cuttings as a whole, particularly the rooting zone, should be possible by the use of radiotracer techniques, photosynthetic labeling, and immunochemical techniques. In particular, we need to know more of the metabolism in cells that initiate primordia. A better cellular understanding could probably be gained by applying cytochemical methods and electron microscopy. Emphasis on cells that initiate primordia would perhaps lead on to biophysical approaches to increasing our understanding of rooting.

Overall, progress would probably benefit from varying the type of plant material used in experiments. In the past, many studies have attempted to define important metabolic aspects of rooting by chemical or environmental treatments of single genotypes. Little advantage has been taken of the fact that metabolism is largely a function of genetics. Studies could profitably become more genetically based through use of mutants of agronomic and other plants that have little genetic variation, and use of full- and half-sib families of species such as forest trees that have wild-type genetic variation. Greater emphasis on the genetic constitution of test plants would aid in identifying genes (and, therefore, individual steps in metabolism) on which rooting specifically depends. Without the genetic approach, it has proven too difficult to separate the metabolic specifics of rooting from the metabolic generalities of whole plant development.

180

14. Acknowledgments

I thank Drs. Patrick J. Breen, Lennart Eliasson, Thomas Gaspar, John R. Potter, and Bjarke Veierskov for major suggestions during the preparation of this review. I also thank Ms. Karin Haissig for translation of French and German language papers, and Ms. Sandra Haissig for clerical work.

15. References

1. Lehninger, A.L. (1970) Biochemistry. Worth Publishers, Inc., New York.
2. Thorpe, T.A. (1982) Callus organization and de novo formation of shoots, roots and embryos in vitro. In Application of Plant Cell and Tissue Culture to Agriculture & Industry (eds. D.T. Tomes, B.E. Ellis, P.M. Harney, K.J. Kasha & R.L. Peterson), pp. 115–138. Plant Cell Culture Centre, University of Guelph, Ontario, Canada.
3. Haissig, B.E. (1974) Metabolism during adventitious root primordium initiation and development. New Zealand Journal of Forest Science, 4, 324–337.
4. Zobel, R.W. (1975) The genetics of root development. In The Development and Function of Roots (eds J.G. Torrey & D.T. Clarkson), pp. 261–275. Academic Press, London, New York, San Francisco.
5. Haissig, B.E. (1974) Origins of adventitious roots. New Zealand Journal of Forest Science, 4, 299–310.
6. Lewis, D.H. (1980) Boron, lignification and the origin of vascular plants – a unified hypothesis. The New Phytologist, 84, 209–229.
7. Esau, K. (1965) Plant Anatomy. Second Edition, John Wiley & Sons, New York, London, Sydney.
8. Sutton, R.F. & Tinus, R.W. (1983) Root and root system terminology. Monograph 24, Supplement to Forest Science, 29, 1–137.
9. McCully, M.E. (1975) The development of lateral roots. In The Development and Function of Roots (eds J.G. Torrey & D.T. Clarkson), pp. 105–124. Academic Press, London, New York, San Francisco.
10. Thorpe, T.A. (1980) Organogenesis in vitro: structural, physiological, and biochemical aspects. In International Review of Cytology (ed I.K. Vasil), Supplement 11A, pp. 71–111. Academic Press, London, New York, San Francisco.
11. Gaspar, T. (1980) Rooting and flowering, two antagonistic phenomena from a hormonal point of view. In Aspects and Prospects of Plant Growth Regulators, Monograph 6 (ed B. Jeffcoat), pp. 39–49. British Plant Growth Regulator Group, Wantage.
12. Wample, R.L. & Reid, D.M. (1975) Effect of aeration on the flood-induced formation of adventitious roots and other changes in sunflower (*Helianthus annuus* L.). Planta, 127, 263–270.
13. Tang, Z.C. & Kozlowski, T.T. (1982) Physiological, morphological and growth responses of *Platanus occidentalis* seedlings to flooding. Plant and Soil, 66, 243–255.
14. Haissig, B.E. (1983) N-phenyl indolyl-3-butyramide and phenyl indole-3-thiolobutyrate enhance adventitious root primordium development. Physiologia Plantarum, 57, 435–440.
15. Mitsuhashi-Kato, M., Shibaoka, H. & Shimokoriyama, M. (1978) Anatomical and physiological aspects of developmental processes of adventitious root formation in *Azukia* cuttings. Plant & Cell Physiology, 19, 393–400.
16. Mitsuhashi-Kato, M., Shibaoka, H. & Shimokoriyama, M. (1978) The nature of the dual effect of auxin on root formation in *Azukia* cuttings. Plant & Cell Physiology, 19, 1535–1542.
17. Haissig, B.E. (1983) The rooting stimulus in pine cuttings. Proceedings of the International Plant Propagators Society, 32, 625–638.
18. Haddon, L.E. & Northcote, D.H. (1975) Quantitative measurement of the course of bean callus

differentiation. Journal of Cell Science, 17, 11–26.

19. Friedman, R., Altman, A. & Zamski, E. (1979) Adventitious root formation in bean hypocotyl cuttings in relation to IAA translocation and hypocotyl anatomy. Journal of Experimental Botany, 30, 767–777.

20. Montain, C.R., Haissig, B.E. & Curtis, J.D. (1983) Initiation of adventitious root primordia in very young *Pinus banksiana* seedling cuttings. Canadian Journal of Forest Research, 13, 191–195.

21. Montain, C.R., Haissig, B.E. & Curtis, J.D. (1983) Differentiation of adventitious root primordia in callus of *Pinus banksiana* seedling cuttings. Canadian Journal of Forest Research, 13, 195–200.

22. Nanda, K.K. (1975) Physiology of adventitious root formation. Indian Journal of Plant Physiology, 18, 80–89.

23. Nanda, K.K. (1979) Adventitious root formation in stem cuttings in relation to hormones and nutrition. In Recent Researches in Plant Sciences (ed S.S. Bir), pp. 461–492. Kalyani Publishers, New Delhi.

24. Gaspar, T., Smith, D. & Thorpe, T. (1977) Arguments supplémentaries en faveur d'une variation inverse du niveau auxinique endogène au cours des deux premières phases de la rhizogénèse. Comptus Rendus Académie des Sciences (Paris), 285, 327–330.

25. Haissig, B.E. (1970) Influence of indole-3-acetic acid on adventitious root primordia of brittle willow. Planta, 95, 27–35.

26. Fowler, M.W. (1975) Carbohydrate metabolism and differentiation in seedling roots. The New Phytologist, 75, 461–478.

27. Audus, L.J. (1959) Plant Growth Substances. Interscience Publishers, Inc., New York.

28. Hardwick, R.C. (1979) Leaf abscission in varieties of *Phaseolus vulgaris* (L.) and *Glycine max* (L.) Merrill – a correlation with propensity to produce adventitious roots. Journal of Experimental Botany, 30, 795–804.

29. Locy, R.D. (1983) Callus formation and organogenesis by explants of six *Lycopersicon* species. Canadian Journal of Botany, 61, 1072–1079.

30. Keyes, G.J., Collins, G.B. & Taylor, N.L. (1980) Genetic variation in tissue cultures of red clover. Theoretical and Applied Genetics, 58, 265–271.

31. Balestrini, S. & Vartanian, N. (1983) Rhizogenic activity during water stress-induced senescence in *Brassica napus* var. *oleifera*. Physiologie Végétale, 21, 269–277.

32. Kauss, H. (1977) Biochemistry of osmotic regulation. International Review of Biochemistry. Plant Biochemistry II, 13, 119–140.

33. Rajagopal, V. & Andersen, A.S. (1980) Water stress and root formation in pea cuttings. I. Influence of the degree and duration of water stress on stock plants grown under two levels of irradiance. Physiologia Plantarum, 48, 144–149.

34. Fischer, P. & Hansen, J. (1977) Rooting of chrysanthemum cuttings. Influence of irradiance during stock plant growth and of decapitation and disbudding of cuttings. Scientia Horticulturae, 7, 171–178.

35. Rajagopal, V. & Andersen, A.S. (1980) Water stress and root formation in pea cuttings. III. Changes in the endogenous level of abscisic acid and ethylene production in the stock plants under two levels of irradiance. Physiologia Plantarum, 48, 155–160.

36. Baadsmand, S. & Andersen, A.S. (1984) Transport and accumulation of indole-3-acetic acid in pea cuttings under two levels of irradiance. Physiologia Plantarum, 61, 107–113.

37. Strömquist, L.-H. & Eliasson, L. (1979) Light inhibition of rooting in Norway spruce (*Picea abies*) cuttings. Canadian Journal of Botany, 57, 1314–1316.

38. Leakey, R.R.B. (1983) Stockplant factors affecting root initiation in cuttings of *Triplochiton scleroxylon* K. Schum., an indigenous hardwood of West Africa. Journal of Horticultural Science, 58, 277–290.

39. Orton, P.J. (1979) The influence of water stress and abscisic acid on the root development of *Chrysanthemum morifolium* cuttings during propagation. Journal of Horticultural Science, 54, 171–180.

182

40. Gay, A.P. & Loach, K. (1977) Leaf conductance changes on leafy cuttings of *Cornus* and *Rhododendron* during propagation. Journal of Hortcultural Science, 52, 509–516.
41. Eliasson, L. (1978) Effects of nutrients and light on growth and root formation in *Pisum sativum* cuttings. Physiologia Plantarum, 43, 13–18.
42. Loach, K. (1977) Leaf water potential and the rooting of cuttings under mist and polythene. Physiologia Plantarum, 40, 191–197.
43. Davis, T.D. & Potter, J.R. (1983) High CO_2 applied to cuttings: Effects on rooting and subsequent growth in ornamental species. HortScience, 18, 194–196.
44. Eliasson, L. & Brunes, L. (1980) Light effects on root formation in aspen and willow cuttings. Physiologia Plantarum, 48, 261–265.
45. Sivakumaran, S. & Hall, M.A. (1979) Hormones in relation to stress recovery in *Populus robusta* cuttings. Journal of Experimental Botany, 30, 53–63.
46. Davis, T.D. & Potter, J.R. (1981) Current photosynthate as a limiting factor in adventitious root formation on leafy pea cuttings. Journal of the American Society for Horticultural Science, 106, 278–282.
47. Veen, B.W. (1980) Energy cost of ion transport. In Genetic Engineering of Osmoregulation. Impact on Plant Productivity for Food, Chemicals & Energy, Volume 14, Basic Life Sciences (eds D.W. Raines, R.C. Valentine & A. Hollaender), pp. 187–195. Plenum Press, New York.
48. Suzuki, T. & Kohno, K. (1983) Changes in nitrogen levels and free amino acids in rooting cuttings of mulberry (*Morus alba*). Physiologia Plantarum, 59, 455–460.
49. Atkinson, D.E. (1969) Limitation of metabolite concentrations and the conservation of solvent capacity in the living cell. In Current Topics in Cellular Regulation (eds B.L. Horecker & E.R. Stadtman), pp. 29–43. Academic Press, London, New York, San Francisco.
50. Rozema, J., Buizer, A.G. & Fabritius, H.E. (1978) Population dynamics of *Glaux maritima* and ecophysiological adaptations to salinity and inundation. Oikos, 30, 539–548.
51. Boyer, J.S. & Meyer, R.F. (1980) Osmoregulation in plants during drought. In Genetic Engineering of Osmoregulation. Impact on Plant Productivity for Food, Chemicals & Energy, Volume 14, Basic Life Sciences (eds D.W. Raines, R.C. Valentine & A. Hollaender), pp. 199–202. Plenum Press, New York.
52. Steingröver, E. (1983) Storage of osmotically active compounds in the taproot of *Daucus carota* L. Journal of Experimental Botany, 34, 425–433.
53. Jefferies, R.L. (1980) The role of organic solutes in osmoregulation in halophytic higher plants. In Genetic Engineering of Osmoregulation. Impact on Plant Productivity for Food, Chemicals & Energy, Volume 14, Basic Life Sciences (eds D.W. Raines, R.C. Valentine & A. Hollaender), pp. 135–154. Plenum Press, New York.
54. Evans, H. (1952) Physiological aspects of the propagation of cacao from cuttings. Report of the Thirteenth International Horticultural Congress, 2, 1179–1190.
55. Zimmerman, P.W. (1930) Oxygen requirements for root growth of cuttings in water. American Journal of Botany, 17, 842–861.
56. Gislerød, H. (1982) Physical conditions of propagation media and their influence on the rooting of cuttings. I. Air content and oxygen diffusion at different moisture tensions. Plant and Soil, 69, 445–456.
57. Gislerød, H.R. (1983) Physical conditions of propagation media and their influence on the rooting of cuttings. III. The effect of air content and temperature in different propagation media on the rooting of cuttings. Plant and Soil, 75, 1–14.
58. Armstrong, W. (1979) Aeration in higher plants. Advances in Botanical Research, 7, 225–332.
59. Alberts, B., Bray, D., Lewis, J., Raff, M., Roberts, K. & Watson, J.D. (1983) Molecular Biology of the Cell. Garland Publishing Co., New York.
60. Pingel, U. (1976) Der Einfluss phenolischer Aktivatoren und Inhibitoren der IES-Oxydase-Aktivität auf die Adventivbewurzelung bei *Tradescantia albiflora*. Zeitschrift für Pflanzenphysiologie, 79, 109–120.

61. Amoore, J.E. (1961) Arrest of mitosis in roots by oxygen-lack or cyanide. Proceedings of the Royal Society B, 154, 95–108.
62. Amoore, J.E. (1961) Dependence of mitosis and respiration in roots upon oxygen tension. Proceedings of the Royal Society B, 154, 109–129.
63. Eliasson L. (1981) Factors affecting the inhibitory effect of indolylacetic acid on root formation in pea cuttings. Physiologia Plantarum, 51, 23–26.
64. Kordan, H.A. (1976) Adventitious root initiation and growth in relation to oxygen supply in germinating rice seedlings. The New Phytologist, 76, 81–86.
65. Pearse, H.L. (1946) Rooting of vine and plum cuttings a (sic.) affected by nutrition of the parent plant and by treatment with phytohormones. Science Bulletin No. 249, Fruit Research: Technical Series No. 6, Department of Agriculture, Union of South Africa. 13 pp.
66. Haun, J.R. & Cornell, P.W. (1951) Rooting response of geranium (*Pelargonium hortorum*, Bailey var. Ricard) cuttings as influenced by nitrogen, phosphorus, and potassium nutrition of the stock plant. Proceedings of the American Society for Horticultural Science, 58, 317–323.
67. Pearse, H.L. (1943) The effect of nutrition and phytohormones on the rooting of vine cuttings. Annals of Botany, 7, 123–132.
68. Samish, R.M. & Spiegel, P. (1957) The influence of the nutrition of the mother vine on the rooting of cuttings. *Katvim*. Records of the Agricultural Research Station, Jewish Agency for Palestine, 8, 93–100.
69. Blazich, F.A., Wright, R.D. & Schaffer, H.E. (1983) Mineral nutrient status of 'Convexa' holly cuttings during intermittent mist propagation as influenced by exogenous auxin application. Journal of the American Society for Horticultural Science, 108, 425–429.
70. White, J.W. & Biernbaum, J.A. (1984) Effects of root-zone heating on elemental composition of *Calceolaria*. Journal of the American Society for Horticultural Science, 109, 350–355.
71. Good, G.L. & Tukey, H.B., Jr. (1967) Redistribution of mineral nutrients in *Chrysanthemum morifolium* during propagation. Proceedings of the American Society for Horticultural Science, 90, 384–388.
72. Blazich, F.A. & Wright, R.D. (1979) Non-mobilization of nutrients during rooting of *Ilex crenata* Thunb. cv. Convexa stem cuttings. HortScience, 14, 242.
73. Gaspar, T., Penel, C., Thorpe, T. & Greppin, H. (1982) *Peroxidases*. Université de Genève, Centre de Botanique, Genève.
74. Reuveni, O. & Raviv, M. (1980) Importance of leaf retention to rooting of avocado cuttings. Journal of the American Society for Horticultural Science, 106, 127–130.
75. Middleton, W., Jarvis, B.C. & Booth, A. (1978) The boron requirement for root development in stem cuttings of *Phaseolus aureus* Roxb. The New Phytologist, 81, 287–297.
76. Middleton, W., Jarvis, B.C. & Booth, A. (1980) The role of leaves in auxin and boron-dependent rooting of stem cuttings of *Phaseolus aureus* Roxb. The New Phytologist, 84, 251–259.
77. Jarvis, B.C. & Booth, A. (1981) Influence of indole-butyric acid, boron, myo-inositol, vitamin D_2 and seedling age on adventitious root development in cuttings of *Phaseolus aureus*. Physiologia Plantarum, 53, 213–218.
78. Lewis, D.H. (1980) Are there inter-relations between the metabolic role of boron, synthesis of phenolic phytoalexins and the germination of pollen? The New Phytologist, 84, 261–270.
79. Veierskov, B., Andersen, A.S., Stummann, B.M. & Henningsen, K.W. (1982) Dynamics of extractable carbohydrates in *Pisum sativum*. II. Carbohydrate content and photosynthesis of pea cuttings in relation to irradiance and stock plant temperature and genotype. Physiologia Plantarum, 55, 174–178.
80. Moore, K.G., Illsley, A. & Lovell, P.H. (1975) The effects of temperature on root initiation in detached cotyledons of *Sinapis alba* L. Annals of Botany, 39, 657–669.
81. Veierskov, B. & Andersen, A.S. (1982) Dynamics of extractable carbohydrates in *Pisum sativum*. III. The effect of IAA and temperature on content and translocation of carbohydrates

in pea cuttings during rooting. Physiologia Plantarum, 55, 179–182.

82. Ooishi, A., Machida, H., Hosoi, T. & Komatsu, H. (1978) Root formation and respiration of the cuttings under different temperatures. Journal of the Japanese Society of Horticultural Science, 47, 243–247.

83. Dykeman, B. (1976) Temperature relationship in root initiation and development of cuttings. Proceedings of the International Plant Propagators Society, 26, 201–207.

84. Haissig, B.E. (1984) Carbohydrate accumulation and partitioning in *Pinus banksiana* seedlings and seedling cuttings. Physiologia Plantarum, 61, 13–19.

85. Strydom, D.K. & Hartmann, H.T. (1960) Effect of indolebutyric acid on respiration and nitrogen metabolism in Marianna 2624 plum softwood stem cuttings. Proceedings of the American Society for Horticultural Science, 76, 124–133.

86. Nanda, K.K., Bansal, G.L., Kochhar, V.K. & Bhattacharya, N.C. (1978) Effect of some metabolic inhibitors of oxidative phosphorylation on rooting of cuttings of *Phaseolus mungo*. Annals of Botany, 42, 659–663.

87. Nanda, K.K., Sethi, R. & Kumar, S. (1982) Some paradoxical effects of metabolic inhibitors in root initiation and floral bud initiation. Indian Journal of Plant Physiology, 25, 1–26.

88. Čiamporová, M. (1983) An ultrastructural study of reserve lipid mobilization in stem root primordia and (sic.) poplar. The New Phytologist, 95, 19–27.

89. Reuveni, O. & Adato, I. (1974) Endogenous carbohydrates, root promoters and root inhibitors in easy- and difficult-to-root date palm (*Phoenix dactylifera* L.) offshoots. Journal of the American Society for Horticultural Science, 99, 361–363.

90. Veierskov, B., Hansen, J. & Andersen, A.S. (1976) Influence of cotyledon excision and sucrose on root formation in pea cuttings. Physiologia Plantarum, 36, 105–109.

91. Champagnol, F. (1981) Relation entre la formation de pousse et de racines par une bouture de vigne et la quantité d'amidon initialement présente. Comptus Rendus Académie des Sciences (Paris), 67, 1398–1405.

92. Nanda, K.K., Kochhar, V.K. & Gupta, S. (1972) Effects of auxins, sucrose and morphactin in the rooting of hypocotyl cuttings of *Impatiens balsamina* during different seasons. Biology Land Plant, 1972, 181–187.

93. Patrick, J.W. & Wareing, P.F. (1973) Auxin-promoted transport of metabolites in stems of *Phaseolus vulgaris* L. Some characteristics of the experimental transport systems. Journal of Experimental Botany, 24, 1158–1171.

94. Patrick, J.W. & Wareing, P.F. (1976) Auxin-promoted transport of metabolites in stems of *Phaseolus vulgaris* L. Effects at the site of hormone application. Journal of Experimental Botany, 27, 969–982.

95. Altman, A. & Wareing, P.F. (1975) The effect of IAA on sugar accumulation and basipetal transport of ^{14}C-labelled assimilates in relation to root formation in *Phaseolus vulgaris* cuttings. Physiologia Plantarum, 33, 32–38.

96. Andersen, A.S., Hansen, J., Veierskov, B. & Eriksen, E.N. (1975) Stock plant conditions and root initiation on cuttings. Acta Horticulturae, 54, 33–37.

97. Haissig, B.E. (1982) Carbohydrate and amino acid concentrations during adventitious root primordium development in *Pinus banksiana* Lamb. cuttings. Forest Science, 28, 813–821.

98. Hansen, J. & Eriksen, E.N. (1974) Root formation of pea cuttings in relation to the irradiance of the stock plants. Physiologia Plantarum, 32, 170–173.

99. Veierskov, B. (1978) A relationship between length of basis and adventitious root formation in pea cuttings. Physiologia Plantarum, 42, 146–150.

100. Veierskov, B., Andersen, A.S. & Eriksen, E.N. (1982) Dynamics of extractable carbohydrates in *Pisum sativum*. I. Carbohydrate and nitrogen content of pea plants and cuttings grown at two different irradiances. Physiologia Plantarum, 55, 167–173.

101. Hansen, J., Strömquist, L.-H. & Ericsson, A. (1978) Influence of the irradiance on carbohydrate content and rooting of cuttings of pine seedlings (*Pinus sylvestris* L.). Plant Physiology, 61, 975–979.

102. Welander, T. (1978) Influence of nitrogen and sucrose in the medium and of irradiance of the stock plants on root formation in *Pelargonium* petioles grown *in vitro*. Physiologia Plantarum, 43, 136–141.

103. Lovell, P.H., Illsley, A. & Moore, K.G. (1974) Endogenous sugar levels and their effects on root formation and petiole yellowing of detached mustard cotyledons. Physiologia Plantarum, 31, 231–236.

104. Rozema, J. (1979) Population dynamics and ecophysiological adaptations of some coastal members of the *Juncaceae* and *Gramineae*. In Ecological Processes in Coastal Environments (eds R.L. Jefferies & A.J. Davy), pp. 229–241. Blackwell, Oxford.

105. Faludi, B., Daniel, A.F., Gyurjan, I. & Anda, S. (1963) Sugar antagonisms in plant tumor cells induced by 2,4-dichlorophenoxyacetic acid. Acta Biologica Academiae Scientianum Hunganicae, 14, 183–190.

106. Maretzki, A. & Hiraki, P. (1980) Sucrose promotion of root formation in plantlets regenerated from callus of *Saccharum* spp. Phyton, 38, 85–88.

107. Hyndman, S.E., Hasegawa, P.M. & Bressan, R.A. (1982) The role of sucrose and nitrogen in adventitious root formation on cultured rose shoots. Plant Cell Tissue Organ Culture, 1, 229–238.

108. Huber, S.C. (1983) Relation between photosynthetic starch formation and dry-weight partitioning between the shoot and root. Canadian Journal of Botany, 61, 2709–2716.

109. Ackerson, R.C. (1981) Osmoregulation in cotton in response to water stress. II. Leaf carbohydrate status in relation to osmotic adjustment. Plant Physiology, 67, 489–493.

110. Hartmann, H.T. & Kester, D.E. (1983) Plant Propagation. Prentice-Hall, New Jersey.

111. Fabijan, D., Yeung, E., Mukherjee I. & Reid, D.M. (1981) Adventitious rooting in hypocotyls of sunflower (*Helianthus annuus*) seedlings. I. Correlative influences and developmental sequence. Physiologia Plantarum, 53, 578–588.

112. Breen, P.J. & Muraoka, T. (1973) Effect of indolebutyric acid on distribution of ^{14}C-photosynthate in softwood cuttings of 'Marianna 2624' plum. Journal of the American Society for Horticultural Science, 98, 436–439.

113. Breen, P.J. & Muraoka, T. (1974) Effect of leaves on carbohydrate content and movement of ^{14}C-assimilate in plum cuttings. Journal of the American Society for Horticultural Science, 99, 326–332.

114. Brossard, D. (1977) Root organogenesis from foliar discs of *Crepis capillaris* L. Wallr. cultured *in vitro*: Cytochemical and microspectrophotometric analysis. The New Phytologist, 79, 423–429.

115. Sawada, S., Igarashi, T. & Miyachi, S. (1983) Effects of phosphate nutrition on photosynthesis, starch and total phosphorus levels in single rooted leaf of dwarf bean. Photosynthetica, 17, 484–490.

116. Haissig, B.E. (1982) Activity of some glycolytic and pentose phosphate pathway enzymes during the development of adventitious roots. Physiologia Plantarum, 55, 261–272.

117. Bhattacharya, N.C., Parmar, S.S. & Nanda, K.K. (1976) Isoenzyme polymorphism of amylase and catalase in relation to rooting etiolated stem segments of *Populus nigra*. Biochemie und Physiologie der Pflanzen, 170, 133–142.

118. Punjabi, B. & Basu, R.N. (1978) Amylolytic activity in relation to adventitious root formation on stem cuttings. Indian Biologist, 10, 65–71.

119. Haissig, B.E. (1972) Meristematic activity during adventitious root primordium development. Plant Physiology, 49, 886–892.

120. Hyndman, S.E., Hasegawa, P.M. & Bressan, R.A. (1982) Stimulation of root initiation from cultured rose shoots through use of reduced concentrations of mineral salts. HortScience, 17, 82–83.

121. Welander, T. (1976) Effects of nitrogen, sucrose, IAA and kinetin on explants of *Beta vulgaris* grown *in vitro*. Physiologia Plantarum, 36, 7–10.

186

122. Haissig, B.E. (1974) Influences of auxins and auxin synergists on adventitious root primordium initiation and development. New Zealand Journal of Forest Science, 4, 311–323.
123. Klämbt, D. (1983) Oligopeptides and plant morphogenesis: A working hypothesis. Journal of Theoretical Biology, 100, 435–441.
124. Bagni, N., Serafini Fracassini, D. & Torrigiani, P. (1981) Polyamines and growth in higher plants. Advances in Polyamine Research, 3, 377–388.
125. Tabor, C.W. & Tabor, H. (1984) Polyamines. Annual Review of Biochemistry, 53, 749–790.
126. White, A., Handler, P. & Smith, E.L. (1968) Principles of Biochemistry. McGraw-Hill, New York, Sydney, Toronto, London.
127. Friedman, R., Altman, A. & Bachrach, U. (1982) Polyamines and root formation in mung bean hypocotyl cuttings. I. Effects of exogenous compounds and changes in endogenous polyamine content. Plant Physiology, 70, 844–848.
128. Jarvis, B.C., Shannon, P.R.M. & Yasmin, S. (1983) Involvement of polyamines with adventitious root development in stem cuttings of mung bean. Plant & Cell Physiology, 24, 677–683.
129. Kaur-Sawhney, R., Flores, H.E. & Galston, A.W. (1980) Polyamine-induced DNA synthesis and mitosis in oat leaf protoplasts. Plant Physiology, 65, 368–371.
130. Mahler, H.R. & Cordes, E.H. (1966) Biological Chemistry. Harper & Row, Inc., New York.
131. Nanda, K.K., Bhattacharya, N.C. & Kochhar, V.K. (1974) Biochemical basis of adventitious root formation on etiolated stem segments. New Zealand Journal of Forest Science, 4, 347–358.
132. Ebrahimzadeh, H. & Amide, M. (1980) Évolution des protéines dans les fragments d'entrenoeuds de tige de *Peperomia blanda* H.B. & K. cultivés *in vitro,* au cours de la néoformation de racines et de bourgeons. Physiologie Végétale, 18, 405–410.
133. Malik, C.P. & Usha, K. (1977) Histochemical studies on the localization of metabolic reserves and enzymes during the initiation and formation of adventitious roots in *Impaniens (sic.) balsamina* Linn. New Botanist, 4, 113–124.
134. Bhattacharya, S. & Nanda, K.K. (1978) Stimulatory effect of purine and pyrimidine bases and their role in the mediation of auxin action through the regulation of carbohydrate metabolism during adventitious root formation in hypocotyl cuttings of *Phaseolus mungo.* Zeitschrift für Pflanzenphysiologie, 88, 283–293.
135. Jordan, M., Iturriaga, L. & Feucht, W. (1982) Effects of nitrogenous bases on root formation of hypocotyls from *Prunus avium* L. 'Mericier' and 'Bing' grown *in vitro.* Gartenbauwissenschaft, 47, 46–48.
136. Bhattacharya, S., Bhattacharya, N.C. & Nanda, K.K. (1976) Synergistic effect of ribose and 2-deoxy-ribose with nutrition and auxin in rooting hypocotyl cuttings of *Phaseolus mungo.* Plant & Cell Physiology, 17, 399–402.
137. Tripathi, R.K. & Schlosser, E. (1979) Effects of fungicides on the physiology of plants. II. Inhibition of adventitious root formation by carbendazim and kinetin. Zeitschrift für Pflanzenkrankheiten und Pflanzenschutz, 86, 12–17.
138. Kantharaj, G.R., Mahadevan, S. & Padmanaban, G. (1979) Early biochemical events during adventitious root initiation in the hypocotyl of *Phaseolus vulgaris.* Phytochemistry, 18, 383–387.
139. Mitsuhashi-Kato, M. & Shibaoka, H. (1981) Effects of actinomycin D and 2,4-dinitrophenol on the development of root primordia in azuki bean stem cuttings. Plant & Cell Physiology, 22, 1431–1436.
140. Oppenoorth, J.M. (1979) Influence of cycloheximide and actinomycin D on initiation and early development of adventitious roots. Physiologia Plantarum, 47, 134–138.
141. Blazich, F.A. & Heuser, C.W. (1981) Effects of selected putative inhibitors of ribonucleic acid or protein synthesis on adventitious root formation in mung bean cuttings. Journal of the American Society for Horticultural Science, 106, 8–11.
142. Jarvis, B.C., Shannon, P.R.M. & Yasmin, S. (1983) Influence of IBA and cordycepin on rooting and RNA synthesis in stem cuttings of *Phaseolus aureus* Roxb. Plant & Cell Physiology, 24, 139–146.

143. Sawhney, S., Sawhney, N. & Kohli, R.K. (1977) Synergistic effect of rifampicin and indole-3-acetic acid in root initiation on hypocotyl cuttings of *Phaseolus mungo*. Indian Journal of Plant Physiology, 20, 164–167.

144. Sawhney, S., Sawhney, N. & Kaur, R. (1981) Auxin-rifampicin interaction in adventitious root formation on hypocotyl cuttings of *Phaseolus mungo*. Indian Journal of Plant Physiology, 24, 199–205.

145. Dhaliwal, G., Bhattacharya, N.C. & Nanda, K.K. (1974) Promotion of rooting by cyclohex-imide on hypocotyl cuttings of *Impatiens balsamina* and associated changes in the pattern of isoperoxidases. Indian Journal of Plant Physiology, 17, 73–81.

146. Tripepi, R.R., Heuser, C.W. & Shannon, J.C. (1983) Incorporation of tritiated thymidine and uridine into adventitious-root initial cells of *Vigna radiata*. Journal of the American Society for Horticultural Science, 108, 469–474.

147. Nanda, K.K. & Bhattacharya, N.C. (1973) Electrophoretic separation of ribonucleic acids on polyacrylamide gels in relation to rooting of etiolated stem segments of *Populus nigra*. Biochemie und Physiologie der Pflanzen, 164, 632–635.

148. James, D.J. & Thurbon, I.J. (1981) Phenolic compounds and other factors controlling rhizogenesis *in vitro* in the apple rootstocks M.9 and M.26. Zeitschrift für Pflanzenphysiologie, 105, 11–20.

149. Zimmerman, R.H. & Broome, O.C. (1981) Phloroglucinol and *in vitro* rooting of apple cultivar cuttings. Journal of the American Society for Horticultural Science, 106, 648–652.

150. James, D.J. & Thurbon, I.J. (1981) Shoot and root initiation *in vitro* in the apple rootstock M.9 and the promotive effects of phloroglucinol. Journal of Horticultural Science, 56, 15–20.

151. James, D.J. (1983) Adventitious root formation 'in vitro' in apple rootstocks (*Malus pumila*). I. Factors affecting the length of the auxin-sensitive phase in M.9. Physiologia Plantarum, 57, 149–153.

152. James, D.J. (1983) Adventitious root formation 'in vitro' in apple rootstocks (*Malus pumila*) II. Uptake and distribution of indol-3yl-acetic acid during the auxin-sensitive phase in M.9 and M.26. Physiologia Plantarum, 57, 154–158.

153. James, D.J. (1979) The role of auxins and phloroglucinol in adventitious root formation in *Rubus* and *Fragaria* grown *in vitro*. Journal of Horticultural Science, 54, 273–277.

154. Fadl, M.S., Sari El-Deen, S.A. & El-Mahdy, M.A. (1979) Physiological and chemical factors controlling adventitious root initiation in carob *Ceratonia siligua* L. stem cuttings. Egyptian Journal of Horticulture, 6, 55–68.

155. Druart, P., Kevers, C., Boxus, P. & Gaspar, T. (1982) *In vitro* promotion of root formation by apple shoots through darkness effect on endogenous phenols and peroxidases. Zeitschrift für Pflanzenphysiologie, 108, 429–436.

156. Lee, C.I. (1971) Influence of intermittent mist on the development of anthocyanins and root-inducing substances in *Euonymus alatus* Sieb. 'Compactus'. Ph.D. Dissertation, Cornell University.

157. Bachelard, E.P. & Stowe, B.B. (1962) A possible link between root initiation and anthocyanin formation. Nature, 194, 209–210.

158. Mosella Chancel, L., Macheix, J.-J. & Jonard, R. (1980) Les conditions du microbouturage *in vitro* de Pecher (*Prunus persica* Batsch): influences combinées des substances de croissance et de divers composés phénoliques. Physiologie Végétale, 18, 597–608.

159. Dixon, M. & Webb, E.C. (1979) Enzymes. Third Edition, Academic Press, New York.

160. Vaughn, K.C. & Duke, S.O. (1984) Function of polyphenol oxidase in higher plants. Physiologia Plantarum, 60, 106–112.

161. Kaminski, C. (1971) Variations des activités peroxydasique et phénoloxydasique au cours de la croissance de *Coleus blumeni* Benth. var. Automne. Planta, 99, 63–72.

162. Dhawan, R.S. & Nanda, K.K. (1982) Stimulation of root formation on *Impatiens balsamina* L. cuttings by coumarin and the associated biochemical changes. Biologia Plantarum, 24, 177–182.

188

163. Bassuk, N.L. & Howard, B.H. (1981) Seasonal rooting changes in apple hardwood cuttings and their implications to nurserymen. Proceedings of the International Plant Propagators Society, 30, 289–293.

164. Habaguchi, K. (1977) Alterations in polyphenol oxidase activity during organ redifferentiation in carrot calluses cultured *in vitro*. Plant & Cell Physiology, 18, 181–189.

165. Habaguchi, K. (1977) A possible function of cyclic AMP in the induction of polyphenol oxidase preceding root formation in cultured carrot-root callus. Plant & Cell Physiology, 18, 191–198.

166. Habaguchi, K. (1979) Purification and some properties of polyphenol oxidase from root-forming carrot callus tissue. Plant & Cell Physiology, 20, 9–18.

167. Clark, S.K., Jr. & Conroy, J.M. (1984) Homology of plant peroxidases: Relationships among acidic isoenzymes. Physiologia Plantarum, 60, 294–298.

168. Gaspar, T., Penel, C. & Greppin, H. (1975) Peroxidase and isoperoxidases in relation to root and flower formation. The Plant Biochemistry Journal, 2, 33–47.

169. Legrand, B., Gaspar, T., Penel, C. & Greppin, H. (1976) Light and hormonal control of phenolic inhibitors of peroxidase in *Cichorium intybus* L. The Plant Biochemistry Journal, 3, 119–127.

170. Legrand, B. (1977) Action de la lumière sur les peroxydases et sur la teneur en composés phénoliques de tissus de feuilles de *Cichorium intybus* L. cultivés *in vitro*. Biologia Plantarum, 19, 27–33.

171. Moncousin, C. & Gaspar, T. (1983) Peroxidase as a marker for rooting improvement of *Cynara scolymus* L. cultured *in vitro*. Biochemie und Physiologie der Pflanzen 178, 263–271.

172. Balasimha, D. & Subramonian, N. (1983) Role of phenolics in auxin induced rhizogenesis & isoperoxidases in cacao (*Theobroma cacao* L.) stem cuttings. Indian Journal of Experimental Biology, 21, 65–68.

173. Chandra, G.R., Gregory, L.E. & Worley, J.F. (1971) Studies on the initiation of adventitious roots on mung bean hypocotyl. Plant & Cell Physiology, 12, 317–324.

174. Chandra, G.R., Worley, J.F., Gregory, L.E. & Clark, H.D. (1973) Effect of 6-benzyladenine on the initiation of adventitious roots on mung bean hypocotyl. Plant & Cell Physiology, 14, 1209–1212.

175. Nanda, K.K., Bhattacharya, N.C. & Kaur, N.P. (1973) Effect of morphactin on peroxidases and its relationship to rooting hypocotyl cuttings of *Impatiens balsamina*. Plant & Cell Physiology, 14, 207–211.

176. Gurumurti, K. & Nanda, K.K. (1974) Changes in peroxidase isoenzymes of *Phaseolus mungo* hypocotyl cuttings during rooting. Phytochemistry, 13, 1089–1093.

177. Nanda, K.K., Bhattacharya, N.C. & Kaur, N.P. (1973) Disc electrophoretic studies of IAA oxidases and their relationship with rooting of etiolated stem segments of *Populus nigra*. Physiologia Plantarum, 29, 422–444.

178. Bhattacharya, N.C., Kaur, N.P. & Nanda, K.K. (1975) Transients in isoperoxidases during rooting of etiolated stem segments of *Populus nigra*. Biochemie und Physiologie der Pflanzen, 167, 159–164.

179. Bhattacharya, N.C., Bhattacharya, S. & Nanda, K.K. (1978) Isoenzyme polymorphism of peroxidase, IAA-oxidase, catalase and amylase in rooting etiolated stem segments of *Salix tetrasperma*. Biochemie und Physiologie der Pflanzen, 172, 439–452.

180. Thorpe, T.A., Tran Thanh Van, M. & Gaspar, T. (1978) Isoperoxidases in epidermal layers of tobacco and changes during organ formation *in vitro*. Physiologia Plantarum, 44, 388–394.

181. Ono, H., Masuda, C. & Nagayoshi, T. (1980) The formation of a root-specific isoperoxidase as an indicator of root primordium differentiation in pith of tobacco. Science Report of the Faculty of Agriculture, Kobe University, 14, 85–91.

182. Haissig, B.E. (1983) Influence of phenyl tryptophyl ether on adventitious root development in bean cuttings. Canadian Journal of Botany, 61, 1548–1549.

183. Gurumurti, K., Chibbar, R.N. & Nanda, K.K. (1973) Evidence for the mediation of indole-

3-acetic acid effects through its oxidation products. Experimentia, 30, 997–998.

184. Tang, Y.W. & Bonner, J. (1947) The enzymatic inactivation of indoleacetic acid. I. Some characteristics of the enzyme contained in pea seedlings. Archives of Biochemistry, 13, 11–25.

185. Wagenknecht, A.C. & Burris, R.H. (1950) Indoleacetic acid inactivating enzymes from bean roots and pea seedlings. Archives of Biochemistry, 25, 30–53.

186. Kenten, R.H. (1955) The oxidation of β-(3-indolyl)propionic acid and γ-(3-indolyl)-n-butyric acid by peroxidase and Mn^{2+}. Biochemistry Journal, 61, 353–354.

187. Waygood, E.R., Oaks, A. & Maclachlan, G.A. (1956) The enzymically catalyzed oxidation of indoleacetic acid. Canadian Journal of Botany, 34, 905–926.

188. Reinecke, D.M. & Bandurski, R.S. (1981) Metabolic conversion of ^{14}C-indole-3-acetic acid to ^{14}C-oxindole-3-acetic acid. Biochemical and Biophysical Research Communications, 103, 429–433.

189. Brennan, T. & Frenkel, C. (1983) Nonenzymatic oxidation of indole-3-acetic acid by H_2O_2 and Fe^{2+} ions. Botanical Gazette, 144, 32–36.

190. Gorst, J.R., Slaytor, M. & de Fossard, R.A. (1983) The effect of indole-3-butyric acid and riboflavin on the morphogenesis of adventitious roots of *Eucalyptus ficifolia* F. Muell. grown *in vitro*. Journal of Experimental Botany, 34, 1503–1515.

191. Cohen, J.D. & Bandurski, R.S. (1982) Chemistry and physiology of the bound auxins. Annual Review of Plant Physiology, 33, 403–430.

192. Epstein, E. (1982) Levels of free and conjugated indole-3-acetic acid in ethylene-treated leaves and callus of olive. Physiologia Plantarum, 56, 371–373.

193. Felker, P. & Clark, P.R. (1981) Rooting of mesquite (*Prosopis*) cuttings. Journal of Range Management, 34, 466–468.

194. Epstein, E., Kochba, J. & Neumann, H. (1977) Metabolism of indoleacetic acid by embryonic and non-embryonic callus lines of 'Shamouti' orange (*Citrus sinensis* Osb.). Zeitschrift für Pflanzenphysiologie, 85, 263–268.

195. Haissig, B.E. (1979) Influence of aryl esters of indole-3-acetic acid and indole-3-butyric acids on adventitious root primordium initiation and development. Physiologia Plantarum, 47, 29–33.

196. Frenkel, C. & Hess, C.E. (1974) Isozymic changes in relation to root initiation in mung bean. Canadian Journal of Botany, 52, 295–297.

197. Chibbar, R.N., Gurumurti, K. & Nanda, K.K. (1979) Changes in IAA oxidase activity in rooting hypocotyl cuttings of *Phaseolus mungo* L. Experimentia, 35, 202–203.

198. Brunner, H. (1978) Einfluss verschiedener Wuchsstoffe und Stoffwechselgifte auf wurzelregenerierendes Gewebe von *Phaseolus vulgaris* L. Veränderungen des Wuchsstoffgehaltes sowie der Peroxydase- und der IAA-Oxydase Aktivität. Zeitschrift für Pflanzenphysiologie, 88, 13–23.

199. Bansal, M.P. & Nanda, K.K. (1981) IAA oxidase activity in relation to adventitious root formation on stem cuttings of some forest tree species. Experimentia, 37, 1273–1274.

200. Chibbar, R.N., Gurumurti, K. & Nanda, K.K. (1980) Effect of maleic hydrazide on peroxidase isoenzymes in relation to rooting hypocotyl cuttings of *Phaseolus mungo*. Biologia Plantarum, 22, 1–6.

201. Quoirin, M., Boxus, P. & Gaspar, T. (1974) Root initiation and isoperoxidases of stem tip cuttings from mature *Prunus* plants. Physiologie Végétale, 12, 165–174.

202. Van Hoof, P. & Gaspar, T. (1976) Peroxidase and isoperoxidase changes in relation to root initiation of *Asparagus* cultured *in vitro*. Scientia Horticulturae, 4, 27–31.

203. Lee, T.T. (1980) Effects of phenolic substances on metabolism of exogenous indole-3-acetic acid in maize stems. Physiologia Plantarum, 50, 107–112.

6. Endogenous control of adventitious rooting in non-woody cuttings.

B C JARVIS

Department of Botany, The University, Western Bank, Sheffield S10 2TN, UK

1. Introduction

In order to understand the endogenous control of any regenerative phenomenon it is first necessary to identify those chemicals involved in the processes of dedifferentiation, cell division and expansion, organization and differentiation. Furthermore, it is essential to relate such chemicals to each of these processes and establish any specific regulatory roles they may have. This is a prerequisite to any appreciation of how the processes themselves are integrated on a spatial and temporal basis.

The concept of a chemical with the specific ability to initiate the regeneration of roots was first propounded by Sachs [1]. He explained the polar regeneration of roots in terms of a rhizogenic substance synthesized in the leaves of cuttings and transported basipetally to the region of regeneration. Bouillenne and Went [2] subsequently used the term 'rhizocaline' to describe such a substance, which was seemingly an acidic compound of low molecular weight, was thermo-stable and stored in cotyledons and buds [2, 3]. Even prior to this time it was widely recognized that other factors, of a nutritive nature, could also influence the extent to which roots developed on stem cuttings (see ref. [4]). Such factors are now known to include carbohydrates, nitrogenous compounds, vitamins and in-organics. Shortly after the discovery of auxin, it was established that IAA and synthetic auxins induced root formation [5, 6, 7] and that 'rhizocaline' was physiologically similar to auxin [8]. Nevertheless, the term rhizocaline, or alter-natively rhizocaline complex, has been retained by some workers. For example, it has been used in more recent times to refer to the postulated interaction between an unidentified diphenol, an enzyme, possibly polyphenol oxidase, and auxin [9]. It is inferred that initiation of adventitious root formation necessarily depends upon interaction between these three factors.

In more recent times the general view that plant development may be control-led by interactions between several plant growth regulators, of which auxin is but one, has been applied to root regeneration in stem cuttings. Thus the effects of supplied gibberellic acid, abscisic acid, ethylene and cytokinins on rooting, which are described below, may be taken as limited evidence for their involvement in the natural processes of regeneration.

This article attempts to relate the influence of some of these factors, auxin in particular, to the temporal control of adventitious root development. Attention is focussed on the control of those events which culminate in the formation of the root primordium. Although it is obviously convenient to subdivide the continu-ous process of root development into sequential 'phases', there is no generally accepted convention for this. Only the terms induction (or preparatory) phase and initiation are used here. The former is defined as that period preceding the first cell divisions of root formation. The latter refers to the period commencing with the earliest cell divisions and terminating with the formation of a root primordium. It will become evident later, however, that initiation, as defined here, is better viewed as at least two distinct phases of development.

2. The central role of auxin

It is generally accepted that auxin has a central role in the initiation and development of adventitous roots. However, it has been emphasized previously that acceptance of such a role is largely based on circumstantial evidence [10], albeit fairly convincing. Auxins comprise the only group of chemicals which consistently enhance root formation in naturally responsive, or so-called easy-to-root, cuttings [11]. In such cuttings auxins invariably induce the formation of a greater number of roots per cutting than other chemicals. Furthermore, in order to demonstrate the stimulatory influence of many other chemicals it is often necessary to supply cuttings with auxin. Several other observations indirectly implicate auxin in the control of adventitious root formation. Young leaves and active buds which are sources of auxin enhance rooting in some cuttings. These promotive affects may be completely, or partially, replaced by exogenous auxin. Auxin is transported preferentially in a basipetal direction, an observation consistent with the polarity of root formation. Compounds which interfere with auxin transport or action, such as triiodobenzoic acid (TIBA), inhibit root regeneration, while chemicals, such as phenolics, which are known to influence auxin metabolism often influence the rooting response of stem cuttings.

2.1. Exogenous applications

Non-woody stem cuttings are usually highly responsive to supplied auxins, although the rooting response of any cutting is dependent upon the age of the stock material from which the cutting is taken, the auxin used and its concentration, the duration of the treatment and the time interval between excision of the cuttings and commencement of auxin treatment. Some of these points are illustrated in Fig. 1. Frequently, maximum rooting occurs when high concentrations of auxin are given immediately, or soon after, cuttings are made [12, 13]. Such concentrations are just below those which induce symptoms of toxicity [14, 15]. Of the more commonly used auxins indolebutyric acid (IBA) and naphthaleneacetic acid (NAA) are usually more effective than indoleacetic acid (IAA) (e.g. [16, 17, 18]). This may be related to differences in the rates with which these auxins are metabolised. Interestingly, 2,4-dichlorophenoxyacetic acid (2,4-D), which is not metabolised to any appreciable extent in plant tissues [19], sometimes appears less effective than the other synthetic auxins [16]. However, unpublished data [20] suggests this is not due to a lack of root initiation *per se,* but rather reflects inhibition of the late stages in root development, possibly because of this lack of metabolism.

A progressive loss in the responsiveness of cuttings to exogenous auxin may be evident if they are 'aged' in water prior to treatment with a high doseage of auxin (Fig. 1B). The basis of this declining response is unknown. Nevertheless, during

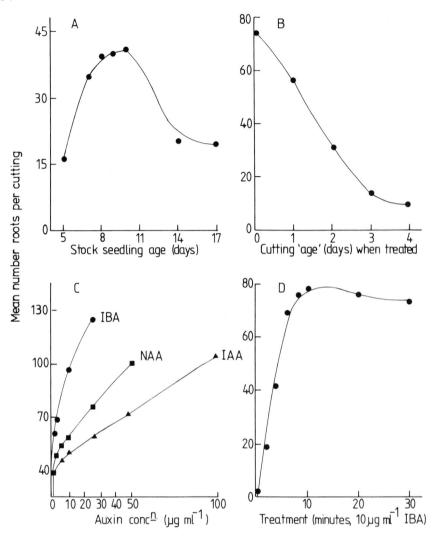

Fig. 1. The rooting response of stem cuttings in relation to stock material used and auxin supplied. A, B and D drawn from data of [52], [12] and [28] respectively. C redrawn, with permission, from [18]. (A) Cuttings, prepared from light-grown seedlings of *Phaseolus aureus* Roxb., were treated with 10^{-4} M IBA for 24 h then transferred to boric acid ($10 \mu g\,ml^{-1}$) for a further 6 d. Cuttings held under continuous fluorescent irradiance. (B) As for (A) except that cuttings were 'aged' in water prior to treatment with IBA. (C) Cuttings, from light-grown seedlings of *Phaseolus vulgaris* L., were treated for 24 h then transferred to tap water for a further 6 d. (D) Cuttings, consisting of 8 cm of hypocotyl only, were from dark-grown seedlings of *Phaseolus vulgaris* L. They were treated with Hoagland's solution containing IBA.

the period of ageing a supply of auxin, at concentrations too low to initiate root formation, delays or partially overcomes the influence of increase age [12]. Treatment of cuttings with low concentrations of supplied auxin evidently results in retention of 'sensitivity' to high doses of auxin. This influence of auxin is also evident in the data of Shibaoka [13]. It could represent an example of auxin-binding activity being induced by exogenous supply of auxin, such has been shown in other tissues [21]. Alternatively, the rate of basipetal transport of auxin, which itself generally declines with increasing age, could be maintained at, or close to, its level in fresh cuttings. This retention of 'sensitivity' to high doses of auxin may be likened to the preparatory action suggested for the so-called auxin-synergists (see Section 3.1).

Went [22] suggested that there are two phases of auxin action during root formation, the first of which can be induced by chemicals other than IAA. Other workers (e.g. [13, 23, 24] have similarly identified phases of adventitious root development which show varying sensitivities to auxins and other chemicals. For example, in cuttings of *Vigna angularis* an early phase, corresponding to early cell divisions, may be induced by IAA,p-chlorophenoxyisobutyric acid (PCIB) or 2,4,6-trichlorophenoxyacetic acid (2,4-T) [13]. This presumably corresponds to the first phase of auxin action described by Went [22]. The second phase, which culminates in the formation of the primordium, is seemingly dependent upon the presence of auxin, at least in its later stages. In both woody [25, 26] and non-woody cuttings [24, 27] there is a need for a constant supply of auxin throughout the initiation phase. High concentrations of auxin, even when applied for periods as short as 30 min (e.g. [28]), must therefore ensure an adequate dosage to initiate cell division and control organisation of the primordium. However, exogenously supplied IAA does not persist at high concentration in the region of regeneration throughout the entire period required for formation of the root primordium [29, 30]. Furthermore, excessively high concentrations of supplied auxin inhibit growth of primordia or even initiation itself. These observations suggest that the second phase of auxin action may be associated with lower auxin concentration than is the first.

2.2. Endogenous auxins

Surprisingly little work has been done on non-woody cuttings to establish endogenous concentrations of auxins in the region of root regeneration. Nevertheless, several investigations, based solely on bioassay procedures, suggest that the number of roots initiated per cutting may be a function of the amount of auxin-like substances in the zone of regeneration [33]. It is, however, unclear from these reports whether the biological activity shown by extracts was solely due to auxin-like substances. Despite the widely acknowledged limitations of bioassay procedures, and the development of more refined techniques for the identification and estimation of auxins, relatively little progress has been made on this front in recent years.

Enzyme immunoassay has been employed recently to study auxin levels in whole terminal cuttings of *Chrysanthemum morifolium* during root regeneration

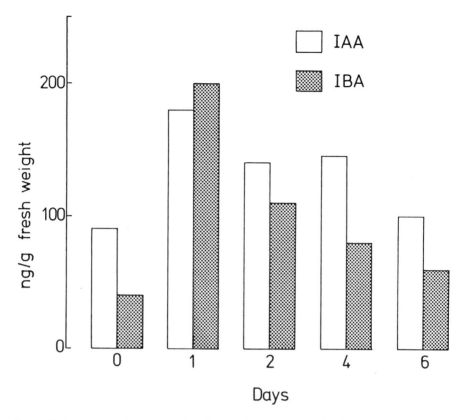

Fig. 2. Endogenous auxin concentration during adventitious root development. Redrawn, with permission, from [29]. Cuttings were taken from light-grown seedlings of *Phaseolus vulgaris* L. They were held in dilute Knop's solution with only the hypocotyl in darkness. Auxin concentration was determined in the basal centimetre of the cuttings.

[34]. Two points emerged from this investigation. Firstly, the greater the auxin content of cuttings at the time they were taken the greater the number of roots which subsequently developed. Secondly, both free and total IAA contents increased markedly from the time cuttings were taken, although declining levels were evident after roots became visible. From the limited samples analysed it is unclear whether maximum auxin content of the cuttings precedes, coincides with, or follows initiation. More detailed analyses of auxin content in specific parts of cuttings will hopefully employ this particular approach. Nevertheless, Brunner [29] has initiated such studies using gas chromatography as the means of assaying endogenous auxins. Within the basal centimetre of the hypocotyl of cuttings of *Phaseolus vulgaris* L. he identified IAA and IBA, the concentrations of which increase markedly within 24 h of taking cuttings (Fig. 2). At this time no root primordia have formed but thereafter concentrations decline. Hypocotyls of cuttings treated with IAA or IBA similarly show greatest accumulation of both

auxins after 24 h. This work clearly identifies high auxin concentrations in the region of root regeneration with the phase of initiation and indicates declining auxin concentration prior to formation of the root primordium, which is not evident until sometime between 48 and 96 h after cuttings are taken. Furthermore, these declining auxin concentrations coincide with increasing activities of peroxidase and IAA oxidase.

2.3. Sources of auxin and/or other factors

The presence of buds on cuttings is beneficial to root formation in some species (e.g. [35] but has little, or no effect in others (e.g. [36, 37]). In pea cuttings it has been shown that decapitation and disbudding during the phase of initiation is particularly detrimental to root formation [38]. In addition, buds may be replaced by auxin [35], suggesting that the influence of buds on root formation is due mainly, if not solely, to supply of auxin. Similarly, in cuttings of young seedlings, removal of the cotyledons may diminish or abolish root formation (e.g. [2, 36, 39]). The influence of removing the cotyledons can be offset to some extent by supply of auxin, suggesting that the latter is at least one of the initiation factors normally supplied by the cotyledons. Consistent with this view is the inhibitory influence of TIBA on root formation when it is placed just below the cotyledonary node [36]. The possibility that other factors involved in root initiation may arise in the cotyledons remains [39], although direct evidence of such factors is lacking.

The influence of leaves on adventitious root formation has been widely interpreted in terms of a supply of both auxin and nutritional factors (for discussion see [4]). In addition many workers have implicated other specific factors in root initiation and some have suggested that such factors may arise in the leaves, stems or buds (e.g. [40, 41]). It should be stressed that evidence for such factors is based on limited, indirect evidence and often questionable interpretation. For example, it has been emphasised that Went's suggestion of such a factor [35], which is initially distributed uniformly throughout the stem but which becomes redistributed in response to auxin, is inconsistent with the data he presents [4]. Moreover, it is claimed that the existence of 'leaf factors' is suggested not only by the influence of leaves on 'normal' regeneration but largely because leaves greatly enhance the response to applied auxin [42]. However, the influence of the leaves in such circumstances may be interpreted in different ways. For example, they may simply enhance uptake of supplied auxin. Alternatively, they may stimulate root production by loading supplied auxin into the appropriate transport system, thereby ensuring a more efficient delivery to those particular cells which are the potential sites for root formation. Data presented in Fig. 3 suggests such interpretations may be valid. Intact and leafless cuttings of mung bean were presented with 3-indolyl [1–14C] acetic acid at two concentrations of known specific activity.

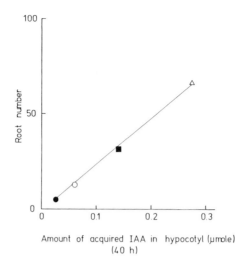

Amount of acquired IAA in hypocotyl (μmole)
(40 h)

Fig. 3. Relationship between adventitious root formation and supplied auxin accumulated in hypoc-otyls of intact and leafless cuttings of mung bean. Amount of auxin was estimated after 40 h, during which time cuttings were held under continuous fluorescent irradiance.
●, Leafless cuttings supplied with 10^{-4} M IAA;
○, Intact cuttings supplied with 10^{-4} M IAA;
■, Leafless cuttings supplied with 5×10^{-4} M IAA;
△, Intact cuttings supplied with 5×10^{-4} M IAA.
(From previously unpublished data of Jarvis *et al.*).

After 40 h, the time at which cell division is initiated in these cuttings [20], radioactivity within the hypocotyl was determined. Chromatography suggested that most of this radioactivity remained as IAA. The eventual rooting response of other cuttings treated with similar dosages of auxin, then transferred to boric acid, was also determined. Irrespective of the presence or absence of leaves the number of roots regenerated was related to the amount of auxin present in the region of root regeneration at the time of the first cell divisions. Leafless cuttings supplied with the higher concentration of auxin produced more roots than intact cuttings treated with the lower auxin concentration. More extensive studies comparing the responsiveness of such intact and leafless cuttings have shown that the supplied auxin dosage required for optimum rooting are quite different in the two types [20]. Other work has similarly led to the suggestions that the concept of cofactors from leaves or stem is not tenable with respect to this material and that rooting response is controlled by auxin [15]. Nevertheless, adherents of the 'rhizocaline' theory (e.g. [9 and 40]) maintain that leaves supply a specific factor, suggested to be an ortho-dihydroxy phenol. Since this compound has not been identified and evidence of its transport is lacking it is possible that during normal regeneration the leaves are simply sources of auxin, and any other specific factors required for root formation already exist, or are synthesised, locally.

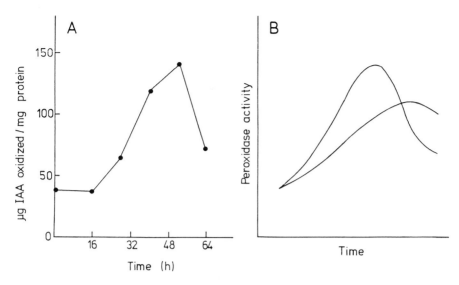

Fig. 4. (A) Activity of IAA oxidase during adventitious root development. Redrawn, with permission, from [44].
Cuttings, consisting of 3 cm each of epicotyl and hypocotyl, were from dark-grown seedlings of mung bean. The hypocotyl, which was submerged in IAA (5 μg ml^{-1}), was used for the assay of IAA oxidase.
(B) The 'ideal' curve of peroxidase activity during adventitious root development. Redrawn, with permission, from [47].

2.4. Auxin metabolism

Auxin catabolism may be attributed to the activity of IAA oxidase, which is most likely part of the peroxidase complex (see [43]). During adventitious root development changes in IAA oxidase activity and peroxidase zymograms have been described. Several reports clearly establish increased activity of IAA oxidase and/ or peroxidase prior to formation of root primordia [44, 45, 46, 47, 48]. Isozymic changes in the peroxidase complex are also evident during this period [48, 49, 50].

Fig. 4A shows the changing activity of IAA oxidase in hypocotyls of cuttings of *Phaseolus mungo* L. when treated with IAA. Initially activity is relatively low but increases within 28 h and reaches maximum activity 52 h after cuttings are made and IAA treatment commenced. Although IAA oxidase activity is lower in the absence of applied IAA it still increases between 16 and 52 h and thereafter declines [44]. Such activity of IAA oxidase most likely accounts for the diminished auxin content observed prior to formation of root primordium (see Fig. 2). Gaspar has suggested that root formation occurs only after total peroxidase has reached and passed a peak of maximum activity and when activity of basic isoperoxidases is declining [48]. He proposed that rooting is improved by conditions which first hasten and increase peroxidase activity then induce a sharp decrease, as shown in the 'ideal' curve of Fig. 4B. Since basic isoperoxidases are of particular importance in the degradation of IAA the data presented in Fig. 4A agree with this suggestion. Those high concentrations of auxin which initiate the rooting response would be inhibitory to subsequent growth, which demands low auxin concentration [12, 14]. This implies a need to regulate such concentration between root initiation and growth of the primordium.

Auxin levels at the site of regeneration could be regulated by means other than peroxidative decarboxylation since oxidation to oxindole-3-acetic acid and conjugates of IAA have been demonstrated in plant tissues [51]. Levels of IAA may be regulated by formation and hydrolysis of conjugates, amongst which *myo*-inositol is a common constituent. Interestingly, *myo*-inositol has been shown to enhance adventitious root formation and could be implicated in the processes of initiation [52]. However, investigation of auxin metabolism in relation to root formation has so far been restricted to consideration of IAA oxidase/peroxidase activity.

2.5. Summary

Supplied auxins are effective in stimulating root formation in many non-woody cuttings. There is limited evidence to show accumulation of auxin in the region of root regeneration under normal conditions. Furthermore, much indirect evidence implicates auxin in such regeneration. Nevertheless, the high concentrations of supplied auxin which initiate root formation are inhibitory to later stages of development, thereby implicating auxin metabolism during the course of primordium formation. Studies of IAA oxidase/peroxidase activity during the course of root development suggest that this enzyme complex has a role in such metabolism.

3. 'Auxin-synergists', 'co-factors' and phenolics

A wide range of chemicals, both natural and synthetic, enhance the rooting response of cuttings to applied auxins. The terms 'auxin-synergists' and rooting 'co-factors' have been used to describe some of these chemicals. The latter term implies natural occurrence and has frequently been used with reference to phenolic compounds (e.g. [15, 41, 53, 54]). Such a term, and particularly its implications, is clearly to be preferred where appropriate, although either of these terms can be readily applied to much previous work. This is particularly so since not all co-factors, or at least those occurring in non-woody cuttings, are phenolics (e.g. [10]).

3.1. 'Auxin-synergists'

The work of van Raalte [55] demonstrated synergistic effects of indole, phenylacetic acid and phenylbutyric acid on IAA-induced rooting of petioles. Subsequent work [56, 57] confirmed the synergism between indole and IAA but showed application of indole alone to enhance rooting. The latter result was interpreted in terms of an interaction between indole and endogenous auxin. The chemical nature of such substances suggests that they might act via IAA oxidase, particularly in view of the known effect of indole in diminishing the activity of IAA oxidase [55, 58]. Whereas indole acts synergistically with IAA, NAA, or 2,4-D, α- and β-naphthol show synergism only with IAA or NAA, while phenol interacts only with IAA [59]. Amongst the 'synergists' can be listed compounds as unrelated as dihydroasparagusic acid [60], actinomycin D [61] and sodium metabisulphite [62], all of which interact with IAA. In addition, although the term 'auxin-synergist' is not generally applied to sucrose it does act synergistically with IAA when supplied to cuttings of light grown seedlings (e.g. [44]). It is difficult, therefore, to envisage that all 'synergists' act in the same way. However, Gorter [59] has suggested that the more active synergists, such as indole and naphthol, may have a 'preparatory action' which makes more cells responsive to auxin. She likens this to the way that sulphuric acid, mercuric chloride and potassium hydroxide, at doses which are 'just sub-lethal', cause prolific rooting in *Coleus* [63]. Such an action, induced by a wide variety of 'auxin-synergists', may be equated with the first phase of auxin action described by Went (see Section 2.1).

3.2. Phenolic compounds and rooting

Bouillenne and Bouillenne-Walrand [9] suggested that root initiation is dependent on at least three factors which constitute the 'rhizocaline' complex. They viewed auxin as a non-specific component of this complex and proposed that it accelerates cell mitosis but does not determine the characteristic structure of the root. The other two components were viewed as a specific orthodiphenolic compound, transported from the leaves, and an enzyme, probably polyphenol oxidase, located in those cells which eventually give rise to the root initials. Direct evidence for transport of specific phenolic compounds into the region of regeneration, or any involvement of an auxin-phenol complex in root initiation, is lacking. Nevertheless, the investigations and suggestions of these authors, and their co-workers, underlie much of the more recent work on the possible involvement of phenolic compounds in the control of root regeneration. This has included attempts to account for the differences between easy- and difficult-to-root woody cuttings (e.g. [41, 45, 54, 64, 65, 66, 67, 68]). Most of these investigations have involved the use of procedures whereby extracts from woody cuttings have been assayed solely by means of the rooting response of non-woody cuttings. Unfortunately, such studies provide little evidence concerning the involvement of phenolic 'co-factors' in non-woody cuttings which are usually highly responsive to auxin. Notwithstanding this, the application of many phenolic compounds to non-woody cuttings, can markedly affect their rooting response, particularly that in response to supplied auxin.

Hess [41] showed catechol and pyrogallol to act synergistically with IAA. Since resorcinol, hydroquinone, phloroglucinol and caffeic acid at similar concentration failed to show such synergism he suggested that for maximum biological activity the structural requirement for a phenolic to act as a co-factor or synergist was an ortho positioning of hydroxy groups and a free para position. The ortho-dihydroxy requirement was seen as similar to that reported to inhibit IAA oxidase [69]. Numerous other reports confirm the activity of ortho-dihydroxy compounds. However, some conflicting results question the significance of this structural requirement. For example, Fernqvist [18] reported the dihdroxy-phenols resorcinol and hydroquinone to be highly promotory in cuttings of mung bean. Of the benzoic acids only para-substituted acids stimulated rooting, with salicylic and benzoic acids showing no effect. Furthermore p-coumaric, caffeic and chlorogenic acids all enhanced rooting when supplied alone and increased the effect of IAA and IBA applications. In cuttings of *Eranthemum,* on the other hand, salicylic and p-hydroxybenzoic acids enhanced the rooting response to auxin applications and ortho or para positioning of the hydroxy group of benzoic acids had no obvious effect on the rooting response [70]. However, none of the phenolics tested promoted root formation when supplied alone. Evidently mono-phenols are effective in some circumstances. This could be taken as evidence that phenolic compounds need not act specifically via IAA oxidase. However, considerable caution should be exercised in attempting to compare data from different plant materials since the effect of any supplied phenolic is likely to depend on the endogenous phenolics present, as well as phenolic metabolism. The latter point is emphasized by the fact that synergistic interactions between auxin and catechol or pyrogallol have been confirmed in cuttings taken from material grown in the light but not the dark [18]. It is possible, therefore, that synergism between supplied phenolics and auxin may only occur when cuttings are rooted in the light or taken from stock material raised in the light.

3.3. Endogenous phenolics

An increase in total phenol content within the region of root regeneration has been demonstrated in cuttings of several species. In *Phaseolus mungo* such an increase was detected 16 h after cuttings were taken [46]. During this period, treatment of cuttings with IAA and/or sucrose had no major effect on total phenol content. Presumably the accumulation was in response to injury. Similarly, total phenolic content of cuttings of *Phaseolus aureus* increased 45% during the first 24 hours of rooting [18]. Thereafter it remained constant until primordia became evident 72 h later. Although there has been little detailed study of the identity and metabolism of phenolic compounds during the early stages of root formation, a correlation between the extent of rooting and the concentration of diphenols has been demonstrated in cuttings of *Impatiens* seedlings [33]. Increased rooting is associated with increased content of anthocyanin-3-glucoside and 2-hydroxy-1,4-dihydroxynaphthalene-4-glucoside, the two major inhibitors of IAA oxidase found in these cuttings. Enhanced rooting of these cuttings is also associated with an increase in auxin-like material, as evidenced by a coleoptile bioassay. Such work clearly needs to be confirmed and extended, preferably using the more advanced techniques currently available. The diminishing rooting response associated with increasing time after decapitation of hypocotyls of mung bean also correlates with a decreasing content of phenolics [71], although this does not necessarily imply a causal relationship. It should be mentioned, however, that rutin, which has been shown to enhance root formation [47], has been identified in both hypocotyl and leaves of this material [72].

The activity of polyphenol oxidase increases in the early stages of root regeneration [46, 73]. Indeed, from work on *Impatiens* it is suggested that sites for potential primordium formation may be detected by histological study of polyphenol oxidase activity [74]. Such work, however, has been very limited. Nevertheless, given its implication in the synthesis of ortho-hydroxylated compounds, increased activity of this enzyme, as a result of injury, could help to ensure increased auxin levels during the early stages of root formation. More detailed investigations of polyphenol oxidase activity in relation to root formation are therefore desirable.

3.4. Summary

Many diverse compounds, which are generally thought to influence auxin levels, have been shown to influence root formation. Phenolic compounds are generally regarded as being of importance in this context. The evidence on which this is based is largely indirect, although there is limited supporting evidence from correlative studies of phenolic levels and rooting. Despite some conflicting results from physiological investigations, the possibility that *o*-dihydroxy compounds may influence root formation via an inhibitory influence on IAA oxidase is widely acknowledged.

4. Nutritional factors

The importance of nutrients in adventitious root development is emphasized by several lines of investigation. Reference may be made to some of these in previous articles (e.g. [4, 10, 18, 41]). What is unclear, however, is the extent to which most nutrients, endogenous or supplied, may specifically influence the phases of induction and initiation. For this reason the discussion which follows is restricted to a consideration of soluble carbohydrate and borate.

4.1. Carbohydrates

Numerous studies have considered the possibility, first postulated by Kraus and Kraybill [75], that a relationship exists between carbohydrate content and rooting of cuttings. Many of these have demonstrated that sugars, including sucrose, glucose, fructose, ribose, deoxyribose, myo-inositol and dextrose, can enhance the rooting response of cuttings in the presence, or absence, of supplied auxin (e.g. [14, 18, 52, 76]). Such observations, however, cannot be taken as evidence of a direct involvement of carbohydrate in root initiation. This is particularly so in cuttings taken from young seedlings where, for example, soluble sugar content may be up to 20% of the dry weight of the hypocotyl [77]. Similarly correlation of rooting response with irradiance is, in itself, inadequate evidence of an involvement of any carbohydrate in adventitious root development. Alternative interpretation of such influence of irradiance is clearly possible [78].

The most detailed studies to date, conducted on cuttings of pea and including two mutants deficient in Photosystem II activity and chlorophyll, have failed to demonstrate any obvious relationship between irradiance, carbohydrate content and rooting response [79, 80, 81]. Less detailed studies on mung bean similarly failed to establish such a relationship [37]. However, cuttings do generally accumulate soluble carbohydrates at their bases prior to formation of root primordia and, even when root initiation is not obviously limited by local availability of sugars, enhanced transport of sucrose from leaves to the rooting region

Table 1. The influence of auxin and sucrose on the rooting response of mung bean cuttings.

Treatment 0–24 h	24 h–48 h	Intact	Leafless
IAA	Water	25.5 ± 4.3	16.5 ± 4.5
IAA	Sucrose	35.5 ± 5.3	20.0 ± 3.5
Water	Sucrose	5.3 ± 4.1	2.0 ± 0.8
Water	IAA	22.5 ± 7.4	10.8 ± 2.5
Sucrose	Water	3.0 ± 1.3	1.3 ± 0.9
Sucrose	IAA	35.8 ± 9.9	10.3 ± 4.0

IAA (5×10^{-4} M) and sucrose (1%) were supplied to intact or leafless cuttings of light-grown seedlings. After 48 h leaves were removed from intact cuttings and all cuttings transferred to boric acid (10 μg ml^{-1}) for a further 5 d. Seedlings were raised, and cuttings treated, under continuous fluorescent irradiance. Data cited are mean number of roots per cutting ± 95% confidence limits.

has been demonstrated during the early stages of initiation [37, 77]. The possibility arises, therefore, that adventitious root formation is dependent upon sugar transported via the stele, irrespective of the sugar content of the surrounding tissues. Auxin application readily enhances transport of sucrose from leaves to the base of the cutting. This it does at least partly, by its influence on the sugar content of the leaves [37, 77].

Although the role of carbohydrates in adventitious root development has yet to be defined, under some circumstances sugar availability can limit adventitious root development in its early stages. Such an early influence of sugars is hardly surprising and is most evident in leafless cuttings from etiolated material (e.g. [18]). In intact cuttings of light-grown material the stimulatory effect of supplied sugars is particularly marked when cuttings are also treated with exogenous auxin. In such circumstances the relationship between supplied carbohydrate and auxin has not been investigated with respect to the possible transport of root-inducing factors out of the leaves. The influence of this carbohydrate could be largely to enhance transport of the supplied auxin to the site of regeneration. The data presented in Table 1 may be interpreted in this way. They indicate that the rooting response is controlled by auxin but that sugar availability is not limiting, a point previously made after carbohydrate analysis of similar cuttings [37]. Rooting of both intact and leafless cuttings is enhanced by supply of IAA whereas a stimulatory effect of sucrose is evident only when intact cuttings are also supplied with auxin. Supplied sugar may therefore enhance the loading/transport of auxin out of leaves in the appropriate transport system. Such interpretation is consistent with the proposition that the rooting response is controlled primarily by the amount of auxin in the rooting region. Nevertheless, the stimulatory effect of sugar on the rooting response of etiolated, leafless cuttings emphasizes the importance of an adequate supply of carbohydrate within the region of regeneration.

4.2. Boron – an essential micronutrient

An essential role for boron in the development of adventitious roots in stem cuttings was first described by Hemberg [82] for cuttings of *Phaseolus vulgaris* L. This was subsequently confirmed by Gorter [83] and both workers concluded that whereas roots were initiated in response to auxin, boron was essential for growth of primordia. Despite these, and subsequent reports, the involvement of boron in adventitious root development has been largely ignored by most investigators. This may be because cuttings contain sufficient boron for some root development at least, or rooting is facilitated by contaminating boron from water source, glassware, chromatography paper or other chemicals supplied to cuttings [14, 20]. Indeed the beneficial effect of tap, as opposed to deionised, water on rooting is related to its boron content.

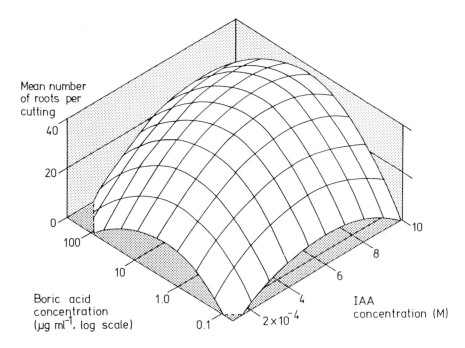

Fig. 5. A fitted response curve depicting the influence of supplied borate and IAA on adventitious root formation. Redrawn, with permission, from [94].

Stem cuttings, from light-grown seedlings of *Phaseolus aureus* Roxb., were treated with IAA for 24 h then transferred to boric acid for a further 6 d. They were held under continuous fluorescent irradiance.

Stem cuttings of light-grown mung bean seedling (*Phaseolus aureus* Roxb. cv. Berkin), which are widely used in studies on adventitious root development, regenerate no visible roots, nor root primordia, when placed in deionised water. Several roots per cutting, however, develop in response to boric acid. Treatment of cuttings with auxin leads to the development of large numbers of primordia and roots only when boron is also supplied [84]. Therefore events initiated by auxin *only* lead to formation of root primordia in the presence of sufficient boron. However, varying sensitivity of mung bean cuttings to supplied boron is evidenced by the results of other workers (e.g. [15, 85, 86].

In mung bean the requirement for supplied boron in primordium development is induced by growing stock seedlings in the light. Cuttings taken from seedlings raised in darkness develop primordia without an exogenous supply of boron, although subsequent root growth may be limited [14, 84]. The boron requirement for root development in intact plants grown in solution culture may also be diminished under conditions of low light intensity [87], while long, as opposed to short, daylengths have been reported to enhance symptoms of boron deficiency [88, 89, 90]. However, the relationship between light treatment and the boron requirement of plant tissues is not understood. Stimulatory effects of boron on rooting have been reported for cuttings of other non-woody stock materials (e.g. [76]) as well as some woody cuttings (e.g. [91, 92, 93]) although whether, or not, the boron requirement of these cuttings is dependent upon the light regime under which stock plants are raised has not been investigated.

Where boron is essential for development of the root primordium it may be provided during growth of stock plants or supplied to the cuttings themselves [94]. Despite this, such boron is not required until between 24 and 48 h after treatment of cuttings with auxin, at which time auxin-stimulated cell division has already begun. It appears, therefore, that either auxin itself is responsible for initiating early cell divisions or it does so in conjunction with a limited amount of boron already available within the cutting. It is evident, however, that boron is essential for the sustained cell division and organization necessary for formation of the root primordium.

Boron influences the *number* of primordia developing whether rooting is in response to auxin treatment or due solely to endogenous factors [12, 84, 94]. Fig. 5. shows the interaction between supplied boron and IAA. The optimum rooting response to either of these variables is dependent on the concentration of the other, such that rooting is enhanced by high levels of auxin only when the boron concentration is also relatively high. Whereas a high concentration of boric acid diminishes root *number*, even if supplied for a relatively short period immediately following treatment of cuttings with auxin [94], it enhances root *growth*. These, and other observations [12, 94], have led to the proposal that boron has a promotive influence on the oxidative destruction of auxin. Since the high concentrations of auxin which initiate the rooting response would be highly inhibitory to later root growth [14], such regulation of endogenous auxin concentration during root development may involve borate (see Section 6).

4.3. Summary

Transport of sugars to the site of regeneration is evident during the early stages of root regeneration. This is so even where root initiation is unlikely to be limited by the local availability of carbohydrate in the region of root formation. The stimulatory effect of supplied sugars on root formation is particularly evident when auxin is also supplied to leafy cuttings. This raises the possibility that supplied sugar enhances the loading and transport of auxin such that it more readily reaches potential sites of initiation.

An adequate supply of boron is essential for the formation of the root primordium as well as its subsequent growth. Cuttings taken from stock material grown in the light are more dependent on supplied boron for development of the primordium than are those from stock material raised in darkness. Borate may be necessary for regulation of auxin levels.

5. 'Phytohormones' other than auxin

5.1. Cytokinins

Adventitious root formation is generally inhibited by apical or basal application of kinetin or 6-benzyladenine (BA) to stem cuttings. Such inhibitory effects have been reported for *Phaseolus aureus* [18], *Pisum sativum* [95] and *Phaseolus vulgaris* [96]. However, the response of pea cuttings is dependent upon the type of cutting used and the time when such application is made [95]. High concentrations of 6-benzyladenine (BA), applied to decapitated and disbudded cuttings during the early stages of initiation, are particularly inhibitory to rooting. On the other hand, similar application of BA to cuttings which are only decapitated leads to enhanced rooting. Since cytokinin application does not replace the influence of the apex it is tempting to explain such results in terms of relative concentrations of auxin and cytokinin in the rooting region. If buds supply auxin to the rooting region their removal could lead to an unfavourably high ratio of cytokinin: auxin when BA is supplied. Such interpretation is consistent with work on callus culture which indicates that a high auxin:cytokinin ratio favours root initiation [97]. Whereas kinetin inhibits root formation it induces development of callus on petioles of *Phaseolus vulgaris*. Transfer of petioles, which have formed callus, to a solution without kinetin results in the emergence of roots from the callus itself [96]. In epicotyls, kinetin inhibits development of hyperplastic tissue normally induced by auxin and ensures production of regular and functional xylem vessels. In this respect it has been suggested that kinetin effectively changes the stem from herbaceous to woody [98]. Whether this is the general basis of the inhibitory effects of cytokinins is unknown.

Despite the inhibitory effects mentioned above, it appears that some cytokinin may be necessary for root formation since addition of small amounts of cytokinin is necessary for formation of roots in tissue culture. In hypocotyl tissue of sunflower rooting is also enhanced by supply of low concentrations of BA or zeatin, although high concentrations are inhibitory [99]. Furthermore, root initiation in cuttings of *Coleus* is inhibited by 3-methyl-7-(3-methylbutylamino) pyrazolo [4,3-d] pyrimidine, a cytokinin antagonist [100]. It seems likely, therefore, that the requirement for root initiation is for a low cytokinin level *and* favourable auxin:cytokinin ratio.

5.2. Gibberellins

Gibberellic acid (GA_3) has been widely reported to inhibit adventitious root formation in cuttings of a variety of species [18, 99, 101, 102, 103, 104]. Such inhibition is particularly evident when gibberellin is supplied before, or soon after, cuttings are made [102]. This coupled to the greater efficacy of basal, as opposed to apical, application has led to the suggestion that gibberellin in the stem base inhibits cell divisions associated with the phase of initiation. Experiments with gibberellin antagonists or inhibitors of gibberellin synthesis have given results consistent with such a suggestion. For example EL531 (α-cyclopropyl-α-(4-methoxyphenyl)-5-pyrimidine methanol), CCC (2-chloroethyl-trimethylammonium chloride) and AMO-1618 (2-isopropyl-4-dimethylamino-5-methylphenyl-1-piperidinecarboxylate methyl chloride) have all been shown to enhance adventitious root formation [99, 105, 106].

Root formation may also be stimulated by application of GA_3 [78, 107, 108]. Such stimulation is dependent upon conditions of irradiance during growth of the stock plants from which cuttings are taken. In cuttings of pea, for example, stimulated rooting in response to gibberellin application is dependent upon growing stock pea seedlings under relatively low irradiance, which in itself is beneficial to root formation [78, 107]. Such an effect of gibberellin may be indirect, a suggestion supported by other observations. In leaf cuttings, root formation is stimulated by treatment of primary leaves with GA_3 prior to excision of the cuttings, but application of TIBA below the pulvinus inhibits the effect of the GA_3 [109]. Given also that tryptophan enhances the effect of GA_3 in such cuttings, the role of gibberellin may be to enhance the supply of auxin from the leaves to the site of regeneration.

5.3. Ethylene

It has been suggested that auxin-stimulated rooting may result from stimulated ethylene production by the treated tissues [110]. There exist, however, numerous contradictory reports concerning ethylene production and the influence of ethylene on adventitious root development. Various workers have shown ethylene or ethephon to stimulate [111, 112, 113], to inhibit [114, 115], or to have no influence [116, 117, 118] on the rooting response of stem cuttings. Furthermore, such contradictory findings cannot always be ascribed to the use of different plant materials, as evidenced by the variable results obtained with cuttings of mung bean (e.g. [111, 115, 118]).

There appears to be no correlation between total ethylene production, or the concentration of ethylene in the tissue, and the number of roots formed in response to IAA and IBA [16, 114, 118]. Moreover, in the absence of exogenous auxin numerous roots may develop with no detectable [16], or relatively little [116], ethylene production. However, since adventitious root formation is enhanced where high concentrations of applied auxin are associated with relatively low rates of ethylene production, auxin and ethylene may be antagonists in the early stages of root formation [114]. This suggestion is supported by the observation that inhibitors of ethylene synthesis or action stimulate rooting of sunflower cuttings only when supplied during the early stages of regeneration [99]. Inhibitors such as aminoethoxyvinyl glycine (AVG), supplied for relatively long periods of time, however, generally reduce the number of visible roots developing [111]. It appears, therefore, that ethylene may be required for adventitious root development, albeit possibly between the early events of initiation and emergence of the roots. Evidence from cuttings of *Phaseolus vulgaris* suggests a crucial role for ethylene in the early cellular organization of the root primordium [119]. Auxin induces cell division, but in the absence of ethylene there appears to be no order to the divisions and no primordia appear. Ethylene supplied in the absence of exogenous auxin initiates little, if any, cell division but leads to swelling and separation of cortical and pith cells. It seems that cell division initiated by auxin and subsequent cell breakdown caused by ethylene are both necessary for root formation. The possibility that endogenous ethylene is inhibitory to the early stages of regeneration but stimulatory or essential to a later stage of development is also suggested by work on organogenesis in cultured tobacco cells [120].

5.4. Abscisic acid

Leafy stem cuttings vary considerably in their response to basal application of abscisic acid (ABA). For example those of *Phaseolus aureus* show enhanced rooting when supplied with relatively high concentrations [121, 122] whereas those of *Phaseolus vulgaris* show no response over a range of concentrations [121]. In other species rooting is reportedly inhibited by ABA (for references see [104]). Given the dose-response curves produced for cuttings of pea [123] and runner bean [104] it may be that apparently contradictory results simply reflect the use of limited concentration ranges and differences in sensitivity between different materials.

Stimulatory effects of supplied ABA may be interpreted either in terms of a direct promotion of cell division or via interaction with other growth regulators. Such regulators in themselves could be inhibitory to root initiation and therefore might include gibberellins or cytokinins. Indeed, the inhibitory effect of gibberellin on rooting of mung bean cuttings is significantly reduced by subsequent treatment with ABA, whereas that of kinetin is not [122]. On the other hand, ABA may promote rooting indirectly since lateral bud break may be influenced during the rooting period [104, 123]. Consistent with this suggestion are the observations that, in cuttings of runner bean, ABA is transported to the apex and its action may depend on the presence of illuminated leaves [104].

5.5. Summary

In the absence of data concerning endogenous concentrations and transport into the region of regeneration, it is unclear whether gibberellins, cytokinins, and abscisic acid play a major role in root initiation, at least under natural conditions. The use of putative inhibitors of their synthesis or action suggests that low concentrations of gibberellin and cytokinin are necessary to facilitate auxin-induced rooting. There is limited evidence for production of an inhibitory factor within roots, which could prevent root formation in the intact plant [124]. It has been suggested that this substance could be a gibberellin or cytokinin [10]. The contradictory reports concerning the influence of supplied ethylene or ethephon may be related to the dual role [99] suggested for endogenous ethylene. Thus 'wound-ethylene' may be inhibitory to root development whereas, at a later time, ethylene may be essential for organization of the root primordium. Confirmation of the solitary report that ethylene is essential for this organization is needed.

Given the central role of auxin in the control of root regeneration it is not surprising that interactions between supplied plant growth regulators and auxin have been widely reported. Examples of these are seen in Table 2. Concentrations of GA_3, ABA and kinetin, which, in themselves, are either inhibitory to root formation or without significant effect, actually enhance rooting when supplied with IAA. This underlines the importance of investigating the influence of these regulators on the transport and metabolism of auxin. In this context it is disappointing that no one has followed, and extended, the simple approach of Varga and Humphries [109], whose results suggest that GA_3 influences rooting via an effect on auxin transport out of the leaves.

Table 2. The influence of GA_3, ABA and kinetin on adventitious root formation of mung bean cuttings.

Plant growth regulator	Auxin (3×10^{-4} M)	
	Absent	Present
None	13.2 ± 2.1	9.7 ± 1.0
10^{-8} M GA_3	13.0 ± 1.7	14.6 ± 3.3
10^{-5} M GA_3	5.1 ± 0.9	17.9 ± 2.8
10^{-6} M ABA	13.7 ± 1.5	12.0 ± 2.5
5×10^{-4} M ABA	19.1 ± 2.1	30.1 ± 2.7
10^{-8} M kinetin	14.9 ± 2.2	11.4 ± 2.4
10^{-6} M kinetin	4.9 ± 0.5	16.7 ± 3.8

Stem cuttings were prepared from light grown seedlings. They were treated for 24 h ± IAA then transferred to boric acid (10 μg ml⁻¹) for a further six d. Seedlings were raised, and cuttings treated, under continuous fluorescent irradiance. Data presented are mean root number per cutting ± 95% confidence limits. (Taken from unpublished data of S. Yasmin.)

6. Discussion

To implicate any chemical in the control of rooting it should be ascertained, amongst other things, that the chemical occurs naturally at the site of normal regeneration. Furthermore, the concentration should, at an appropriate time, reach a level which, when also attained by exogenous supply, induces a specific response. The latter has yet to be established for any factor. Consequently, current acceptance of an involvement in root development must rely on alternative, and essentially indirect, evidence. It must be stressed, therefore, that currently of the five recognised groups of plant growth regulators only auxins can be confidently implicated in the initiation of root regeneration. In addition two other factors, carbohydrate and borate, are important in initiation processes, although any specific regulatory role of carbohydrate is obscure.

Although an essential role for auxin in the regeneration of roots seems reasonably well established, it is still unclear whether, or not, auxin is the sole regulatory agent for initiation *per se*. Nevertheless, it is proposed that auxin may be the only initiation factor which must be *transported* into the region of regeneration from the leaves. This, of course, does not preclude the possibility that other essential factors may be present, or synthesized, locally. The widespread efficacy with which supplied auxin induces root formation may be taken as evidence that it is often the sole limiting factor, at least for initiation. Such evidence, however, is only acceptable in view of other information which implicates auxin, such as that concerning basipetal transport and changes in endogenous concentration.

The fact that so many other chemicals, of diverse nature, stimulate rooting raises the important question of whether they act by influencing the availability of auxin at the sites of initiation. It is difficult to envisage a common means by which all such chemicals have this effect. Some, such as GA_3, may influence the amount of endogenous auxin moving from the leaves, others, such as phenolics, may influence the activity of IAA oxidase/peroxidase (e.g. [125]) at the sites of initiation. It has been suggested that injurious treatment of cuttings with non-specific chemicals, such as sulphuric acid and potassium hydroxide, enhances root formation via an indirect inhibition of IAA oxidase [126]. Clearly there is a need to investigate the relationship between some of these various chemicals and auxin metabolism and transport.

Despite the requirement for application of high concentrations of auxin in order to initiate a good rooting response, such concentrations are inhibitory to the essential organization of the primordium and its subsequent growth [12, 14]. Furthermore, levels of both endogenous and supplied auxin decline prior to formation of the primordium [19, 30]. Clearly, regulation of auxin concentration is of fundamental importance during adventitious root development and numerous studies on IAA oxidase/peroxidase activity implicate this enzyme complex in such control. Fig. 6, which is the basis of the discussion which follows, describes three phases of development which culminate in the formation of the root

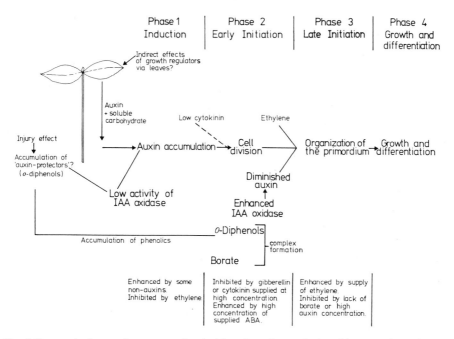

| Phase 1 Induction | Phase 2 Early Initiation | Phase 3 Late Initiation | Phase 4 Growth and differentiation |

Fig. 6. Proposed scheme of events associated with various phases of adventitious root formation.

primordium. It affords a central role to IAA oxidase and its control by an interaction between borate and phenolics. It is necessarily speculative, not least because the extent to which auxin concentration is controlled *in vivo* by IAA oxidase has yet to be resolved.

The first phase of the regeneration process is essentially characterised by lack of cell division at the potential sites of root formation. It has previously been referred to as the inductive or preparatory phase (e.g. [48]). It presumably corresponds to the IAA-less-sensitive phase described for *Vigna angularis* [23], which itself is in agreement with the earlier suggestion of Went [22] that auxin has no specific role in the earliest stage of rhizogenesis. Indeed it appears that this phase of regeneration is induced or enhanced by a variety of chemicals which are not auxins [13, 22, 23]. During this first phase of rhizogenesis auxin accumulates in the region of root formation. Accumulation results from basipetal transport and is probably aided by the relatively low activity of IAA oxidase/peroxidase [44, 45, 50] at least in the early stages. It is tempting to speculate that this low activity is related to an altered phenolic metabolism since auxin protectors show a transient increase following wounding [127] and in stem cuttings increased phenolic content has been observed during this phase [18]. There is also evidence to suggest that in the later stages of this induction phase there are changes in the isoenzyme pattern of peroxidase (e.g. [49, 50]).

The second phase, or early initiation, is characterised by cell division. This cell division may be triggered by the auxin accumulation of the first phase. Several lines of evidence suggest this may be so. Continued supply of auxin at high concentration up to, and just beyond, the earliest divisions results in the production of large numbers of roots [20]. Other evidence also indicates the need for supply of endogenous auxin into the phase of initiation [24]. Furthermore, those hypocotyls containing the greatest amounts of auxin at the time of the earliest divisions eventually produce the largest number of roots. The early divisions of this phase are probably dependent on low endogenous cytokinin levels and are inhibited by supply of gibberellin. Excision of the roots is likely to remove a source of both cytokinin and gibberellin. This, together with accumulation of auxin and soluble carbohydrate, would therefore ensure favourable conditions for these early events of initiation.

Gaspar has suggested that it is the induction phase which is characterised by a decrease in endogenous auxin concentration and has therefore associated high root-forming capacity with low auxin content [48]. This difference of opinion is one of interpretation and merely emphasizes the need for detailed investigations, in which changes in auxin concentration and the precise time when cell division begins are monitored simultaneously. However, his proposition is not entirely at variance with that made here since the essence of the third phase, outlined in Fig. 6, is a reduction in auxin level associated with high peroxidase/IAA oxidase activity. Such a drop in auxin content within the region of root formation has been observed during natural regeneration (see Fig. 2) and is also evident in auxin-treated cuttings [29, 30]. Marked changes in the isoenzyme patterns of peroxidase and IAA oxidase seem characteristic of this part of the induction phase and it has previously been suggested that primordium formation occurs after maximum peroxidase activity, when basic iso-peroxidases, which destroy auxin, are declining in activity [47, 48]. The final phase of adventitious root development, which has not been discussed here, would involve subsequent growth and differentiation of the primordium.

Work concerning the boron requirement of mung bean cuttings [12, 94] suggests that it may have a role in the control of auxin levels. In the absence of boron phases 1 and 2 proceed but no primordia appear [12, 14, 20, 94]. Several observations indicate that boron may control the effective auxin concentrations at sites of initiation. The interaction shown in Fig. 5 shows that rooting is enhanced by high concentrations of supplied auxin only when the concentration of supplied boron is also relatively high. In addition, cuttings taken from stock plants supplied with boron show a smaller rooting response to supplied auxin than do those taken from stock plants grown in its absence [12]. Furthermore, high concentrations of boron diminish the number of roots which regenerate, particularly when supplied immediately after auxin treatment. Nevertheless such treatments enhance root growth. Finally, boron has been shown to enhance the activity of IAA oxidase (e.g. [128, 129]). If the primary role of boron in plants is to enhance metabolism of IAA such observations would enforce the view propounded here that root initiation is related to high auxin levels, but low auxin concentration is essential for organization and growth of the primordium. Interestingly, borate readily complexes with compounds containing *cis*-hydroxyl groups [130], amongst which are *o*-diphenols. Given that *o*-diphenols inhibit IAA oxidase activity [69], the borate required for organization of the primordium could complex with such phenolics and thereby enhance IAA oxidase activity [94]. The diminished auxin level resulting from this, together with endogenous ethylene, may be two of the prerequisites for organization of the root primordium.

It has been previously suggested that increased activity of IAA oxidase during root initiation is at variance with the proposition that polyphenol oxidase generates diphenols which inhibit IAA oxidase activity [73]. The proposals made here, concerning auxin levels during early and late stages of initiation, would suggest that the relative importance of these enzymes may be temporally separated and that borate may well be important in controlling the ultimate effects of these two enzymes.

7. Suggestions for future studies

In 1965 Dore [4] stressed that little direct work had been done on relations between regeneration and auxins, whereas considerable efforts had been made to prove the existence of other, presumably specific, root-forming substances. This is still the case. Detailed studies on the transport of auxin to the sites of root initiation and its metabolism at those sites are urgently needed. In such work use of modern physico-chemical methods for identification and estimation of auxins will be essential. Given that the relative concentrations of various growth regulators are important in controlling morphogenesis, these studies on auxin must be made alongside investigations concerning the nature, amounts and metabolism of other natural regulators. In such work the possible need to identify novel growth regulators should be borne in mind, a point emphasized by recent work suggesting that oligosaccharins may regulate various aspects of morphogenesis, including root formation [131]. The relatively ill-defined, but seemingly important, relationships between auxin, phenolics, boron and IAA oxidase/peroxidase need further study. Basic information concerning phenolic metabolism in the region of regeneration will be a pre-requisite for such work. Of relevance in this context could be the 'auxin-protectors'. These compounds, the active sites of which are probably *o*-dihydroxyphenols, show a transient increase after injury and exhibit several other features which suggest that they might act as rooting co-factors (e.g. [127]).

Previous work has been largely concerned with the influence of supplied chemicals on the rooting response as evidenced solely by the number of external roots which are so readily recorded. Little, if anything, will be gained by pursuing such studies further. Future work should concentrate on those chemicals shown to occur naturally in regenerating tissues and endeavour to establish their role in *normal* regeneration, rather than study the gross 'pharmacological' effects. The possible roles of growth regulators, other than auxin, would surely be better studied in this way. In addition more attention should be focussed on establishing precisely when in root development such chemicals play regulatory roles.

Current speculation concerning the endogenous control of root formation is necessarily based on gross changes determined within the region of root regeneration. However, those cells from which primordia develop are but a small fraction of the total cell number in this region and, as yet, they are not readily identifiable. It is of considerable importance, therefore, that future studies should include the coordinated use of biochemical, histochemical and anatomical techniques.

8. References

1. Sachs, R.M. (1880, 1882) Stoff und Form der Pflanzenorgane I & II. Arbeiten aus dem botanischen Institut des Wurzburg, 2, 452–488; 689–718.
2. Bouillenne, R. & Went, F. (1933) Recherches experimentales sur la neoformation des racines dans les plantules et les boutures des plantes superieures. Annales du Jardin botanique de Buitenzorg, 43, 25–202.
3. Went, F.W. (1934) A test method for rhizocaline, the root-forming substance. Proceedings Koninklijke Nederlandse Akademie van Wetenschappen, 37, 445–455.
4. Dore, J. (1965) Physiology of regeneration in cormophytes. In Encyclopedia of Plant Physiology (ed. W. Ruhland), Vol. XV/2, pp. 1–91. Springer-Verlag, Berlin.
5. Thimann, K.V. & Went, F.W. (1934) On the chemical nature of the root-forming hormone. Proceedings Koninklijke Nederlandse Akadamie van Wetenschappen, 37, 456–459.
6. Zimmerman, P.W. & Wilcoxon, F. (1935) Several chemical growth substances which cause initiation of roots and other responses in plants. Contributions from the Boyce Thompson Institute for Plant Research, 7, 209–229.
7. Cooper, W.C. (1935) Hormones in relation to root formation on stem cuttings. Plant Physiology, 10, 789–794.
8. Thimann, K.V. & Koepfli, J.B. (1935) Identity of the growth-promoting and root-forming substances of plants. Nature, 135, 101–102.
9. Bouillenne, R. & Bouillenne-Walrand, M. (1955) Auxines Bouturage. 14th International Horticultural Congress Schweninger I, 231–238.
10. Batten, D.J. & Goodwin, P.B. (1978) Phytohormones and the induction of adventitious roots. In Phytohormones and Related Compounds: A Comprehensive Treatise (Ed. D.S. Letham, P.B. Goodwin and T.J.V. Higgins). Vol. II, pp. 137–173. Elsevier/North Holland, Amsterdam.
11. Haissig, B.E. (1974) Influences of auxins and auxin synergists on adventitious root primordium initiation and development. New Zealand Journal of Forestry Science, 2, 311–323.
12. Jarvis, B.C., Ali, A.H.N. & Shaheed, A.I. (1983) Auxin and boron in relation to the rooting response and ageing of mung bean cuttings. The New Phytologist, 95, 509–518.
13. Shibaoka, H. (1971) Effects of indole-acetic acid, p-chlorophenoxyisobutyric acid and 2-4-6 trichlorophenoxyacetic acids on three phases of rooting *Azukia* cuttings. Plant and Cell Physiology, 12, 193–200.
14. Middleton, W. (1977) Root development in cuttings of *Phaseolus aureus* Roxb. Ph.D. thesis. University of Sheffield, UK.
15. Jackson, M.B. & Harney, P.M. (1970) Rooting cofactors, indoleacetic acid and adventitious

root initiation in mung bean cuttings, (*Phaseolus aureus*). Canadian Journal of Botany, 48, 943–946.

16. Geneve, R.L. & Heuser, C.W. (1982) The effect of IAA, IBA, NAA, and 2-4-D on root promotion and ethylene evolution in *Vigna radiata* cuttings. Journal of the American Society for Horticultural Science, 107, 202–205.

17. Kawase, M. & Matsui, H. (1980) Role of auxin in root primordium formation in etiolated 'red kidney' bean stems. Journal of the American Society for Horticultural Science, 105, 898–902.

18. Fernqvist, I. (1966) Studies on factors in adventitious root formation. Lantbruskhögskolans Annaler, 32, 109–244.

19. Andraea, W.A. (1967) Uptake and metabolism of indoleacetic acid, naphthalenacetic acid, and 2,4-dichlorophenoxyacetic acid by pea root segments in relation to growth inhibition during and after auxin application. Canadian Journal of Botany, 45, 737–753.

20. Jarvis, B.C., et al. Unpublished data.

21. Trewavas, A. (1980) An auxin induces the appearance of auxin-binding activity in artichoke tubers. Phytochemistry, 19, 1303–1308.

22. Went, F.W. (1939) The dual effect of auxin on root formation. American Journal of Botany, 26, 24–29.

23. Mitsuhashi, M., Shibaoka, H. & Shimokoriyama, M. (1969) Morphological and physiological characterization of IAA-less-sensitive and IAA-sensitive phases in rooting of *Azukia* cuttings. Plant and Cell Physiology, 10, 867–874.

24. Mohammed, S. & Eriksen, E.N. (1974) Root formation in pea cuttings. IV. Further studies on the influence of indole-3-acetic acid at different developmental stages. Physiologia Plantarum, 32, 94–96.

25. Haissig, B.E. (1970) Influence of indole-3-acetic acid on adventitious root primordia of brittle willow. Planta, 95, 27–35.

26. Haissig, B.E. (1972) Meristematic activity during adventitious root primordium development. Plant Physiology, 49, 886–892.

27. Eriksen, E.N. & Mohammed, S. (1974) Root formation in pea cuttings. II. The influence of indole-3-acetic acid at different developmental stages. Physiologia Plantarum, 30, 158–162.

28. Kantharaj, G.R., Mahadevan, S. & Padmanaban, G. (1979) Early biochemical events during adventitious root initiation in the hypocotyl of *Phaseolus vulgaris*. Phytochemistry, 18, 383–387.

29. Brunner, H. (1978) Influence of various growth substances and metabolic inhibitors on root regenerating tissue of *Phaseolus vulgaris* L. Changes in the contents of growth substances and in peroxidase and IAA oxidase activities. Zeitschrift für Pflanzenphysiologie, 88, 13–23.

30. Kefeli, V.I. (1978) Principles of analysis of phytohormones and natural growth inhibitors. In Natural Plant Growth Inhibitors and Phytohormones. Dr W. Junk, Dordrecht.

31. Hemberg, T. (1954) The relation between the occurrence of auxin and the rooting of hypocotyls in *Phaseolus vulgaris* L. Physiologia Plantarum, 7, 323–331.

32. Blahova, M. (1969) Changes in the level of endogenous gibberellins and auxins preceding the formation of adventitious roots in isolated epicotyls of pea plants. Flora, 160, 493–499.

33. Bastin, M. (1966) Root initiation, auxin level and biosynthesis of phenolic compounds. Photochemistry and Photobiology, 5, 423–429.

34. Weigel, U., Horn, W. & Hock, B. (1984) Endogenous auxin levels in terminal stem cuttings of *Chrysanthemum morifolium* during adventitious rooting. Physiologia Plantarum, 61, 422–428.

35. Went, F.W. (1938) Specific factors other than auxin affecting growth and root formation. Plant Physiology, 13, 55–80.

36. Katsumi, M., Chiba, Y. & Fukuyama, M. (1969) The roles of the cotyledons and auxin in the adventitious root formation of hypocotyl cuttings of light grown cucumber seedlings. Physiologia Plantarum, 22, 993–1000.

37. Middleton, W., Jarvis, B.C. & Booth, A. (1980) The role of leaves in auxin and boron dependent rooting of stem cuttings of *Phaseolus aureus* Roxb. The New Phytologist, 84, 251–259.

218

38. Eriksen, E.N. (1973) Root formation in pea cuttings. I. Effects of decapitation and disbudding at different developmental stages. Physiologia Plantarum, 28, 503–506.
39. Fabijan, D., Yeung, E., Mukherjee, I. & Reid, D.M. (1981) Adventitious rooting in hypocotyls of sunflower seedlings (*Helianthus annus*). I. Correlative influence and developmental sequence. Physiologia Plantarum, 53, 578–588.
40. Bastin, M. (1966) Metabolisme auxinique et rhizogenese: III. Interactions entre l'auxine et quelques substances phenoliques présentes dans les boutures. In Les Phytohormones et l'organogenese, pp. 123–140. Congres et Colloques de l'Université de Liège.
41. Hess, C.E. (1969) Internal and external factors regulating root initiation. In Root Growth (ed. W.J. Whittington) pp. 42–64. Butterworths, London.
42. Audus, L.J. (1963) Plant Growth Substances, p. 150. Leonard Hill Books, London.
43. Gaspar, Th., Penel, C., Thorpe, T. & Greppin, H. (1982) In Peroxidases 1970–1980. A Survey of their Biochemical and Physiological Roles in Higher Plants. Université de Genève, Centre de Botanique.
44. Chibbar, R.N., Gurumurti, K. & Nanda, K.K. (1979) Changes in IAA oxidase activity in rooting hypocotyl cuttings of *Phaseolus mungo* L. Experientia, 35, 202–203.
45. Foong, T.W. & Barnes, M.F. (1981) Rooting 'cofactors' in Rhododendron: The fractionation and activity of components from an easy-to-root and a difficult-to-root variety. Biochemie und Physiologie der Pflanzen, 176, 507–523.
46. Bhattacharya, N.C. & Kumar, A. (1980) Physiological and biochemical studies associated with adventitious root formation in *Phaseolus mungo* L. in relation to auxin-phenol synergism. Biochemie und Physiologie der Pflanzen, 175, 421–435.
47. Moncousin, Ch. & Gaspar, Th. (1983) Peroxidase as a marker for rooting improvement of *Cynara scolymus* L. cultured *in vitro*. Biochemie und Physiologie der Pflanzen, 178, 263–271.
48. Gaspar, Th. (1981) Rooting and flowering, two antagonistic phenomena from a hormonal point of view. In Aspects and Prospects of Plant Growth Regulators (ed. B. Jeffcoat), pp. 39–49. British Plant Growth Regulator Group, Wantage.
49. Gurumurti, K. & Nanda, K.K. (1974) Changes in peroxidase isoenzymes of *Phaseolus mungo* hypocotyl cuttings during rooting. Phytochemistry, 13, 1089–1093.
50. Chandra, G.R., Gregory, L.E. & Worley, J.F. (1971) Studies on mung bean hypocotyl. Plant and Cell Physiology, 12, 317–324.
51. Bandurski, R.S. & Nonhebel, H. (1984) Auxins. In Advanced Plant Physiology (ed M.B. Wilkins), pp. 1–20. Pitman, London.
52. Jarvis, B.C. & Booth, A.B. (1981) Influence of indole-butyric acid, boron, myo-inositol, vitamin D_2 and seedling age on adventitious root development in cuttings of *Phaseolus aureus*. Physiologia Plantarum, 53, 213–218.
53. Hess, C.E. (1962) A physiological analysis of root initiation in easy and difficult to root cuttings. XVIth International Horticultural Congress, 4, 375–387.
54. Hess, C.E. (1964) Naturally occurring substances which stimulate root initiation. In Régulateurs Naturels de la Croissance Végétale (ed. J.P. Nitsch) pp. 517–527. C.N.R.S., Paris.
55. Van Raalte, M.H. (1954) On the synergism of indole and indole acetic acid in root production. Annales Bogarienses, 1, 167.
56. Gorter, C.J. (1958) Synergism of indole and indole-3-acetic acid in the root production of *Phaseolus* cuttings. Physiologia Plantarum, 11, 1–9.
57. Gorter, C.J. (1962) Further experiments on auxin synergists. Physiologia Plantarum, 15, 88–95.
58. Pilet, P-E. (1958) Action de l'indole sur la destruction des auxines en relation avec la senescence cellulaire. Comptes Rendus, 246, 1896–1958.
59. Gorter, C.J. (1969) Auxin synergists in the rooting of cuttings. Physiologia Plantarum, 22, 497–502.
60. Kuhnle, J.A., Corse, J. & Chan, B.G. (1975) Promotion of rooting of mung bean cuttings by dihydroasparagusic acid synergistic interaction with indole acetic acid. Biochemie und Phys-

iologie der Pflanzen, 167, 563–556.

61. Mitsuhashi-Kato, M. & Shibaoka, H. (1981) Effects of actinomycin D and 2.4,dinitrophenol on the development of root primordia in *Azuki* bean stem cuttings. Plant and Cell Physiology, 22, 1431–1436.

62. Gurumurti, K., Chibbar, R.N. & Nanda, K.K. (1974) Evidence for the mediation of indole-3-acetic acid effects through its oxidation products. Experientia, 30, 997–998.

63. Soekarjo, R. (1966) On the formation of adventitious roots in cuttings of Coleus in relation to the effect of indole acetic acid on the epinastic curvature of isolated petioles. Acta Botanica Neerlandica, 14, 373–399.

64. Hess, C.E. (1957) A physiological analysis of rooting in cuttings of juvenile and mature *Hedera helix* L. Ph.D. Thesis, Cornell University.

65. Basu, R.N., Ghosh, B. & Sen, P.K. (1968) Naturally occurring rooting factors in mango (*Mangifera indica* L.). Indian Agriculturalist, 12, 194–196.

66. Kawase, M. (1964) Centrifugation, rhizocaline and rooting in *Salix alba* L. Physiologia Plantarum, 17, 855–965.

67. Fadl, M.S. & Hartman, H.T. (1967) Relationship between seasonal changes in endogenous promoters and inhibitors in pear buds and cutting bases and the rooting of pear hardwood cuttings. Proceedings of the American Society for Horticultural Science, 91, 96–112.

68. Roy, B.N., Roychoudhury, N., Bose, T.K. & Basu, R.N. (1972) Endogenous phenolic compounds as regulators of rooting in cuttings. Phyton, 30, 147–151.

69. Zenk, M.H. & Muller, G. (1963) In vivo destruction of exogenously applied indolyl-3-acetic acid as influenced by naturally occurring phenolic acids. Nature, 200, 761–763.

70. Basu, R.N., Bose, T.K., Roy, B.N. & Mukhopadhyay, A. (1969) Auxin synergists in rooting of cuttings. Physiologia Plantarum, 22, 649–652.

71. Heuser, C.W. & Hess, C.E. (1972) Isolation of three lipid root-initiating substances from juvenile *Hedera helix* shoot tissue. Journal of the American Society of Horticultural Science, 97, 571–574.

72. Duke, S.O. & Vaughn, K.C. (1982) Lack of involvement of polyphenol oxidase in ortho-hydroxylation of phenolic compounds in mung bean seedlings. Physiologia Plantarum, 54, 381–385.

73. Frenkel, C. & Hess, C.E. (1973) Isozymic changes in relation to root initiation in mung bean. Canadian Journal of Botany, 52, 295–297.

74. Kaminski, C. (1959) Recherches sur les phenoloxydases dans les hypocotyles de *Impatiens balsamina* L. Bulletin de l'Academie royal de Belgique. Classe des Sciences, 65, 154–168, 299–315.

75. Kraus, E.J. & Kraybill, H.R. (1918) Vegetation and reproduction with special reference to the tomato. Oregon Agricultural College Experimental Station Bulletin, p. 149.

76. Eliasson, L. (1978) Effects of nutrients and light on growth and root formation in *Pisum sativum* cuttings. Physiologia Plantarum, 43, 13–18.

77. Altman, A. & Wareing, P.F. (1975) The effect of IAA on sugar accumulation and basipetal transport of ^{14}C labelled assimilates in relation to root formation in *Phaseolus vulgaris* cuttings. Physiologia Plantarum, 33, 30–38.

78. Hansen, J. (1976) Adventitious root formation induced by gibberellic acid and regulated by the irradiance to the stock plants. Physiologia Plantarum, 36, 77–81.

79. Veirskov, B., Andersen, A.S. & Eriksen, E.N. (1982) Dynamics of extractable carbohydrates in *Pisum sativum*. I. Carbohydrate and nitrogen content in pea plants and cuttings grown at two different irradiances. Physiologia Plantarum, 55, 167–173.

80. Veirskov, B., Andersen, A.S., Stummann, B.M. & Henningsen, K.W. (1982) Dynamics of extractable carbohydrates in *Pisum sativum*. II. Carbohydrate content and photosynthesis of pea cuttings in relation to irradiance and stock plant temperature and genotype. Physiologia Plantarum, 55, 174–178.

81. Veirskov, B. & Andersen, A.S. (1982) Dynamics of extractable carbohydrates in *Pisum sativum* III. The effect of IAA and temperature on content and translocation of carbohydrates in pea cuttings during rooting. Physiologia Plantarum, 56, 179–182.
82. Hemberg, T. (1951) Rooting experiments with hypocotyls of *Phaseolus vulgaris* L. Physiologia Plantarum, 4, 358–369.
83. Gorter, C.J. (1958) Synergism of indole and indole-3-acetic acid in the root production of *Phaseolus* cuttings. Physiologia Plantarum, 11, 1–9.
84. Middleton, W., Jarvis, B.C. & Booth, A. (1978) The boron requirement for root development in stem cuttings of *Phaseolus aureus* Roxb. The New Phytologist, 81, 287–297.
85. Blazich, F.A. & Heuser, C.W. (1978) The mung bean rooting bioassay – a re-examination. Journal of the American Society of Horticultural Science, 104, 117–120.
86. Girouard, R.M. (1969) Physiological and biochemical studies on adventitious root formation. Extractable co-factors from *Hedera helix*. Canadian Journal of Botany, 47, 687–699.
87. Eaton, F.M. (1940) Interrelations in the effects of boron and indoleacetic acid on plant growth. Botanical Gazette, 101, 700–705.
88. MacVicar, R. & Struckmeyer, B.E. (1946) The relation of photoperiod to the boron requirement of plants. Botanical Gazette, 107, 454–461.
89. Skok, J. (1941) Effect of boron on growth and development of the radish. Botanical Gazette, 103, 280–294.
90. Warington, K. (1933) The influence of length of day on the response of plants to boron. Annals of Botany, 47, 429–458.
91. Weiser, C.J. & Blaney, L.T. (1960) The effects of boron on the rooting of English Holly cuttings. Proceedings of the American Society of Horticultural Science, 75, 704–710.
92. Murray, H.R., Taper, C.D., Pickup, T. & Nussey, A.N. (1957) Boron nutrition of softwood cuttings of geranium and currant in relation to root development. Proceedings of the American Society of Horticultural Science, 69, 498–501.
93. Weiser, C.J. (1959) Effect of boron on the rooting of clematis cuttings. Nature, 183, 559–560.
94. Jarvis, B.C., Yasmin, S., Ali, A.H.N. & Hunt, R. (1984) The interaction between auxin and boron in adventitious root development. The New Phytologist, 97, 197–204.
95. Eriksen, E.N. (1974) Root formation in pea cuttings. III. The influence of cytokinins at different developmental stages. Physiologia Plantarum, 30, 163–167.
96. Humphries, E.C. (1960) Inhibition of root development on petioles and hypocotyls of dwarf bean (*Phaseolus vulgaris*) by kinetin. Physiologia Plantarum, 12, 659–663.
97. Skoog, F. & Miller, C.O. (1957) Chemical regulation of growth and organ formation in plant tissues cultured in vitro. Symposia of the Society for Experimental Biology, 11, 118–131.
98. Sorokin, H.P., Mathur, S.N. & Thimann, K.V. (1962) The effects of auxins and kinetin on xylem differentiation in the pea epicotyl. American Journal of Botany, 49, 444–454.
99. Fabijan, D., Taylor, J.S. & Reid, D.M. (1981) Adventitious rooting in hypocotyls of sunflower (*Helianthus annus*) seedlings. II. Action of gibberellins, cytokinins, auxins and ethylene. Physiologia Plantarum, 53, 589–597.
100. Skoog, F., Schmitz, R.Y., Bock, R.M. & Hecht, S.M. (1973) Cytokinin antagonists:synthesis and physiological effects of 7-substituted 3-methylpyrazolo [4,3-d] pyrimidines. Phytochemistry, 12, 25–37.
101. Brian, P.W. & Radley, M. (1955) A physiological comparison of gibberellic acid with some auxins. Physiologia Plantarum, 8, 899–912.
102. Brian, P.W., Hemming, H.G. & Lowe, D. (1960) Inhibition of rooting of cuttings by gibberellic acid. Annals of Botany, 24, 407–419.
103. Kato, J. (1958) Studies on the physiological effect of gibberellin. II. On the interaction of gibberellin with auxins and growth inhibitors. Physiologia Plantarum, 11, 10–15.
104. Hartung, W., Ohl, B. & Kummer, V. (1980) Abscisic acid and the rooting of Runner Bean cuttings. Zeitschrift für Pflanzenphysiologie, 98, 95–103.

105. Kefford, N.P. (1973) Effect of a hormone antagonist on the rooting of shoot cuttings. Plant Physiology, 51, 214–216.

106. Libbert, E. & Krelle, E. (1966) Die Wirkung des 'Gibberellin-antagonisten' 2-chlorathyltrimethylammoniumchlorid (CCC) auf die Stecklings-bewurzelung windender und nicht windender pflanzen. Planta, 70, 95–98.

107. Hansen, J. (1975) Light dependent promotion and inhibition of adventitious root formation by gibberellic acid. Planta, 123, 203–205.

108. Nanda, K.K., Purohit, A.N. & Bala, A. (1967) Effect of photoperiod, auxins and gibberellic acid on rooting of stem cuttings of Bryophyllum tubiflorum. Physiologia Plantarum, 20, 1096–1102.

109. Varga, M. & Humphries, E.C. (1974) Root formation on petioles of detached primary leaves of dwarf beans (Phaseolus vulgaris) pretreated with gibberellic acid, triiodobenzoic acid and cytokinins. Annals of Botany, 38, 803–807.

110. Crocker, W., Hitchcock, A.E. & Zimmerman, P.W. (1935) Similarities in the effects of ethylene and the plant auxins. Contributions from the Boyce Thompson Institute for Plant Research, 7, 231–248.

111. Robbins, J.A., Kays, S.J. & Dirr, M.A. (1983) Enhanced rooting of wounded mung bean cuttings by wounding and ethephon. Journal of American Society for Horticultural Science, 108, 325–239.

112. Andersen, A.S. (1977) Ethylene and root initiation in cuttings. In Plant Growth Regulators (ed. T. Kudrev et al.), pp. 524–530, Bulgarian Academy of Sciences, Sofia.

113. Krishnamoorthy, H.N. (1970) Promotion of rooting in mung bean hypocotyl cuttings with ethrel, an ethylene releasing compound. Plant and Cell Physiology, 11, 979–982.

114. Mullins, M.G. (1972) Auxin and ethylene in adventitious root formation in Phaseolus aureus (Roxb.). In Plant Growth Substances 1970, (ed. D.J. Carr), pp. 526–533, Springer Verlag, Heidelburg, Berlin, New York, London.

115. Geneve, R.L. & Heuser, C.W. (1983) The relationship between ethephon and auxin on adventitious root initiation in cuttings of Vigna radiata (L.) R. Wilcz. Journal of the American Society for Horticultural Science, 108, 330–333.

116. Batten, D.J. & Mullins, M.G. (1978) Ethylene and adventitious root formation in hypocotyl segments of etiolated mung bean (Vigna radiata (L.) Willczek) seedlings. Planta, 138, 193–197.

117. Michener, H.D. (1935) Effects of ethylene on plant growth hormone. Science, 88, 551–552.

118. Mudge, K.W. & Swanson, B.T. (1978) Effect of ethephon, indole-butyric acid and treatment solution pH on rooting and ethylene levels within mung bean cuttings. Plant Physiology, 61, 271–273.

119. Linkens, A.E., Lewis, L.N. & Palmer, R.L. (1973) Hormonally induced changes in the stem and petiole anatomy and cellulase enzyme patterns in Phaseolus vulgaris L. Plant Physiology, 52, 554–560.

120. Huxter, T.J., Thorpe, T.A. & Reid, D.M. (1981) Shoot initiation in light- and dark-grown tobacco callus: the role of ethylene. Physiologia Plantarum, 53, 319–326.

121. Basu, R.N., Roy, B.N. & Bose, T.K. (1970) Interaction of abscisic acid and auxins in rooting of cuttings. Plant and Cell Physiology, 11, 681–684.

122. Chin, T.Y., Meyer, M.M. Jr. & Beevers, L. (1969) Abscisic acid stimulated rooting of stem cuttings. Planta, 88, 192–196.

123. Rasmussen, S. & Andersen, A.S. (1980) Water stress and root formation in pea cuttings. II. Effect of abscisic acid treatment of cuttings from stock plants grown under two levels of irradiance. Physiologia Plantarum, 48, 150–154.

124. Libbert, E. (1957) Untersuchungen über die Physiologie der Adventivwurzelbildung. III. Untersuchung der Hemstoffe, mittels derer eine Wurzel die Adventivwurzelbildung beeinflukt. Zeitschrift für Botanik, 45, 57–76.

125. Sembdner, G., Gross, D., Liebisch, H.-W. & Schneider, G. (1980) Biosynthesis and metabo-

lism of plant hormones. In Encyclopedia of Plant Physiology, New Series, 9, (ed. J. MacMillan) pp. 281–444. Springer-Verlag, Heidelburg, Berlin, New York, London.

126. Soekarjo, R. & Janssen, M.G.H. (1969) The liberation of inhibitors of indole acetic acid oxidase activity out of Coleus internodes treated with potassium hydroxide and sulphuric acid. Acta Botanica Neerlandica, 18, 651–653.

127. Stonier, T. (1971) The role of auxin protectors in autonomous growth. In Les cultures de tissue de plantes, pp. 423–435. Colloques internationaux, C.N.R.S., Paris (No. 193).

128. Shkol'nik, M.J., Krupnikova, T.A. & Dimitrieva, N.N. (1964). Influence of boron deficiency on some aspects of auxin metabolism in the sunflower and corn. Soviet Plant Physiology, 11, 164–169.

129. Parish, R.W. (1968) In vitro studies on the relationship between boron and peroxidase. Enzymologia, 35, 239–252.

130. Lewis, D.H (1980) Boron, lignification and the origin of vascular plants – a unified hypothesis. The New Phytologist, 84, 209–230.

131. Tran Thanh Van, K., Toubart, P., Cousson, A., Darvill, A.G., Gollin, D.J., Chelf, P. & Albersheim, P. (1985) Manipulation of the morphogenetic pathways of tobacco explants by oligosaccharins. Nature, 314, 615–617.

7. Environmental influences on adventitious rooting in cuttings of non-woody species

ARNE SKYTT ANDERSEN[1]
Royal Veterinary and Agricultural University, Copenhagen, Denmark

1. Introduction

The environment which supports and facilitates rooting of cuttings has a profound influence on the success or failure of the establishment of a new plant. The purpose of this chapter is to assess the types of variation in environment that are

[1] Present address: Department of Horticulture, Rolighedsvej, Frederiksberg, Denmark.

224

important. The intention is to focus on the entire process of adventitious rooting i.e. to include the preceding pre-cutting period, hereafter designated stock plant, as well as the subsequent post-rooting period which will be called establishment. These two marginal periods have all too often been neglected in the literature concerned with the effects of environmental conditions on adventitious rooting and the effects of various hormonal treatments on this process. Frequently there is a lack of concern for the environmental conditions of stock plants. Also many physiologically oriented papers neglect the fact that adventitious rooting is a means of propagation i.e. the establishment of a new plant. The environment which has received the most attention, judging from the number of research papers, is the rooting period itself. It is also the period where changes in environment have the most pronounced effect on the percentage of cuttings that succesfully form a root system and on the number and mass of roots formed. These effects are, however, greatly influenced by the physiological condition of the stock plant and this in turn affects the extent to which cuttings respond to their immediate environment. The latter defines the 'quality' of the cutting.

2. Environment of stock plants

If the environment of stock plants has been neglected it is not a consequence of any lack of early concern for this parameter. Sachs [1], one of the founders of plant science, observed that the absence of light which caused etiolation of the stems of intact plants would promote the rooting of cuttings made from these plants. In the more recent literature there is frequently a nonspecific reference to the environmental conditions of stock plants [2]. Nevertheless one still sees recent papers that lack concern for this important aspect e.g. [3].

What then are the important environmental conditions surrounding stock plants which may affect subsequent rooting of the cuttings? The answer is simply all of them. However, difficulties arise when trying to rank them in order of importance and to ascribing causal mechanisms. An additional problem is the difficulty of transferring information obtained with one species to others. The problem is further confounded by marked varietal differences. Therefore it is important not to generalize.

2.1. Temperature

Since the rate of many plant metabolic processes is controlled by temperature it is not surprising that temperature also influences the condition of stock plants. It is more surprising, that relatively few investigations have attempted to determine the optimal temperature for stock plants. Heide has carried out one of the first thorough experiments ([4], and references therein). It was evident that in *Be-*

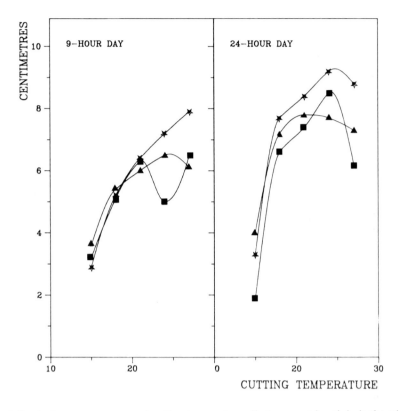

Fig. 1. Stock plant temperature and cutting temperature effects on root length in leaf cuttings of *Begonia × cheimantha* cv. 'Mørk Marina'. The cuttings were rooted at either 9 or 24 h day length (from [4]). △, 12–14° C; ■, 16–18° C; ★, 20–22° C.

gonia, temperatures within the range 12 to 22° C had relatively little influence on the mass of roots produced on single leaf cuttings (Fig. 1). However, bud initiation was severely inhibited in cuttings from high temperature stock plants. Hilding [5] has also worked with *Begonia* 'Schwabenland' and confirmed Heide's results. A paper by Kaminek [5A] suggests that cool temperatures (3 to 4° C) given to stock plants of pea can increase the number of roots formed on cuttings from these stock plants. Similarly, in leaf disks of *Streptocarpus* there was a clear promotion of root number and bud formation if the stock plants were kept at 12° C rather than at 24° C [6]. Leafy stem cuttings of *Campanula isophylla* had a higher dry weight if taken from stock plants grown at warmer (18 to 22° C) temperatures [7], but there was no difference in the rooting of the cuttings. Fischer has carried out a series of temperature experiments with the leafy pea cutting system [8]. Through these we have become aware of the following pitfalls when working with these problems: (a). *Changes in the stage of development.* The development of similar cuttings may take longer at low temperatures, therefore the question arises whether to compare plants (cuttings) of different age, but at a similar stage of development or vice versa? In the work of Fischer [8] there was a delay in development of four to five days if the temperature of the stock plants was decreased from 25 to 15° C. (b). *Air temperature versus leaf temperature.* Experience in the writer's laboratory has shown a need to standardise irradiance conditions. We found that leaf temperature could rise 12 to 15° C above air temperatures in the light as compared to darkness under growth room conditions [9]. (c). *Interactions between stock plant and cutting temperatures.* An abrupt change in temperature at the time of cutting affects subsequent metabolism [10], but complete factorial experiments showed no interaction on root formation or on the length of the rooting period produced by moving cuttings from stock plants grown at one temperature to a rooting medium at another temperature (see [8] and Table 1.).

The available evidence, based on experiments with very few species, points to temperature during stock plant growth having only a minor effect on the rooting of cuttings taken from these stock plants. There is a suggestion that cool temperatures may be beneficial in some species.

2.2. Irradiance

In recent years the trade in cuttings between the tropics and higher latitudes has increased interest in this subject, since stock plants may have been exposed to much higher irradiances than the cuttings experience later. Formerly, the major practical interest was annual differences in cutting quality, some of which could be related to changing light conditions [11]. However, reviewing the literature on this matter is extremely difficult because so many different and in reality non-comparable systems of light measurement have been used. Many reports in this area adopt photometric units (e.g. lux, foot candle) and try to compare different light sources on this basis. Generally, such reports and papers are excluded from consideration in the following discussion. The term 'irradiance' which is largely synonomous with light intensity, is adopted and expressed in radiometric units ($W\,m^{-2}$, cal, J) or photosynthetic photon flux densities (PPFD, $\mu mol\,m^{-2}\,s^{-1}$) as suggested by Incoll *et al.* [12]. The irradiance environment of stock plants received some attention throughout the last century, starting with Sachs [1], but only during the last two decades have intensive and well controlled experiments been made with a number of species – model systems as well as economically important stock plant species [13 to 20].

Experiments investigating the effects of light on rooting fall into four main, if overlapping categories. Firstly there is the etiolation type where darkness is contrasted with light of some kind [13]; secondly, the screening of natural light to stock plants [14, 15]; thirdly, various types of growth chamber lighting [16, 17, 18]. These three have all been used to assess the energy aspect of light by varying the irradiance levels. The fourth category is concerned mainly with day length [19] or

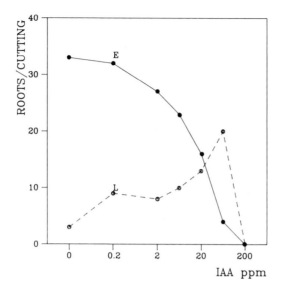

Fig. 2. Root formation in cuttings from etiolated or light grown mung bean stock plants treated with IAA for 20 h (from [21]). E = etiolated. L = light grown.

light quality [20] although this type of experiment often also involves quantity.

Etiolated stems usually form more roots than green stems ([1, 13] and references in the latter). This explains the practical importance of 'stooling', a method of propagation in which the bases of shoot are excluded from light for some considerable time before cuttings are taken. The background for the effectiveness of etiolating the part of the stem which later forms the base of the cutting has been described well by Herman & Hess [13]. They studied the anatomy, auxins and 'cofactors' (see Chapter 6) in green and etiolated bean and *Hibiscus* stems in which both rooting percentages and number of roots were increased by several weeks of etiolation. The anatomical differences due to etiolation were not, in the authors' opinion, extensive enough to account for the doubling of root numbers. Differences in the content of auxin or rooting cofactors between the two types of cutting as measured by bioassay were also unconvincing [13, 21]. However, Tillberg [22] found a higher concentration of auxins in light grown bean plants than in etiolated plants, but did not examine the possibility of an opposite effect on the content of growth or rooting inhibitors [22]. As shown by Kawase [21] the root promoting effects of auxin treatments may be reversed if cuttings are etiolated (Fig. 2). Etiolation of the part of the stem which later forms the base of the cutting is thus an effective way of increasing the root initiation and may even explain data [23] which indicate that the length of the basal internode of the cuttings is positively correlated with rooting (Fig. 3).

An obvious extension of the etiolation procedure is to shade the entire stock plant. This has proved to be a pratical means for improving rooting in a number of species (e.g. [14, 15]) as well as explaining some of the seasonal differences in

228

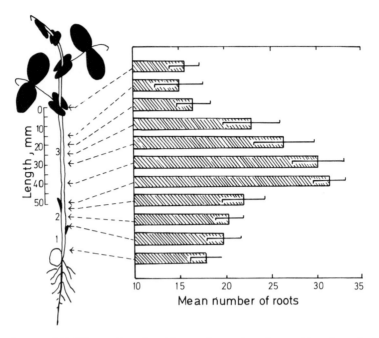

Fig. 3. Influence of 11 different sites of excision on the rooting of pea cuttings from stock plants irradiated for 11 d at 16 W m^{-2}. Numbers along stem axis indicate internode number. Means of three replicates +−95% confidence limits (from [23]).

rooting not related to daylength, dormancy or temperature [24]. Explanation of the physiological background has been sought mainly through growth room experiments using both model systems [11, 16, 17] and plants of commercial interest [24, 25, 26]. The main experimental approach has been to subject the stock plants to several levels of irradiance rather than simply to impose 'on-off' treatments of dark and light. Such studies were initiated by Hansen & Eriksen [17] who utilised a model system consisting of leafy pea seedlings which were made into cuttings. This system has a number of advantages, but also some drawbacks. The young pea plant is physiologically and genetically well described and easy to manipulate, but large storage reserves in the hypogeal cotyledons pose a problem since mobile sugars may circumvent the effects of irradiance treatments [27]. Furthermore the third internode which is used as the base of the cuttings is in a juvenile stage and possesses preformed root initials [30]. The problem thus arises as to whether it is root initiation or root growth that is being studied. Nevertheless the system has yielded a great many results that may or may not be applicable to other genera and species. Decreasing the irradiance to stock plants of peas [17] increases the number of roots produced by the cuttings (Fig. 4). This was initially explained as a simple carbohydrate deficiency that might encourage root formation. There are some indications from other model studies, notably those of Lovell [29] with *Sinapis* leaf cuttings, that the contrary applies

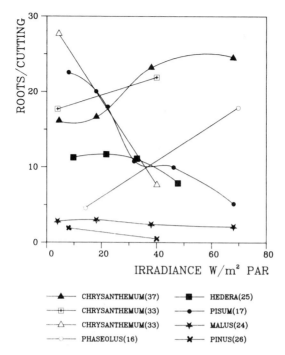

Fig. 4. The influence of irradiance to stock plants on the rooting of cuttings of several different genera.
Redrawn from source papers cited in the figure. It should be noted that light quality was not the same
in all experiments although measurements of irradiance are assumed to be comparable. In most cases
the cuttings of one genus were rooted at a single irradiance that was below the most intense stock plant
irradiance.

i.e. that feeding with carbohydrates reduces rooting. However, a thorough
investigation of the changes in carbohydrate contents of pea plants and cuttings
failed to support the hypothesis [30]. A large accumulation of various carbohydrates in the cuttings was found during the rooting period, but rooting could not
be correlated with any of the minor changes in the stock plants brought about by
the differences in irradiance. Also the old theory of a high carbon: nitrogen
balance of cuttings as an indication of prospects for good rooting [2, 28] was
disproved at least as far as the pea system goes [18, 30].

Since auxin transport is known to be influenced by light [9] the possibility of a
change in auxin transport could conceivably explain the phenomenon because it is
well known that cuttings without an auxin supply form few if any roots [31, 32].
Most auxin transport studies concerned with light effects have focused on the
extreme situations of light or darkness [22]. Few authors have attempted to vary
the light intensity or quality, but when this has been done the results have been
unexpected. Higher irradiances were found to increase auxin transport and
accumulation of auxin at the base of the pea plants [9]. Analyses of endogenous
auxin content have been carried out with this problem in mind. The new tech-

nique of immunoassay has yielded data for *Chrysanthemum* cuttings which indicates more auxin to be present in plants grown under $40\,W\,m^{-2}$ than under $4\,W\,m^{-2}$ [33].

It should be emphasised that findings from the pea system exemplified in Fig. 3 also apply to other genera e.g. *Pinus* [26], *Hedera* [25], *Malus* [24], *Rhododendron* [14], *Sinapis* [29], *Dahlia* [34], *Hibiscus* [15], *Ligustrum* [35], *Berberis* [35], and *Triplochiton* [36]. The opposite response has been observed in three genera so far: *Chrysanthemum* [37], *Campanula* [6] and *Phaseolus* [16]. Thus there appears to be a variable response to irradiance that is related to genetic differences. Some of the data suggest an optimum curve for irradiance of stock plants [35] and of course all plants will fail to form roots at very low (almost darkness) levels of irradiance. Similarly, a reduction at extremely high irradiance may be found for those genera which exhibit the *Chrysanthemum* type response. These levels have probably not been achieved in the growth chamber experiments depicted in Fig. 4 since 60 or $80\,W\,m^{-2}$ PAR [16, 37] is not a very high irradiance compared with natural summertime sunshine conditions. The type of response may depend on the photosynthetic adaptation characteristics of the genus, species or perhaps even variety under investigation. A number of the 'pea type' genera have a fairly low light compensation point and saturation irradiance level, although there are exceptions, while *Chrysanthemum* and *Phaseolus* both have high irradiance requirements [74].

The effects of day-length on rooting of cuttings has been little investigated. According to an earlier review [38] short days (SD) to stock plants always inhibited rooting of cuttings while long days (LD) promoted it. Most of the evidence cited was related to rooting of woody cuttings which also exhibit LD-sensitivity in respect to several other morphogenetic responses. Consequently the effects of daylength on rooting may have been confused with an induction of bud dormancy or senescence. These reservations were eliminated in the experiments of Smith & Wareing [39], who concluded that SD given to stock plants do indeed reduce the number of roots in *Populus* cuttings. Even in the SD-plant *Begonia,* there were more roots on cuttings from stock plants grown in LD than from those grown in SD [40]. In *Chrysanthemum* another SD-plant, Lesham & Schwarz [19] found the opposite effect i.e. SD-treated stock plants yielded cuttings which rooted better than those from LD. In a complicated experiment by Heins *et al.* [20], red light night-breaks given to *Chrysanthemum* stock plants increase root number, a result which seems to contradict the finding of Lesham and Schwarz [19]. In *Rhododendron* and *Acer* soft wood cuttings where SD induce dormancy, rather ambigous results have been obtained. These may have been because the stock plants were not grown under well defined environmental conditions [41, 42].

2.3. Water relations

Too much or too little of anything may be harmful to the rooting process. This is certainly true for the watering of stock plants. The emphasis has generally been on 'too little' which I shall call drought stress. The opposite i.e. waterlogging has received only scant attention.

When pea stock plants were subjected to osmotic stress by growing them in nutrient solutions containing different amounts of polyethylene glycol (PEG 6000), the number of roots on cuttings from these stock plants was decreased, but only if the plants had been grown under low irradiance (16 W m^{-2} PAR). Osmotic stress applied to plants growing under relatively high irradiance (38 W m^{-2} PAR) apparently resulted in more roots [43]. However, when the results were recalculated on a 'stress-hour factor' basis these differences disappeared and a decrease in root numbers was found in response to PEG 6000. Stress-hour factor incorporates the duration as well as the severity of osmotic stress. To some extent these results can be explained by changes in the content of abscisic acid [44, 45], which has a different effect on cuttings from stock plants grown under high or low irradiance [46].

Waterlogging has the distinction of being one of the few natural causes of adventitious root formation which has an ecological function. Kawase [21] has suggested that the ethylene produced in the shoots above waterlogged root systems may be the cause of this root formation. In view of the rather ambiguous role of ethylene in root formation, this explanation should be regarded as no more than a suggestion [15, 45].

2.4. Nutrients

Mineral nutrient relationships of stock plants have been regarded as an area capable of optimization [2, 47], but it is rarely mentioned that the optimum for shoot growth may not be optimal for production of well-rooted cuttings. Stock plant nutrition has been examined thoroughly in certain species, but neglected in other commercially important groups. For *Chrysanthemum* and *Pelargonium* a research group in Geisenheim, Germany has established the optimal nitrogen supply to stock plants [47, 48]. The largest number of cuttings per stock plant of *Chrysanthemum* was obtained with 8 mM ammonium sulphate. Higher or lower amounts gave poorer results. Ammonium alone was directly inhibitory while nitrate alone gave an intermediate response. Unfortunately these authors failed to provide any rooting data although they made the unsubstantiated claim that root number can be calculated from a formula containing the N-concentration and the sugar concentration of the cuttings multiplied by some unidentified factors.

In *Vitis* stock plants, a decrease in the concentration of supplied minerals

232

produced cuttings which all rooted successfully and with more and longer roots per cutting than those taken from stock plants grown in higher concentrations of minerals [49]. Also cuttings from nursery stock plants of *Hypericum, Rosa* and *Rhododendron* which had been grown in stone wool (Grodan) with a lower concentration of the nutrient solution (0.05%) rooted better than those from strongly fertilized stock plants (0.2%) [50]. Thus available evidence suggests that lowering the mineral nutrition of stock plants encourages rooting.

The supply of carbon dioxide to stock plants as a means of influencing rooting of cuttings has not been investigated extensively. Walla & Kristoffersen [51] grew *Chrysanthemum* and *Euphorbia* stock plants at different CO_2 concentrations and found that for some cultivars at least, most cuttings were produced at 2000 ppm CO_2. Unfortunately these authors did not give data for rooting. However, Moe [7] found that CO_2 enrichment of *Campanula* stock plants produced more and higher 'quality' cuttings. The CO_2 gave a significantly greater number of longer roots and 900 ppm CO_2 was more effective than either 300 or 1800 ppm CO_2. With the leafy pea cutting system developed in the writer's laboratory, negligible effects on the rooting of cuttings were obtained with 900 ppm CO_2 supplied to stock plants.

2.5. Pollutants

Although stock plants and cuttings have presumably been subjected accidentally to a number of air and other pollutants this has not been recorded in the literature. However, there is now evidence that oxides of nitrogen, generated as unwanted contaminants in the production of CO_2 for atmospheric enrichment in greenhouses can affect plant growth adversely. This pollutant may be one of the causes of variable success with CO_2-enrichment [52]. Also ethylene may occur as a contaminant in 'pure' CO_2 produced in breweries and sold to growers. The effect of this on stock plants of *Hibiscus* has been the cause of a large sum being paid as compensation to one Danish grower in an out of court settlement for damages.

3. Physiological age

That the physiological stage of a stock plant or some part of it may affect rooting of cuttings from that plant has been known for a long time [53], although this problem has mainly been of concern to propagators of woody cuttings. In genera where there is a clear morphological distinction between the juvenile stage and a later mature stage it is well known that there is a clear difference between the rootability of cuttings from the two shoot types [2]. Cuttings from the juvenile plants or plant parts root much more easily than those from more mature parts.

Examples of genera behaving in this way are *Hedera* [25], *Vacinia* [54], *Ficus* [55], *Castanea* [56], *Triplochiton* [36]. The reason for this difference has been ascribed to auxin content [56], but the explanation is probably less simple. In *Hedera* at least, it has been possible to change the growth habit from mature to juvenile by gibberellin treatment and *vice versa* with abscisic acid [57]. Considering the contrary effects of the two growth regulators on root formation [23, 44, 45], it seems doubtful whether this reversal is accompanied by a concomittant change in rootability, but it has been reported [2] that gibberellin increases the rooting of mature *Hedera* cuttings. Heuser & Hess [58] found a number of lipid 'rooting cofactors' present in higher concentrations in the juvenile *Hedera,* but the chemical nature of these was not further elucidated. There may be differences in the density of stem sclerenchyma tissue between juvenile and mature stems that would pose a greater resistance to the growing roots in mature stems. However, this possibility has been rejected by Davies *et al.,* at least for *Ficus pumila* stems [55]. Leakey also found a connection between the generally lower irradiance which reaches the juvenile plant parts and the previously mentioned better rooting of cuttings from plants grown under a lower irradiance [36]. A logical test of this idea would be to decrease irradiance of mature plant parts and see if this increases the rooting of mature cuttings. Severe cutting-back of the stock plants for continued production of 'juvenile' shoots, for example in blueberry, is a means of utilizing juvenility for propagation purposes [54]. That the degree of juvenility determines the rooting of cuttings is beyond doubt. However, there are several positional (topophysical) effects on rooting which are not explained by a simple statement that the tissue is juvenile. Even in the juvenile form of *Hedera* there are positional or age related differences between the basal internodes which form many roots and the more apical ones which form fewer roots if either are used as cuttings [25]. But since this is to some extent paralleled by a difference in internode length which furthermore is accentuated by irradiance of the stock plants, one can question the conclusions of a number of reports where no account of these important parameters has been taken e.g. [54, 55, 56].

4. Storage of cuttings

In the past, cuttings may have been accidentally left on the bench, perhaps over a weekend, and still have rooted when inserted in the rooting medium some days after excision from the stock plant. In more recent times we have witnessed an explosive increase in the number of cuttings which are transported over great distances, even between continents before they are rooted. In these situations a need for controlled storage is obvious. There is also an increasing requirement for on-site storage in nurseries as a means of levelling-out cutting production to match propagation facilities and sales peaks [59]. Most investigations have concentrated on woody cuttings, but leafy cuttings of *Rhododendron* [60] and *Di-*

234

anthus [61] have been shown to withstand storage for extended periods at low temperatures and high humidity. *Chrysanthemum* cuttings can also be stored for extended periods merely by cooling and restricting evaporation e.g. by wrapping in polyethylene or similar water impermeable material. The introduction of low pressure or hypobaric storage of plant materials has also been of benefit for cutting storage [62]. Experiments with a number of different genera have revealed great differences in their ability to withstand extended periods (several months) of hypobaric storage [64]. Some genera such as *Epipremnum, Begonia, Hibiscus, Cissus* and *Nemathanthus* can be stored successfully at 6.3 kPa (about 1/16 Atm), 5 to 12° C and high humidity. Other genera, noteably *Euphorbia* cannot be stored at all under these conditions. Furthermore it has been found that some other genera such as *Hedera* and *Schlumbergera* can be stored for several months at normal pressure provided that the atmosphere is kept saturated with water. Storage of cuttings for up to seven months for use in micropropagation has been achieved with *Chrysanthemum, Dianthus,* and *Pelargonium* by placing the small (virus free and sterile) cuttings in test tubes with an agar medium in low pressure chambers [62]. The theory behind hypobaric storage was originally that the removal of ethylene from the plant material would prevent the onset of senescence [63]. Apparently this is not the only explanation. Reduced respiration due to low oxygen partial pressure, stomatal opening as a result of the water saturated atmosphere and a low temperature are probably of far greater importance (personal communication, H.G. Kirk, Department of Plant Physiolgy and Anatomy, Royal Veterinary and Agricultural University, Copenhagen). There is however, a paucity of trustworthy data in this area – too much is of the enthusiastic 'advertisement type' with insufficient data to evaluate the claims. Especially lacking are reliable data on survival of cuttings after prolonged low pressure storage. Work in my own laboratory has shown extreme difficulty in repeating some of the results. For some genera there may be a future in the use of hypobaric storage of cuttings since it is already employed commercially in some areas. Major obstacles when scaling up from the laboratory dessicator to a 30 m³ container or warehouse include the circulation of water vapour and maintenance of a tight seal to retain high humidity.

5. Environment of cuttings during rooting

Having produced a cutting of the appropriate quality by proper stock plant maintenance and treatment, it is then necessary to perfect the environment for the rooting of these cuttings. A general problem is the large number of reports that fail to consider the importance of the stock plant on the subsequent reactions of the cuttings to environmental conditions. Such papers have generally been excluded from the following discussion. It is also necessary to clarify some general principles guiding the reporting of rooting. Several papers report only the percen-

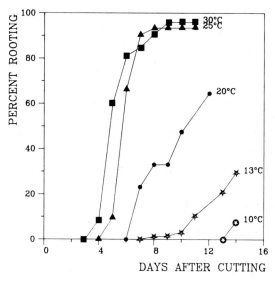

Fig. 5. The effect of rooting temperature on the length of the rooting period of *Sinapis* cotyledon cuttings. There was no rooting at 35 or 40° C. (from [65]).

tage of cuttings that form a root system and fail to address the question of the timing of root formation especially in temperature experiments where the speed of rooting is often a dependent variable. Data from Moore *et al.* [65] illustrate this aspect clearly (Fig. 5). Almost any desired relationship could be deduced from this work if roots had been counted only once at an arbitrary time. Moore's data simply show that at lower temperatures (10 to 13° C) there is a longer lag phase before roots are visible than there is at the higher temperatures (25 to 30° C). Although in some species there is a clear-cut 'initiation period' after which roots are no longer initiated (e.g. [32]) this is not the case for all herbaceous cuttings [66]. Thus, we have several interconnected parameters; the speed of root initiation; number of initials; number of roots; growth of new roots; percentage of cuttings rooted. The extent to which they are expressed will of course be determined at different times after the cuttings are removed from the parent plant.

5.1. Temperature effects during rooting

There are a few investigations of temperature effects where the above mentioned shortcomings are minimal. The first is Heide's work with *Begonia* leaf cuttings [4]. He tested a range of temperatures for stock plants and for cuttings. It is unfortunately difficult to distinguish between the effects on root initiation and on root growth since only maximum root lengths were reported. Generally longer roots were formed with higher propagation temperature. The optimal temperatures for rooting were considerably higher for those cuttings which came from high temperature stock plants. If stock plants had been grown at 14° C there was in some combinations of cultivar/daylength an inhibition of root length (growth?) at 27° C compared to the optimum of 18 to 21° C.

With *Streptocarpus* leaf discs [6] an entirely different pattern was found, although a procedure similar to that of Heide was employed [6]. Low temperature stock plants yielded cuttings which rooted better at low temperatures and even cuttings from stock plants grown at the higher temperature of 24° C had formed more roots at 12° C. Thus more roots are induced at low temperatures, but subsequently they elongate and develop only slowly [65, 66], and this of course may be more important in practical propagation than the number of roots on each cutting [36]. In hydroponic systems for root initiation studies it is possible to monitor root initiation and growth throughout the propagation period [8]. With leafy pea cuttings from stock plants grown at medium irradiance (38 W m^{-2} PAR) and at either 15, 20 or 25° C air temperature, a much longer period was required at 15° C to produce shoots suitable for making cuttings than at the two warmer temperatures (Table 1). Interestingly, the internode length was almost doubled at 15° C, which may be important for rooting [23, 25]. As previously mentioned, different stock plant temperatures had only a minor effect, but the different temperatures during rooting influenced rooting time clearly and also modified the number of roots. Lower temperatures increased the time required for rooting to a

Table 1. The influence of stock plant growing temperature and cutting temperature on days to root emergence (D) and number of roots (R.N.) on pea cuttings. Stock plants and cuttings were grown in controlled environment rooms under an irradiance of 38 W m^{-2} (from [8]).

Stock plant temperature (°C)	Cutting temperature (°C)							
	15		20		25		Means	
	D	RN	D	RN	D	RN	D	RN
15	17.5	34.2	13.0	18.2	11.5	6.5	14.0	19.6
20	17.3	33.4	12.0	24.3	10.3	12.1	13.2	23.3
25	17.3	29.1	11.5	22.2	10.0	10.3	12.9	20.5
Means	17.3	32.2	12.2	21.6	10.6	9.6		

RN = Number of roots per cutting.
D = Days to root emergence.

		D	RN
P. Values	Stock plant temperature	0.24	0.54
	Cutting temperature	0.03	0.03
	Interaction	0.21	0.29
	95% significance		0.05

pre-determined stage, but if the cuttings were allowed to continue growth, root number was tripled at 15°C compared to 25°C, an effect which was almost independent of the stock plant temperature. This combination of slow rooting and increased root number could be a result of a slow-down in metabolic activity and a prolongation of the cell division cycle [67]. That more roots are also produced at the lower temperatures in genera other than pea (reviewed in [8]) suggests it may well be a consequence of a longer period available for initiation. Roots in pea cuttings are initiated in an acropetal succession until the most basal roots emerge [32]. If the lower temperature restricts the growth of the earliest initiated roots, the initiation period will be extended and more roots produced. Recent unpublished results show it is the temperature of the rooting medium that is important. That auxin treatments to the base of cuttings in the form of a quick dip are less effective in promoting root numbers at high or low temperature (F. Okkels, personal communications) may be a reason for the limited practical use of 'rooting hormones' [68]. Changing the temperature during the propagation period may be beneficial for the establishment of some cuttings most notably in *Begonia* leaf cuttings. In this species two weeks of low temperature (18°C) followed by a longer growing period with a higher temperature (27°C) increased root length, (but not number) as well as the number of adventitious buds [69].

5.2. Light

5.2.1. Irradiance. A large body of information describes the relationship between the rooting of cuttings and irradiance during propagation. Unfortunately much of this literature is weak. Many papers show a lack of concern for stock plant conditions (e.g. 'cuttings were taken from plants growing on the campus'), provide unsuitable irradiance measurements or adopt non comparable methods. Where stringent experimental procedures have been employed it has usually been found that increasing irradiance to the leaves of cuttings also enhances the number of roots formed on the cuttings [8, 16, 18, 70]. Such result may indicate that rooting is limited by the availability of current photosynthate [70]. This suggestion had some support from experiments with the photosynthesis inhibitor diuron (DCMU) [29]. However, it is evident that at least in leafy pea cuttings, large quantities of soluble and transportable carbohydrates accumulate soon after the cuttings are excised from the plant [18, 26, 30]. Up to 5 or 6% of sucrose on dry weight basis has been measured. Furthermore, the addition of carbohydrates has no effect in green cuttings of pea [71] and in some cases exogenous sucrose even inhibits root formation [29, 72]. Consequently, other effects of increasing irradiance have been drawn into the discussion of light effects on rooting. These include phenolic rooting inhibitors [73], the accumulation of soluble carbohydrates at the base of auxin treated cuttings [74] and that higher irradiance promotes the basipetal transport of auxin [9, 22]. These possibilities tend to

238

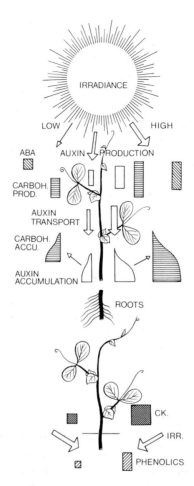

Fig. 6. Generalized interpretation of the effects of irradiance (IRR) of stock plants and cuttings on factors influencing new root formation in cuttings.

complicate the situation and do little to clarify the connection between irradiance, photosynthesis, carbohydrates and rooting of cuttings. It may be added that, under certain circumstances greater irradiance induces the formation of more abscisic acid (at least in stock plants [45]) which may be translocated to the base and promote rooting [44]. To further complicate the picture, the topographical distribution of light to the cutting must be considered. Any irradiance to the base of cuttings will inhibit rooting [70], much more so if the tops of the cuttings are exposed to low irradiance [75]. The complex physiological effects of irradiance during rooting are summarized in Fig. 6.

5.2.2. Light quality. There is evidence that part of the light effect on root formation is mediated through phytochrome [20, 40, 76, 77]. Red light can inhibit root primordia formation in comparison with white or far red irradiation [20, 76]. Preliminary experiments with red or far red light directed towards the base of cuttings through fibre obtics indicate that it is the wavelength of light absorbed by the base that is important (H.F. Wilkins, St. Paul, MN, U.S.A., personal communication). There is also evidence that the alleged phytochrome involvement in rooting has no connection with growth regulator-induced rooting or inhibition of this process [77]. In view of the extremely complicated experimental procedures necessary to elucidate this problem it is not surprising that insufficient evidence has been presented to make it entirely convincing that phytochrome is involved. Obviously, some of the previously mentioned irradiance effects on stock plants (c.f. [103] with [17]) may have different proportions of red/far red irradiance as their cause.

5.2.3. Photoperiodic effects. The photoperiodic control of rooting has received much attention [4, 7, 11, 18, 19, 38, 39, 40, 41, 42, 69, 78, 79]. With pea stock plants grown at 14 h daylength, rooting improved with increasing daylength (from 8 to 22 h) but the difference was small. Decreasing the day to 4 h depressed root number by 50% [18], but since there was no compensation for the decrease in total incident irradiance the result may reflect photosynthetic production rather than day length. Most experiments on photoperiodic effects on rooting of herbaceous cuttings have been performed with plants that are also showing other morphogenetic responses to short days, chiefly flowering [4, 7, 19, 38, 40, 78]. Unfortunately some of these experiments were performed under conditions of variable irradiance (e.g. [4, 19, 38]). With *Begonia,* no marked influence of daylength on rooting was found [4], even when precautions were taken to equalize the energy supplied during short or long days. The largest effect was that of continuous light which gave 25% more roots than an 8 h daylength [40]. In *Chrysanthemum,* the opposite (but unconvincing) results were obtained [19]. *Clematis* cultivars which are long day plants with regard to growth and flowering also root better in long days, especially if the stock plants are also grown under long day conditions [79]. In leafy herbaceous cuttings of *Populus* taken from stock plants grown during long days, a long day treatment was much superior to short days (Fig. 7 and [39]). Similarly, night breaks with weak incandescent light were claimed to promote rooting of leafy *Acer* cuttings [42], but the data were rather inconclusive. In *Rhododendron,* four out of five cultivars responded to supplementary low intensity light with better rooting [41]. In conclusion there appears to be reasonable agreement that long days promote the rooting of cuttings, but the effect is not spectacular.

240

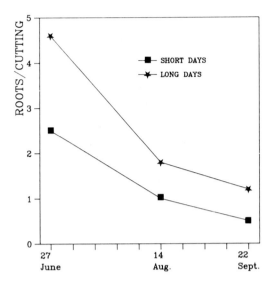

Fig. 7. Rooting response of leafy *Populus* cuttings exposed to short (9 h) or long (18 h) days during rooting. All cuttings received the same total daily irradiance. (from [39]).

5.3. Rooting media

Since the days of Sachs [1], there has been keen interest in different kinds of media for the rooting of cuttings. Laboratory or plant physiologically oriented investigators have often used water or hydroponic cultures [8, 70]. More horticulturally oriented workers have preferred various soil or chemically inert mixtures. However, even strictly practical and commercial propagators have become more and more concerned to standardize the rooting media. Such media may be based on organic or inorganic materials such as compressed peat, perlite, vermiculite, stone wool (Grodan) or nutrient solution. Hydroponics has the advantage of being easily controllable and enables roots to be counted during the propagation period but there are some problems, most notably lack of oxygen [2, 80]. Futhermore, roots produced in water may not be well suited for transplanting into soil since they often lack root hairs. Hydroponic systems are however helpful in the study of inorganic nutrition [8], and for long term treatments, for instance with auxins [70, 78]. One of the first thorough investigations of hydroponic propagation was in 1930 by Zimmerman [80], whose pictures continue to appear in textbooks (2). A criticism of Zimmerman's experiments is his use of undefined tap water as control, an oversight which continues to be made by others. But this may be better than those all too numerous reports where distilled or deionized water are used as controls against which it is much too easy to obtain 'results'. It is surprising that so many cuttings survive the often week-long deprivation of the essential minerals of life. The main feature to which Zimmerman drew attention was the necessity of aeration of the solution [80]. This problem also continues to be of interest with the newer substrates [81] since availability of oxygen and removal of excess CO_2 and other gases is a prerequisite for good rooting. Interestingly the above mentioned deleterious effect of light on the basal parts of the cutting can to some extent be reversed if no oxygen is available [80].

More usually, rooting is achieved by placing the base of the cuttings in some form of solid substrate. It is important that the substrate has the appropriate physical characterisitics i.e. an optimal volume of gas-filled porespace, an oxygen diffusion rate adequate for the needs of respiration, and a reasonable water retention capacity. For gas volume, values between 15 and 20% have been suggested [81], but this parameter is dependent on the water tension under which the substrate is placed. With similar tensions, the rooting of poinsettia cuttings in two substrates with different proportions of gas-filled space (Jiffy 7 and Jiffy 9), however, resulted in considerably more roots at the lowest gas volume (approx. 7.5%) than the highest gas volume (approx. 15%) [82]. But other factors may have been involved since the two substrates also differed with respect to nutrient content. In addition the difficulties of measuring partial gas volumes lower than about 20% may contribute to the unexpected results with these two otherwise very similar substrates.

Considerable differences exist among the many available completely inorganic and artificial media (e.g. perlite, grodan, vermiculite and styrofoam) with respect to gas-filled volume and water retention capacity. Perlite has the highest and vermiculite the lowest gas content while contrariwise, vermiculite retains the greatest volume of water. Nevertheless all these substrates have been used with reasonable success as media for both experimental and practical propagation [17, 32, 42, 50, 82, 83]. A lack of systematic investigation is evident in this area, especially regarding variations in gas-filled pore space of substrate without including concomitant measurements in other parameters such as mineral nutrients [82] or pH [83].

5.4. Water relations

The water relations of cuttings have always been of concern to propagators and this has given rise to a number of inventions designed to maintain turgidity and to reduce transpiration by establishing a constant water film on the leaves of the cuttings [2]. However, a worry here is that a covering of water will impede diffusion of gasses in and out of the leaves [85]. In most systems some screening of excess light is necessary, not because of detrimental effects of light, but as a means of reducing the water loss from the cuttings which may be unable to take up sufficient water to replace that lost by rapid transpiration. Even cuttings with their bases in water develop a water deficit if the irradiance exceeds $100 \, W \, m^{-2}$ PAR. This is remarkable since intact plants of many species will not suffer from water deficit until irradiance is about four times this intensity. Until quite recently these rather obvious water requirements of leafy, herbaceous cuttings had not been subjected to rigorous experimental testing. Lately, however, the Glass-house Crops Research Institute in England has carried out a series of very penetrating investigations of this subject and the following discussion is largely based on their work ([84, 85] and references cited therein).

When a leafy cutting is severed from the stock plant a severe stress situation is created [43]. The cutting loses water at a rate which is mainly dependent on the water vapour pressure difference between the leaf and the air surrounding it. This difference further depends to a considerable extent on the incoming and absorbed radiation which creates a temperature gradient. Thus water loss can be restricted by shading to reduce irradiance; by enclosing the cuttings in plastic tents to increase the water content of the surrounding air; by misting to cool the leaves and by applying anti-transpirants to depress the rate of diffusive water loss. The first three methods, either separately or in various combinations are usually the most satisfactory. If a polyethylene tent system is used alone it appears to be most advantageous to leave as little air space as possible above the cuttings. This lowers the temperature difference between leaf and air, partly due to the direct transfer back to the leaves of the cuttings of water condensing on the plastic. Shading may still be necessary at high irradiance ($>100\,\mathrm{W\,m^{-2}}$ PAR). Of course problems of disease control and limitations to shoot growth may lead to a rejection of these so called 'contact tents', on practical grounds.

During the first week of rooting, water uptake by cuttings is reduced to half of its initial value. This is a result of xylem blockage in the basal part of the stem. In some species the blockage can be alleviated by slitting the stem base. Naturally the rate of water uptake is dependent upon the availability of liquid water in the medium and upon direct contact being established with the cutting base. In this respect some interesting differences among media have been revealed [85]. In perlite, Grange and Loach found that the water uptake by cuttings was nil at a water content of $0.11\,\mathrm{m^3m^{-3}}$, but it rose sharply with a small increase in water content [85]. By contrast, water uptake from media containing peat was possible even when the water content was as small as $0.05\,\mathrm{m^3m^{-3}}$. Further wetting of these peat mixtures increased water absorbtion by cuttings only slightly [85].

Water tension of the medium has already been mentioned briefly. It seems generally to be true that some tension is necessary, especially for media with high water holding capacity placed under a mist system. Gislerød [82] mentions 4–8 cm of tension for peat blocks (Jiffy) and Grodan blocks when these were placed under an open mist system. The data of Gislerød indicate complex relationships between water tension, gas volume, temperature and oxygen diffusion rate. The relationship between oxygen diffusion rate and percentage of gas-filled space is given in Fig. 8 for three propagation media. The rooting of poinsettia cuttings was found not to be directly correlated with the gas or water content of the different media. A relationship of this kind was probably confounded by large temperature

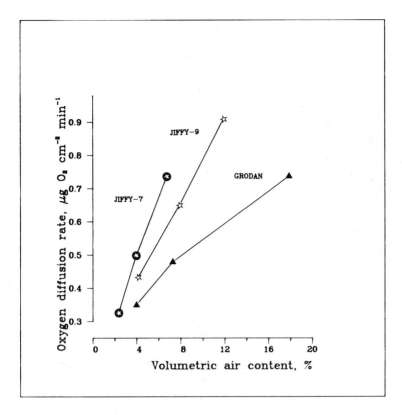

Fig. 8. Oxygen diffusion rates for three different propagation blocks of contrasting structure as a function of their volumetric gas-filled (air) content at increasing water tensions (0.4 or 8 cm). Means of 8 replicates and standard errors. (from [81]).

differences and the effect of this on rooting in this species. In *Hydrangea* cuttings there was a clear promotion of rooting when the peat blocks were subjected to 16 cm tension as compared to zero under conditions of intermittent mist [81]. It is evident that the water relationships of herbaceous leafy cuttings is one of the most, if not the most important environmental factor affecting rooting. Obviously the cutting which looses its turgidity has very limited chance of survival, yet surprisingly there are reports of better rooting in certain genera (e.g. *Pelargonium,* cacti) if they are allowed to desiccate partially before insertion into the rooting medium [86]. This may be related to a similar effect obtained by droughting stock plants prior to excision of cuttings. The resulting improved rooting is probably the consequence of an accumulation of abscisic acid rather than direct positive effect of the drying *per se* [43, 44, 45, 46]. Ensuring that the surface of the foliage is dry may of course be beneficial for pathological reasons.

244

5.5. Nutrients

Interest in the nutrition of cuttings has been intensified along with the increased commercial use of inert and nutrient-poor substrates for rooting of cuttings, often in combination with mist systems where a considerable leaching of mineral nutrients occurs [8, 47]. The carbon dioxide and organic nutrition of cuttings have also received some attention [6, 18, 29, 71].

The resolution of the complex nutrient requirements of cuttings has been approached in two ways. Firstly, by using different concentrations of 'complete' nutrient solutions [8, 50] or by the omission or addition of single elements believed to be of particular importance [47, 87]. In Fig. 9 data are compiled for experiments with the leafy pea cutting system with respect to the concentration of the nutrient solution in which the cuttings rooted for two weeks [8]. This solution described by Nielsen [88] is based on the mineral content of the leaves of a great number of normal, healthy pea plants. It is clear from Fig. 9 that deionized water inhibits rooting. The data also show the need to distinguish between root number, weight and rooting percentage as indicators of optimal rooting conditions. Furthermore, it is evident that less concentrated nutrient solution than that normally used to secure optimal growth of intact plants is beneficial for rooting. The '100-concentration' contained about $0.8\,g\,l^{-1}$ of salt which is dilute compared with a nutrient solution such as the familiar Hoaglands. In a study with mung bean cuttings, a similar effect of concentration of nutrients was observed [89]. Furthermore it was found that the initial pH of the solution was more important than the actual composition. A pH of 6.5 was optimal regardless of the source of minerals. There was very poor rooting in 'control', which presumably was distilled water. Decreasing ionic strength is a trick commonly used in tissue culture to induce rooting [90].

A number of elements have been suggested to be active in the root induction process. These include nitrogen in the form of either nitrate or ammonium [89], Mg [91], P [92], Mn [93], Zn [2] and B [87]. That nitrogen is necessary is hardly surprising, although one may not yet be ready to accept the suggestion by Trewavas [94] that nitrate should be regarded as a hormone. Relevant to this claim is evidence cited by Trewavas that small amounts of nitrate can favour lateral root formation. Other studies [89, 90], however, seem to contradict this interpretation, at least for adventitious roots on cuttings. The suggestion that the actual effect of nitrate is on the proton motive force across membranes and its role in expelling calcium ions [94] may not be so easy to apply to the work of Hyndman et al. [90] with cuttings. These authors made no distinction between root initiation and root growth and the media (peat/sand) were not fully defined in terms of nutrients and waterholding capacity [90]. Nevertheless the investigators have made a contribution which may tie-in with the Trewavas hypothesis as follows. The balance between Mg and Ca is important. If all Ca is replaced by Mg, roots are not initiated on *Chrysanthemum* cuttings. Anything above 40% exchangeable

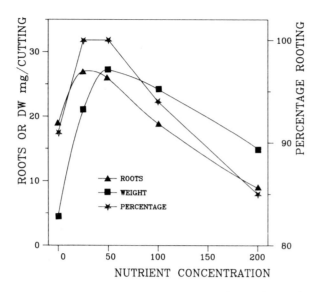

Fig. 9. Rooting of leafy pea cuttings in different concentrations of a complete nutrient solution. On an arbitrary scale, a nutrient solution concentration of 100 comprised: 1.575 mM Ca $(NO3)_2$, $4H_2O$; 0.39 mM KH_2PO_4; 1.84 mM KNO_3; 0.305 mM $MgSO_4$, $7H_2O$; 0.27 mM Mg $(NO_3)_2$, $6H_2O$; 0.7 mM NH_4NO_3; 75 μM NaCl; 1.0 μM $MnSO_4$, H_2O; 0.5 μM $CuSO_4$, $5H_2O$; 2.5 μM $ZnSO_4$, $7H_2O$; 3.5 μM H_3BO_3; 0.65 μM Na_2MoO_4, $2H_2O$; 4.5 μM Fe-EDTA. The cuttings were rooted in a growth chamber without mist. Data are means from a multifactorial experiment and the results are expressed on the basis of total number of cuttings, i.e. rooted and unrooted. There was a highly significant effect of nutrient concentration (P = 0.0001) on all three parameters. [from [8]].

$Mg \left(\dfrac{Mg}{Ca + Mg + Na} \times 100 \right)$ in the medium inhibits rooting. Thus it is probably a lack of Ca due to its replacement by Mg that is responsible. Such a situation may occur if mist water is high in Mg and the medium low in Ca as is the case for a number of media. One role of Ca is thought to be the stimulation of cell division, a function which cannot be reproduced by Mg.

Manganese is also inhibitory to root initiation. This can be surmised from the report that difficult-to-root avocado cultivars contain much more Mn than easy-to-root cultivars [93]. The physiological explanation could be the well-known promoting effect of Mn on the activity of auxin degrading enzymes. Unfortunately no evidence is available showing that decreasing the Mn-concentration in the medium increases rooting. The necessity for Zn in rooting processes could be explained in terms of the activation of tryptophan and indole-acetic acid synthesis, but no recent investigation of the involvement of this element in rooting has been published, due perhaps to the rather infrequent occurrence of Zn deficiency.

Another minor element which has been implicated in root initiation or early root growth is boron (see Chapter 6). B.C. Jarvis at Sheffield University, U.K.,

246

has addressed this problem in work with mung bean cuttings ([87] and references therein). Unfortunately, various B-treatments of cuttings or stock plants have been compared to distilled water treatments. This has probably been necessary due to problems with contaminations from macro nutrients, but nevertheless as mentioned previously, distilled water is an unsuitable control medium. In the pea cutting system it has not been possible to show any inhibiting effect of omitting or adding B to the nutrient solution that is described in Fig. 8. If, however, the major objection of using distilled water controls is ignored, a case for B as an activator of indole-acetic acid oxidase has been built up, leaving the paradox that more active auxin degrading appears to promote rather than inhibit rooting [87]. This is in accord with Philbeam & Kirkby [95] who stated that *'auxin concentrations are higher in B-deficient plants'*. Therefore the role of boron in rooting may be more related to the out-growth of primordia rather than to the initiation phase *per se*. A definite role for B in plants has not so far been established, but several possibilities have been explored. Among these is a suggestion that B 'is a key element in phytochrome action' [96]. This is intriguing in relation to the previously mentioned effects of phytochrome-related events on rooting [19, 39] and also the mounting evidence for boron involvement in stabilizing membrane functions [95].

In conclusion: The mineral nutrient requirements during rooting are low, but real and there must be enough Ca, N, P, (and probably K), but Mg is not necessary in most cases. The pH should be close to 6.5. Anything more than trace ammounts of Mn should be avoided, but some Zn and B seems to be necessary.

Influencing the organic nutrition of cuttings has been attempted by feeding with sugars or amino acids through solutions entering the transpiration stream at the cut base [29, 71] or by elevating the carbon dioxide concentration in the atmosphere [6, 18]. The former method has been used mainly to probe the functions and physiological mechanisms underlying rooting, while the latter often has a practical and economic motivation. The addition of sugars [29] or amino acids [97] to media has largely been done to help evaluate the effects of light [29, 71] on the quality of cuttings. In general it has not been possible to mimic the effect of light on rooting by feeding sugars to cuttings [71]. Investigations which have shown various interactions between sugars and auxins or inhibitors e.g. [98] all lack a theoretical explanation of why such interactions should influence rooting over and above the gross requirement for energy and carbon skeletons [29].

That basally applied auxin should attract soluble carbohydrates to the rooting zone of cuttings and thereby increase metabolism is contradicted by the fact that the accumulation occurs several days after auxin treatment [74]. Cuttings generally accumulate much higher concentrations of sugars during the rooting period even when no auxin is given than uncut plants do under similar conditions. The idea that current photosyntate under certain low irradiance conditions limits rooting [18] is confounded by the finding that cool temperatures strongly increase root formation [8, 74], but have little or no effect on carbohydrate accumulation [74].

The addition of CO_2 to the atmosphere surrounding the cuttings may have several effects. These include increased photosynthesis, reduced transpiration, counteracting the action of ethylene, altering auxin levels [18]. Whether it is beneficial for the rooting process or not will depend on species, other environmental factors and condition of the stock plant. *Chrysanthemum* and *Weigela* can respond with about twice as many roots when the CO_2 concentration is raised to 2000 ppm [99]. In an experiment with a number of different genera Davies & Potter [100] found slightly better rooting in *Pelargonium, Peperomia* and *Hemigraphis* in response to 1200 ppm CO_2. Carbonated water in the form of a mist has been used as a means of supplying CO_2 to cuttings, some in experiments [101] where increased rooting resulted. This may be a quite promising treatment especially if combined with dilute ethanol, which also promotes rooting [102]. Who would not like to be a propagator with whisky and soda in the mist ?

5.6. Pollutants

Pollutants either in the air, in the water or in the substrate may affect the rooting of cuttings. Ethylene, nitrogen oxides, ozone or sulphur dioxide may enter the propagation rooms unnoticed and reduce growth at subvisual levels i.e. a 10 to 15% decrease with no other apparent symptoms [54]. As previously mentioned, such pollution may be the unwanted result of addition of CO_2 either through burning of petroleum products or from the manufacture of compressed CO_2. Also the more limited presence of ethylene in badly aerated media can be considered pollution [103]. The specific effect of ethylene on rooting is controversial [45, 104, 105], but most investigators now favour the view that concentrations of ethylene above 1 ppm (v/v) are inhibitory to root initiation. Such concentrations can occur in enclosed spaces when gasoline motors or other polluting equipment has been operated there. This should be kept in mind when evaluating propagation failures. As far as I am aware no work has been done specifically to evaluate the effects of other air pollutants as possible inhibitors (or promotors?) of rooting.

Interference with rooting of cuttings through pollution of water or media with Mg has been suggested [91]. Of greater significance is a build-up of salts in certain media because of misting with water containing high salt concentrations [9, 90, 92], which according to previously cited evidence [90] would limit root formation.

Few publications have addressed the problem of fungicide interaction with rooting of cuttings [3, 106]. This is surprising since it is often necessary to include fungicide treatments of cuttings to maintain disease-free propagation. Most of the fungicides which have been tested inhibit rooting greatly [20 to 75%]. Clearly it is therefore better to eliminate pathogens by using healthy stock plants and clean propagation facilities than to try to repair the damage by subsequent applications of fungicide that may inhibit rooting.

248

6. Conclusions

It should be evident from the foregoing that the environmental condition of stock plants and cuttings is of the utmost importance for the initiation and growth of adventitious roots. Unfortunately the presently available information makes general conclusions difficult. Findings from experiments with one species are not applicable to another and laboratory results are definitely not always a safe basis for large scale commercial production. Nevertheless it is appropriate to conclude that close attention must be directed towards LIGHT – WATER – NUTRIENTS – TEMPERATURE. Evidence to-hand suggests that for many of these factors, moderation gives better rooting performance by most cuttings.

7. Acknowledgements

Without the generous support during the last 10 years from the Danish State Agricultural Science Foundation to a number of projects concerned with root initiation, the present work could not have been covered. I also thank the numerous colleagues and students who have been active in pursuing the problems and creating more along the way. I hope they all have been cited properly and thus acknowledged. Mr. and Mrs. E. Steele have given invaluable help with the language.

8. References

1. Sachs, J. (1865) Über die Neubildung von Adventivwurzeln durch Dunkelheit. Verhandlungen des Naturhistorischen Vereins Preussen, Rheinland und Westphalen. 21, 110–111.
2. Hartmann, H.T. & Kester, D.E. (1983) Plant Propagation, Principles and Practices. 4th ed. Prentice Hall Inc., Englewood Cliffs, N.J. ISBN 0–13–681007–1.
3. Moorman, G.W. & Woodbridge, W.C. (1983) Effect of fungicide drenches on root initiation by geranium cuttings. Plant Disease, 67, 612–613.
4. Heide, O.M. (1964) Effects of light and temperature on the regeneration ability of *Begonia* leaf cuttings. Physiologia Plantarum, 17, 789–804.
5. Hilding, A. (1974) Inverkan av dagslängd och temperatur vid förökning av høstbegonie (*Begonia* × *hiemalis*) med bladstecklinger. Lantbrukshøgskolans Meddelinger, A 209, 1–15.
5a. Kaminek, M. (1962) Prspeek k fyziologii otuzovani rostlin zakarenovani rostlinnych rizku. Rostlin. Vyro. 8, 959–978.
6. Appelgren M. & Heide, O.M. (1972) Regeneration in *Streptocarpus* leaf and its regulation by temperature and growth substances. Physiologia Plantarum, 27, 417–423.
7. Moe, R. (1977) Effect of light, temperature and CO_2 on the growth of *Campanula isophylla* stockplants and on the subsequent growth and development of their cuttings. Scientia Horticulturae, 6, 129–141
8. Fischer, P. (1981) Temperatur og indstrålingsforholdenes betydning for roddannelse og vækst i ærtestiklinger. Thesis Royal Veterinary & Agricultural University, Copenhagen.
9. Baadsmand, S. & Andersen, A.S. (1984) Transport and accumulation of indole-3-acetic acid in

pea cuttings under two levels of irradiance. Physiologia Plantarum, 61, 107–113.

10. Andersen, Aa. (1979) The influence of temperature on photosynthetic rate of *Dieffenbachia maculata* Lodd. Royal. Veterinary & Agricultural University Yearbook, 78, 15–24.

11. Andersen, A.S., Hansen, J., Veierskov, B. & Eriksen, E.N. (1975) Stock plant conditions and root initiation on cuttings. Acta Horticulturae, 54, 33–38.

12. Incoll, L.D., Long, S.P. & Ashmore, M.R. (1977) SI units in publications in plant science. Current Advances in Plant Science, 9, 331–343.

13. Herman, D.E. & Hess, C.E. (1963) The effect of etiolation upon the rooting of cuttings. Combined Proceedings of the International Plant Propagation Society, 13, 42–62.

14. Johnson, C.R. & Roberts, A.N. (1971) The effect of shading *Rhododendron* stock plants on flowering and rooting. Journal of the American Society for Horticultural Science, 96, 166–168.

15. Johnson, C.R. & Hamilton, D.F. (1977) Rooting of *Hibiscus rosa-sinensis* as influenced by light intensity and ethephon. HortScience, 12, 39–40.

16. Jarvis, B.C. & Ali, A.H.N. (1984) Irradiance and adventitious root formation in stem cuttings of *Phaseolus aureus* Roxb. The New Phytologist, 32, 97, 31–36.

17. Hansen, J. & Eriksen, E.N. (1974) Root formation of pea cuttings in relation to the irradiance of the stock plants. Physiologia Plantarum, 32, 170–173.

18. Davis, T.D. & Potter, J.R. (1981) Current photosyntate as a limiting factor in adventitious root formation on leafy pea cuttings. Journal of the American Society for Horticultural Science, 106, 278–282.

19. Lesham, Y. & Schwarz, M. (1968) Interaction of photoperiod and auxin metabolism in rooting of *Chrysanthemum morifolium* cuttings. Proceedings of the American Society for Horticultural Science, 93, 589–594.

20. Heins, R., Healy, W.E. & Wilkins, H.F. (1980) Influence of night lighting with red, far red and incandescent light on rooting of *Chrysanthemum* cuttings. HortScience, 15, 84–85.

21. Kawase, M. (1965) Etiolation and rooting in cuttings. Physiologia Plantarum, 18, 1066–1076.

22. Tillberg, E. (1974) Levels of indolyl-3-acetic acid and acid inhibitors in green and etiolated bean seedlings (*Phaseolus vulgaris*). Physiologia Plantarum, 31, 106–111.

23. Veierskov, B. (1978) A relationship between length of basis and adventitious root formation in pea cuttings. Physiologia Plantarum, 42, 146–150.

24. Christensen, M.V., Eriksen, E.N. & Andersen, A.S. (1980) Interaction of stock plant irradiance and auxin in the propagation of apple root stocks by cuttings. Scientia Horticulturae, 12, 11–17.

25. Poulsen, A. & Andersen, A.S. (1980) Propagation of *Hedera helix*: Influence of irradiance to stock plants, length of internode and topophysis of cutting. Physiologia Plantarum, 49, 359–365.

26. Hansen, J., Strömquist, L-H. & Ericsson, A. (1978) Influence of the irradiance on carbohydrate content and rooting of cuttings of pine seedlings (*Pinus sylvestris*). Plant Physiology, 61, 975–979.

27. Suttcliffe, J.F. & Pate, J.S. (1977) The Physiology of the Garden Pea. Academic Press, London, ISBN 0–12–677550–8.

28. Kraus, E.J. & Kraybill, H.R. (1918) Vegetation and reproduction with special reference to the tomato. Oregon Agricultural Experiment Station Bulletin, 149.

29. Lovell, P.H., Illsley, A. & Moore, K.G. (1972) The effects of light intensity and sucrose on root formation, photosynthetic activity and senescence in detached cotyledons of *Sinapis alba* L. and *Raphanus sativus* L. Annals of Botany, 36, 126–134.

30. Veierskov, B., Andersen, A.S. & Eriksen, E.N. (1982) Dynamics of extractable carbohydrates in *Pisum sativum*. I. Carbohydrate and nitrogen content in pea plants and cuttings grown at two different irradiances. Physiologia Plantarum, 55, 167–173.

31. Thimann, K.V. & Went, F.W. (1934) On the chemical nature of the rootforming hormone. Proceedings of the Royal Academy of Science, Amsterdam, 37, 456–459.

32. Eriksen, E.N. (1973) Root formation in pea cuttings. I. Effects of decapitation and disbudding at

different developmental stages. Physiologia Plantarum, 28, 503–506.

33. Weigel, U., Horn, W. & Hock, B. (1984) Endogenous auxin levels in terminal cuttings of *Chrysanthemum*. Physiologia Plantarum, 61, 422–428.

34. Biran, I. & Halevy, A.H. (1973) Stock plant shading and rooting of *Dahlia* cuttings. Scientia Horticulturae, 1, 125–131.

35. Knox, G.W. & Hamilton, D.F. (1982) Rooting of *Berberis* and *Ligustrum* cuttings from stock plants grown at selected light intensities. Scientia Horticulturae, 16, 85–90.

36. Leakey, R.R.B. (1983) Stock plant factors affecting root initiation in cuttings of *Triplochiton scleroxylon* K. Schum. an indigenous hardwood of West Africa. Journal of Horticultural Science, 58, 277–290.

37. Fischer, P. & Hansen, J. (1977) Rooting of *Chrysanthemum* cuttings. Influence of irradiance during stock plant growth and of decapitation and disbudding of cuttings. Scientia Horticulturae, 7, 171–178.

38. MacDonald, A.B. (1969) Lighting – its effect on rooting and establishment of cuttings. Combined Proceedings of the International Plant Propagator Society, 9, 241–246.

39. Smith, N.G. & Wareing, P.F. (1972) The rooting of actively growing and dormant leafy cuttings in relation to endogenous hormone levels and photoperiod. The New Phytologist, 71, 483–500.

40. Heide, O.M. (1965) Photoperiodic effects on the regeneration ability of *Begonia* leaf cuttings. Physiologia Plantarum, 8, 185–193.

41. French, C.J. (1983) Stimulation of rooting in *Rhododendron* by increasing natural daylight with low intensity lighting. HortScience, 8, 88–89.

42. Still, S.M. & Lane, B.H. (1984) Influence of extended photoperiod and rooting media fertility on subsequent growth of *Acer rubrum* L'Red Sunset'. Scientia Horticulturae, 22, 129–132.

43. Rajagopal, V. & Andersen, A.S. (1980) Water stress and root formation in pea cuttings. I. Influence of the degree and duration of water stress on stock plants grown under two levels of irradiance. Physiologia Plantarum, 48, 144–149.

44. Rasmusen, S. & Andersen, A.S. (1980) Water stress and root formation in pea cuttings. II. Effect of abscisic acid treatment of cuttings from stock plants grown under two levels of irradiance. Physiologia Plantarum, 48, 150–154.

45. Rajagopal, V. & Andersen, A.S. (1980) Water stress and root formation in pea cuttings. III. Changes in the endogenous level of abscisic acid and ethylene production in the stock plants under two levels of irradiance. Physiologia Plantarum, 48, 155–160.

46. Orton, P.J. (1979) The influence of water stress and abscisic acid on the root development of *Chrysanthemum morifolium* cuttings during propagation. Journal of Horticultural Science, 54, 171–180.

47. Roeber, R. & Reuther, G. (1982) Der Einfluss unterschiedlicher N-Formen und -Koncentrationen auf den Ertrag und die Qualitaet von Chrysanthemen Stecklingen. Gartenbauwissenschaften, 47, 182–188.

48. Reuther, G. & Roeber, R. (1980) Einfluss unterschiedlicher N-Versorgung auf Photosynthese und Ertrag von Pelargonienmutterpflanzen. Gartenbauwissenschaften, 45, 21–29.

49. Pearse, H.L. (1943) The effect of nutrition and phytohormones on the rooting of vine cuttings. Annals of Botany, N.S., 7, 123–132.

50. Knoblauch, F. (1976) Gødskning af *Hypericum Rosa* og *Rhododendron* i containere. Meddelelser fra Statens Forsøgsvirksomhed i Plantekultur, §1275, 1276 & 1279.

51. Walla, J. & Kristoffersen, T. (1974) The effect of CO_2 application under various light conditions on growth and development of some florist crops. Agricultural University of Norway Report, 170, 1–46.

52. Saxe, H. & Christensen, O.V. (1984) Effects of carbon dioxide with and without nitric oxide pollution on growth, morphogenesis and production time of potted plants., Journal of Environmental Science, 38, 159–169.

53. Gardner, F.E. (1929) The relationship between tree age and rooting of cuttings. Procedings of

the American Society of Horticultural Science, 26, 101–104.

54. Lyrene, P.M. (1981) Juvenility and production of fast rooting cuttings from blueberry shoot cultures. Journal of the American Society for Horticultural Science, 106, 396–398.

55. Davies jr., F.T., Lazarte, J.E. & Joiner, J.N. (1982) Initiation and development of roots in juvenile and mature leaf bud cuttings of *Ficus pumila* L., American Journal of Botany, 69, 804–811.

56. Vazquez, A. & Gesto, D.V. (1982) Juvenility and endogenous rooting substances in *Castanea sativa* Mill., Biologia Plantarum, 24, 48–52.

57. Rogler, C.E. & Hackett, W.P. (1982) Phase change in *Hedera helix*. Induction of the mature to juvenile phase change by gibberellin A3. Physiologia Plantarum, 34, 141–147.

58. Heuser, C.W. & Hess, C.E. (1972) Isolation of three lipid root initiating substances from juvenile *Hedera helix* shoot tissue. Journal of the American Society for Horticultural Science, 97, 571–574.

59. Christensen, O.V., ed. (1984) Symposium on production planning in glasshouse floriculture. Acta Horticulturae, 147.

60. Pryor, R.L. & Holley, W.F.D. (1964) Storage of unrooted azalea cuttings. Procedings of the American Society of Horticultural Science, 82, 483–484.

61. Allstadt, R.A. & Holley, W.F.D. (1964) Effects of storage on the performance of carnation cuttings. Colorado. Flower Growers Association Bulletin, 173, 1–2.

62. Paludan, N. (1980) *Chrysanthemum* stunt and chlorotic mottle. Establishment of healthy *Chrysanthemum* plants and storage at low temperature of *Chrysanthemum,* carnation and *Pelargonium* plants. Tidsskrift for Planteavl, 84, 349–360.

63. Eisenberg, B.A., Fretz, T.A. & Staby, G.L. (1978) A comparison of low pressure and common cold storage of unrooted woody ornamental cuttings. Ohio Agriculture Research Circular, 236, 24–29.

64. Jensen, K. (1983) Hypobaric storage of cuttings. 21 International Horticultural Congress Abstract 2, 1804.

65. Moore, K.G., Illsley, A. & Lovell, P.H. (1975) The effects of temperature on root initiation in detached cotyledons of *Sinapis alba* L. Annals of Botany, 39, 1–13.

66. Takahashi, K., Matsuda, S. & Uemoto, S. (1981) Studies on the herbaceous cutting propagation in *Chrysanthemum morifolium.* Influences of temperatures on the increasing rate of adventitious root. Scientific Bulletin of the Faculty of Agriculture, Kyushu University, 35, 1–4.

67. Brown, R. (1951) The effects of temperature on the durations of the different stages of cell division in the root tip. Journal of Experimental Botany, 2, 96–110.

68. Baadsmand, S.B. (1983) Survey of the usage of rooting 'hormones' in Danish nurseries in 1982. (in Danish) Report to Danish National Agricultural Science Foundation

69. Heide, O.M. (1965) Interaction of temperature, auxins, and kinins in the regeneration ability of *Begonia* leaf cuttings. Physiologia Plantarum, 18, 891–920.

70. Eliasson, L. (1980) Interaction of light and auxin in regulation of rooting in pea stem cuttings. Physiologia Plantarum, 48, 78–82.

71. Veierskov, B., Hansen, J. & Andersen, A.S. (1976) Influence of cotyledon excision and sucrose on root formation in pea cuttings. Physiologia Plantarum, 26, 105–109.

72. Larrien, C. & Trippi, V.S. (1979) Etude du viellissement d' *Angallis arvensis* L. Action regulatric de la lumière du saccarose et de l'acide indolyl-acetique sur la capacité rhizogenique des feuilles et des rameaux aicillaires. Biologia Plantarum, 21, 336–344.

73. Druart, P., Kevers, C., Boxus, P. Gaspar, T. (1982) In vitro promotion of root formation by apple shoots through darkness effect on endogenous phenols and peroxidases. Zeitschrift für Pflanzenphysiologie, 108, 429–436.

74. Veierskov, B. & Andersen, A.S. (1982) Dynamics of extractable carbohydrates in *Pisum sativum.* III. The effect of IAA on content and translocation of carbohydrates in pea cuttings during rooting. Physiologia Plantarum, 55, 179–182.

75. Eliasson, L. & Bruns, L. (1980) Light effects on root formation in aspen and willow cuttings. Physiologia Plantarum, 48, 261–265.

76. Pfaff, W. & Schopfer, P. (1974) Phytochrome-induzierte Regeneration von Adventivwurzeln beim Senfkeimling (Sinapis alba). Planta, 117, 269–278.

77. Pfaff, W. & Schopfer, P. (1980) Hormones are no causal links in phytochrome-mediated adventitious root formation in mustard seedlings (Sinapis alba L.). Planta, 150, 321–329.

78. Hansen, J. & Ernstsen, A. (1982) Seasonal changes in adventitious root formation in hypocotyl cuttings of Pinus sylvestris. Influence of photoperiod during stock plant growth and of indole butyric acid treatment of cuttings. Physiologia Plantarum, 54, 99–106.

79. Weyland, H.B. (1978) The effect of photoperiod on Clematis cuttings. American Nurseryman, 148, 13/48/49.

80. Zimmerman, P.W. (1930) Oxygen requirements for root growth of cuttings in water. American Journal of Botany, 17, 842–861.

81. Gislerød, H. (1982) Physical conditions of propagation media and their influence on the rooting of cuttings. I. Air content and oxygen diffusion at different moisture tensions. Plant & Soil, 69, 445–456.

82. Gislerød, H. (1983) Physical conditions of propagation media and their influence on the rooting of cuttings. III. The effect of air content and temperature in different propagation media on the rooting of cuttings. Plant & Soil, 75, 1–14.

83. Altman, A. & Freudenberg, D. (1983) Quality of Pelargonium graveolens cuttings as affected by the rooting medium. Scientia Horticulturae 19, 379–384.

84. Grange, R.I. & Loach, K. (1984) Comparative rooting of eighty-one species of leafy cuttings in open and polyethylene-enclosed mist systems. Journal of Horticultural Science, 59, 15–22.

85. Grange, R.I. & Loach, K. (1983) The water economy of unrooted leafy cuttings. Journal of Horticultural Science 58, 9–17.

86. Wood, H.J. (1966) Pelargoniums, a Complete Guide to Their Cultivation. Faber and Faber, London.

87. Jarvis, B.C., Ali, A.H.N. & Shaheed, A.I. (1983) Auxin and boron in relation to the rooting response and ageing of mung bean cuttings. The New Phytologist, 95, 509–518.

88. Nielsen, N.E. (1976) Om formuleringen af næringsstofopløsninger og dyrkning af planter i vandkultur. Tidsskrift for Planteavl, 80, 175–180.

89. Zucconi, F. & Pera, A. (1978) The influence of nutrients and pH effect on rooting as shown by mung bean cuttings. Acta Horticulturae 79, 57–62.

90. Hyndman, S.E., Hasegawa, P.M. & Bressan, R.A. (1982) Stimulation of root initiation from cultured rose shoots through the use of reduced concentrations of mineral salts. HortScience 17, 82–83.

91. Paul, J.L. & Thornhill, W.H. (1969) Effects of magnesium on rooting of Chrysanthemums. Journal of the American Society for Horticultural Science, 94, 280–282.

92. Kabi, T. & Pujari, R. (1980) Effects of IAA, GA_3 and CCC at different levels of inorganic phosphate on rooting of Impomea batatas L. leaves. Geobios 7, 170–171.

93. Reuveni, O. & Raviv, M. (1981) Importance of leaf retention to rooting of avocado cuttings. Journal of the American Society for Horticultural Science, 106, 127–130.

94. Trewavas, A.J. (1983) Nitrate as a plant hormone. In Interactions between Nitrogen and Growth Regulators in the Control of Plant Development, Monograph 9. (ed M.B. Jackson), pp 97–110. British Plant Growth Regulator Group, Wantage.

95. Philbeam, D.J. & Kirby, E.A. (1983) The physiological role of boron in plants. Journal of Plant Nutrition 6, 563–582.

96. Tanda, T. (1978) Boron – key element in the actions of phytochrome and gravity? Planta, 143, 109–111.

97. Suzuki, T. & Kohno, K. (1983) Changes in nitrogen levels and free amino acids in rooting cuttings of mulberry (Morus alba). Physiologia Plantarum 59, 455–460.

98. Nanda, K.K. & Dhaliwal, G. (1974) Interaction of auxin, nutrition, and morphological nature in rooting hypocotyl cuttings of *Impatiens balsamina*. Indian Journal of Experimental Biology, 1, 82–84.

99. Molnar, J.N. & Cummings, W.A. (1968) Effect of carbon dioxide on propagation of softwood, conifer and herbaceous cuttings. Canadian Journal of Plant Science, 48, 595–599.

100. Davis, T.D. & Potter, J.R. (1983). High CO_2 applied to cuttings: Effect on rooting and subsequent growth in ornamental species. HortScience, 18, 194–196.

101. Lin, W.C. & Molner, J.M. (1980) Carbonated mist and high intensity lighting for propagation of selected woody ornamentals. Proceedings of the International Plant Propagator Society, 30, 104–109.

102. Middelton, W., Jarvis, B.C. Booth, A. (1978) The effects of ethanol on rooting and carbohydrate metabolism in stem cuttings of *Phaseolus aureus* Roxb. The New Phytologist, 81, 279–285.

103. Konings, H. & Jackson, M.B. (1979) A relationship between rates of ethylene production by roots and the promoting or inhibiting effects of exogenous ethylene and water on root elongation. Zeitschrift für Pflanzenphysiologie 92, 385–397.

104. Andersen, A.S. (1977) Ethylene and root initiation in cuttings. In Plant Growth Regulators 2 (eds T. Kudrev, I. Ivanova & E. Karnovov), pp 524–530. Bulgarian Academy of Science, Sofia, Bulgaria.

105. Geneve, R.L. & Heuser, C.W. (1982) The effect of IAA, IBA, NAA and 2,4-D on root promotion and ethylene evolution in *Vigna radiata* cuttings. Journal of the American Society for Horticultural Science, 107, 202–205.

106. Lee, K.W., Sander, K.C. & Williams, J.G. (1983) Effect of fungicides applied to polyurethane propagation blocks on rooting of poinsettia cuttings. HortScience, 18, 359–360.

Subject index

Page numbers in *italics* refer to Figures

Index of plant names

Page numbers in *italics* refer to Figures